15th Annual International Astrophysics Conference 2016

The Science of Ed Stone: Celebrating his 80th Birthday

Journal of Physics: Conference Series Volume 767

Cape Coral, Florida, USA
4 – 8 April 2016

Editors:

G.P. Zank

ISBN: 978-1-5108-3384-5
ISSN: 1742-6588

Printed from e-media with permission by:

Curran Associates, Inc.
57 Morehouse Lane
Red Hook, NY 12571

Some format issues inherent in the e-media version may also appear in this print version.

Printed by Curran Associates, Inc. (2017)

For permission requests, please contact the Institute of Physics
at the address below.

Institute of Physics
Dirac House, Temple Back
Bristol BS1 6BE UK

Phone: 44 1 17 929 7481
Fax: 44 1 17 920 0979

techtracking@iop.org

Additional copies of this publication are available from:

Curran Associates, Inc.
57 Morehouse Lane
Red Hook, NY 12571 USA
Phone: 845-758-0400
Fax: 845-758-2633
Email: curran@proceedings.com
Web: www.proceedings.com

Table of contents

Volume 767

15th Annual International Astrophysics Conference: "The Science of Ed Stone: Celebrating his 80th Birthday"

4–8 April 2016, Cape Coral, Florida, USA

Accepted papers received: 19 October 2016
Published online: 22 November 2016

Preface

Papers

Journal of Physics: Conference Series **767** (2016) 011001 doi:10.1088/1742-6596/767/1/011001

Preface

G P Zank
Editor, *Proceedings of the 15th Annual International Astrophysics Conference,*
Center for Space Plasma and Aeronomic Research (CSPAR) and Department of Space Physics, University of Alabama in Huntsville, Huntsville AL 35805, USA

E-mail: garyp.zank@gmail.com

The 15th Annual International Astrophysics Conference was held at the Westin Cape Coral Resort at Marina Village, Cape Coral, Florida, during the week of 4-8 April 2016. This year the meeting was rather different from previous years in that it honoured the scientific achievements and recognized the extraordinary contribution of Ed Stone to space physics. Entitled "The Science of Ed Stone: Celebrating his 80th Birthday," the meeting drew some 85 participants from all over the world. The number of participants and the range of topics covered paid testament to the enormous range of the science that Ed influenced profoundly. Indeed, Ed Stone has made foundational contributions to our understanding of all facets of the physics of energetic particles in the solar wind and the interstellar medium. Much of our current understanding of the interaction of the solar wind with the local interstellar medium is derived from the Voyager Interstellar Mission, which has operated for 39 years under Ed's leadership. His remarkable career has spanned the space age, and he has contributed not only to science but to the leadership of our community.

It is with great sadness that we note that one of the people greatly influenced by Ed Stone, Harm Moraal, died in the past year. A brief obituary of Prof. Harm Moraal is presented below. He will be missed greatly in our community.

Finally, we thank Adele Corona and ICNS for her continued excellent organization of the AIAC meetings and her help in providing the logistical support for this volume of papers.

Journal of Physics: Conference Series **767** (2016) 011001 doi:10.1088/1742-6596/767/1/011001

Remembering HARM MORAAL
1947 - **November 10, 2015**

(R.A. Burger, lightly edited by G.P. Zank)

Harm Moraal, Emeritus Professor of Physics at North-West University, passed away on 10 November 2015 after a brief illness at the age of 68.

Harm was born in the Netherlands in 1947 and his family relocated to South Africa in 1957. Harm spent his career at the former Potchefstroomse Universiteit vir Christelike Hoër Onderwys, beginning his studies there in 1965, obtaining a DSc degree in 1973, and eventually attaining the position of full professor in 1985. He was Chair of the Department of Physics from 1991 to 2002 and retired in 2012.

Harm rapidly established himself as a leader in the field of cosmic-ray studies, including their transport in the heliosphere, acceleration at shocks, how and why their intensity vary with changes in solar activity, and the properties of galactic and of solar cosmic rays measured with neutron monitors at Earth. Harm's leadership extended to the local and international physics community. He became Secretary of the Commission on Cosmic Rays (C4) of the International Union for Pure and Applied Physics (IUPAP) in 1990 and was Chair from 1994 to 1997. Harm was instrumental in bringing the 25th International Cosmic Ray Conference to South Africa, held in Durban in 1997.

Harm Moraal was actively involved in the management of the South African Institute of Physics (SAIP). Harm was Secretary of the Sun-Earth and Plasma Physics Specialist Group from 1993 to 1995. He was elected member of the SAIP Council in 1999, Vice-President from 2005 to 2007, and President from 2005 to 2007. As Past-President he was ex officio member of the Council from 2007 to 2009.

Harm also served the South African physics community in various other ways. He was associate editor of the South African Journal of Science from 2004 to 2010. Harm was also co-author of a number of publications related to physics in South Africa: Growing opportunities for international collaboration in physics in South Africa, Shaping the future of physics in South Africa, The state of physics in South Africa, and the comprehensive History of Physics in South Africa.

His contributions to physics and the physics community, only some of which are noted above, led to a number of awards. He was awarded the Havenga Prize for Technical Sciences of the Suid-Afrikaanse Akademie vir Wetenskap en Kuns in 2004, and the MT Steyn Medal of the same Academy in 2014 for his significant contribution to the development and organization of physics. The SAIP awarded him the De Beers Gold Medal for his original research in physics in 2012.

Harm Moraal remained humble despite his achievements and honors. He will remain a role model for all of us on how to balance teaching, research, and service to both the national and the international physics community. Harm is survived by his wife Christina and their four children.

Journal of Physics: Conference Series **767** (2016) 011002 doi:10.1088/1742-6596/767/1/011002

Peer review statement

All papers published in this volume of *Journal of Physics: Conference Series* have been peer reviewed through processes administered by the proceedings Editors. Reviews were conducted by expert referees to the professional and scientific standards expected of a proceedings journal published by IOP Publishing.

The interaction of turbulence with parallel and perpendicular shocks

L Adhikari[1,2], G P Zank[1,2], P Hunana[1,2], and Q Hu[1,2]

[1]Department of Space Science, University of Alabama in Huntsville, Huntsville, AL 35899, USA

[2]Center for Space Plasma and Aeronomic Research (CSPAR), University of Alabama in Huntsville, Huntsville,
AL 35899, USA

E-mail: la0004@uah.edu

Abstract. Interplanetary shocks exist in most astrophysical flows, and modify the properties of the background flow. We apply the Zank et al 2012 six coupled turbulence transport model equations to study the interaction of turbulence with parallel and perpendicular shock waves in the solar wind. We model the 1D structure of a stationary perpendicular or parallel shock wave using a hyperbolic tangent function and the Rankine-Hugoniot conditions. A reduced turbulence transport model (the 4-equation model) is applied to parallel and perpendicular shock waves, and solved using a 4th-order Runge Kutta method. We compare the model results with ACE spacecraft observations. We identify one quasi-parallel and one quasi-perpendicular event in the ACE spacecraft data sets, and compute various turbulent observed values such as the fluctuating magnetic and kinetic energy, the energy in forward and backward propagating modes, the total turbulent energy in the upstream and downstream of the shock. We also calculate the error associated with each turbulent observed value, and fit the observed values by a least square method and use a Fourier series fitting function. We find that the theoretical results are in reasonable agreement with observations. The energy in turbulent fluctuations is enhanced and the correlation length is approximately constant at the shock. Similarly, the normalized cross helicity increases across a perpendicular shock, and decreases across a parallel shock.

1. Introduction

Collisionless shocks exist in any astrophysical flows, and are important structures in the heliosphere. They are formed throughout the heliosphere due to the interaction of solar wind disturbances with the solar wind [1, 2]. Collisionless shocks change the characteristics of the solar wind when the solar wind passes through the shocks, through both heating and compression. Collisionless shocks play an important role in particle acceleration to high energies (several tens of ~MeVs) [3-10] including cosmic rays [11], which causes a significant hazard to satellite technology and human activity in space [7, 10, 12]. Shocks can also be responsible for geomagnetic activity [13, 14].

Interplanetary shocks (IP) can be formed either by the interaction of fast solar wind streams with slow preceding plasma or by fast interplanetary coronal mass ejections (ICMEs). However, to form a shock in the heliosphere, the difference in speed between two solar wind streams or the ICME and the preceding plasma should exceed the magnetosonic speed. The source of shocks depends on solar cycle, and during solar minimum most shocks are driven by stream interactions [15, 16], whereas during solar maximum most shocks are driven by ICMEs [15]. The acceleration characteristics for charged particles accelerated by ICME driven shocks and by stream interaction driven shocks is different for several reasons. The particle acceleration rate is governed by the magnetic field strength, shock strength, turbulence levels (also related to shock obliquity; Zank et al 2006 [17]), as discussed by Zank et al 2001 [5]. The acceleration is therefore determined in part by where the shock forms since the acceleration rate is largest near the Sun [5]. The time available for a particle to be accelerated also depends on where the shock forms, and so the maximum achievable particle energy to which a particle can be accelerated is determined by a balance of the particle acceleration rate and the dynamical time scale of the shocks ([5], [18], [19]). ICME driven shocks form near the Sun so that particles are accelerated for a longer time. Stream interaction driven shocks form at larger distances and so particles experience acceleration for a shorter time [20], and the particle acceleration rate is much slower [5]. Typically, coronal mass ejection driven shocks are faster and stronger, and show a larger distribution of shock parameters than stream interaction driven shocks [21].

The structure of a shock depends on its strength, the magnetic field geometry, and the plasma beta β. The strength is characterized by the magnetosonic Mach number and the compression ratio, and the geometry by the obliquity angle $\theta_{\rm bn}$, where $\theta_{\rm bn}$ is the angle between the magnetic field and the shock normal. Shocks are quasi-parallel if $\theta_{\rm bn} \leq 45°$, and quasi-perpendicular if $\theta_{\rm bn} > 45°$. With increasing Mach number, the shock profile changes and develops a foot and overshoot in association with ion reflection and gyration. [22, 23] found that $\theta_{\rm bn}$ plays an important role in the shock geoeffectiveness in that most perpendicular shocks are more geoeffective than quasi-parallel shocks (see also [24]). The shock normal is another important property of IP shocks. The IP shocks propagate along the shock normal direction. Usually, the normal associated with ICMEs-driven IP shock is close to the Sun-Earth line when the shock passes 1 AU [2]. The shock normal for a stream interaction driven shock typically is at a large angle compared to the Sun-Earth line [25].

Shocks are categorized as fast forward (FF) and fast reverse (FR) shocks with respect to the solar wind frame. Forward shocks propagate away from the Sun, and reverse shocks propagate towards the Sun. Generally, shocks associated with ICMEs are forward shocks. Most observed IP shocks at 1 AU are also found to be fast forward shocks. An increase in the magnitude of the magnetic field and other parameters is observed when a fast forward shock passes over the spacecraft in the spacecraft frame solar wind. On the other hand, a decrease in the magnitude of the magnetic field, the solar wind density and the temperature, and an increase in the solar wind speed are observed when a fast reverse shock passes over the spacecraft.

Shocks are also responsible for the generation of turbulence in the heliosphere. In this manuscript, we study the properties of several turbulent parameters in the upstream and downstream regimes of a shock. We present a turbulence transport model that describes the fluctuations of the solar wind parameters when interacting with quasi-parallel and quasi-perpendicular shocks. We introduce a steady state shock wave in the form of hyperbolic tangent function of x and solve the 1D steady state coupled turbulence transport equations. We compare the numerical solutions of the turbulence transport equations applied to the quasi-parallel shock of day 272 of year 2001 and the quasi-perpendicular shock of day 250 of year 2000, respectively. In so doing, we calculate turbulent parameters such as the energy in forward and backward propagating modes, the residual energy, their corresponding correlation lengths, the cross helicity, and the total turbulent energy from the upstream region to the downstream region of the shock. We also calculate the error associated with each turbulent quantity. We discuss the shock model equations in Section 2. Section 3 presents a comparison between the theoretical results and the observed values. Section 4 contains some conclusions. Finally, Appendix A presents an error analysis associated with several turbulent quantities.

2. Turbulence and Shock model equations

To study the interaction of turbulence with parallel and perpendicular shocks, we apply the Zank et al 2012 [26] (see also [27, 28]) turbulence transport model equations. We reduce the 1D six coupled equations of Zank et al 2012 model equations to four coupled equations. The reduction follows if we assume $\lambda^+ = \lambda^- = \lambda$ and $\lambda_D = 2\lambda$, where λ^+ and λ^- are the correlation length corresponding to backward and forward propagating modes, and λ_D is the correlation length of the residual energy. The Zank et al model equations can be reduced to four coupled equations for both the zero and nonzero cross helicity cases. Here we use a four coupled equation model that corresponds to the nonzero cross helicity case. The equations are expressed in a Cartesian coordinate system x for both parallel and perpendicular shocks. We adopt a coordinate system in which x is parallel to the shock normal.

2.1. Quasi-Parallel Shock Model Equations

For a parallel shock, the magnetic field **B** is parallel to the shock normal. The 1D steady state four coupled turbulence transport equations can be written as

$$(U - V_A)\frac{\partial f}{\partial x} + \left[\frac{f}{2} + \left(2a - \frac{1}{2}\right)E_D\right]\frac{\partial U}{\partial x} + \frac{\partial V_A}{\partial x}(f - E_D) = -2\frac{fg^{1/2}}{\lambda\cos\theta_{\rm bn}}; \quad (1)$$

$$(U + V_A)\frac{\partial g}{\partial x} + \left[\frac{g}{2} + \left(2a - \frac{1}{2}\right)E_D\right]\frac{\partial U}{\partial x} - \frac{\partial V_A}{\partial x}(g - E_D) = -2\frac{gf^{1/2}}{\lambda\cos\theta_{\rm bn}}; \quad (2)$$

$$U\frac{\partial E_D}{\partial x} + \frac{1}{2}\frac{\partial U}{\partial x}\left[(2a - \frac{1}{2})(f+g) + E_D\right] + \frac{V_A}{2\sqrt{fg}}\left[f\frac{\partial g}{\partial x} - g\frac{\partial f}{\partial x}\right] + \frac{\partial V_A}{\partial x}\frac{f-g}{2} = -E_D\left[\frac{f^{1/2} + g^{1/2}}{\lambda\cos\theta_{\rm bn}}\right]; \quad (3)$$

$$U\frac{\partial \lambda}{\partial x} = \frac{f^{1/2} + g^{1/2}}{\cos\theta_{\rm bn}}, \tag{4}$$

where $f \equiv \langle z^{+2} \rangle$ and $g \equiv \langle z^{-2} \rangle$ are the energy in backward and forward propagating modes respectively, E_D is the residual energy, and λ is the correlation length. Here the parameters $\mathbf{z}^{\pm} = \mathbf{u} \pm \mathbf{b}/\sqrt{\mu_0 \rho}$ are Elsässer variables [29] for the turbulent fluctuations $\mathbf{u}$ and $\mathbf{b}$. The parameter U is the solar wind velocity, and V_A is the Alfvén velocity. The parameter a=1/2 or 1/3 for 2D or 3D mixing tensor ([26], [30]). Note that the magnetic field is outwardly directed. The parameter $\theta_{\rm bn}$ is the angle between the normal to an interplanetary shock front and the upstream magnetic field. The value of $\theta_{\rm bn}$ depends on the configuration of the interplanetary magnetic field through which the shock passes. In this manuscript, we solve turbulence transport equations (1)–(4) for $\theta_{\rm bn} = 0$.

2.2. Perpendicular Shock Model Equations

At a perpendicular shock, the magnetic field $\mathbf{B}$ is perpendicular to the shock normal. The 1D steady state four coupled equations in this case become

$$U\frac{\partial f}{\partial x} + \left[\frac{f}{2} + \left(2a - \frac{1}{2}\right)E_D\right]\frac{\partial U}{\partial x} = -2\frac{f g^{1/2}}{\lambda}; \tag{5}$$

$$U\frac{\partial g}{\partial x} + \left[\frac{g}{2} + \left(2a - \frac{1}{2}\right)E_D\right]\frac{\partial U}{\partial x} = -2\frac{g f^{1/2}}{\lambda}; \tag{6}$$

$$U\frac{\partial E_D}{\partial x} + \frac{1}{2}\frac{\partial U}{\partial x}\left[\left(2a - \frac{1}{2}\right)(f+g) + E_D\right] = -E_D\left[\frac{f^{1/2} + g^{1/2}}{\lambda}\right]; \tag{7}$$

$$U\frac{\partial \lambda}{\partial x} = f^{1/2} + g^{1/2}. \tag{8}$$

The coupled turbulence transport equations (5)–(8) are simpler than (1)–(4) in that the Alfvén velocity is absent. The coupled equations (1)–(4) and (5)–(8) are solved using a 4th-order Runge Kutta method for different initial conditions, and the numerical solutions are then compared to ACE observations.

The other turbulent quantities can be calculated as [26]

$$E_T = \frac{f+g}{2}; \quad E_C = \frac{g-f}{2}; \tag{9}$$

$$\sigma_c = \frac{E_C}{E_T}; \quad \sigma_D = \frac{E_D}{E_T}; \tag{10}$$

$$\langle u^2 \rangle = \frac{f+g+2E_D}{2}; \quad \left\langle \frac{b^2}{\mu_0 \rho} \right\rangle = \frac{f+g-2E_D}{2}, \tag{11}$$

where E_T is the total turbulent energy, E_C is the cross helicity, σ_c is the normalized cross helicity, σ_D is the normalized residual energy, $\langle u^2 \rangle$ is the fluctuating kinetic energy, and $\langle b^2 \rangle$ is the fluctuating magnetic energy.

2.3. Solar wind velocity and Alfvén velocity

To study the interaction of turbulence with shock waves, we introduce a stationary shock wave in the form of a hyperbolic tangent function of x [31, 32]

$$U = \frac{1}{2}(U_1 + U_2) - \frac{1}{2}(U_1 - U_2)\tanh\left(\frac{x - x_0}{D_{sh}}\right), \tag{12}$$

where U_1 and U_2 are the upstream and downstream velocities of the shock. The parameter x_0 is the position of the shock, and D_{sh} is the thickness of the shock. The thickness of a shock is determined by the appropriate dissipation mechanism. In the case of perpendicular shocks, the dissipation mechanism is due to reflected ions from the cross shock potential. The reflected ions form a foot in front of the ramp, which roughly determines the thickness of a shock. In the case of a parallel shock, the ions that reflect from the ramp propagate away from the shock as a beam in the upstream region. This introduces instabilities that generate waves that scatter the ions. Wave particle interactions leads to the formation of a diffuse ion population which eventually convects through the shock. For parallel shocks, the dissipation mechanism occurs over an extended range. Hence, the thickness of parallel shock D_{sh} is typically much larger than the thickness of perpendicular shock.

The Alfvén velocity can also be expressed in the form of a hyperbolic tangent function of x. The Alfvén velocity can be written as

$$V_A = \frac{B}{\sqrt{\mu_0 \rho}} = \frac{B}{\sqrt{\mu_0 m}}\sqrt{U}, \tag{13}$$

using mass conservation $\rho U = m$. Here B is the magnetic field, μ_0 is the permeability of free space, ρ is the density, and m is a constant. Equation (13) shows that the Alfvén velocity is proportional to the product of the magnetic field B and the square root of the solar wind velocity. Note that for a parallel shock, the magnetic field is constant across a shock, and for a perpendicular shock the magnetic field in the downstream region is larger than the magnetic field in the upstream region (by the product of the upstream B with the compression ratio r). We use Equations (12) and (13) in Equations (1)–(4) and (5)–(8), and solve the coupled turbulence transport equations.

3. Data Analysis

In this section, we compare our theoretical results with ACE observations. We selected two shock events, the quasi-parallel shock of day 272 of year 2001 and the quasi-perpendicular shock of day 250 of year 2000. We calculate various turbulent values such as the fluctuating magnetic energy, the fluctuating kinetic energy, the energy in forward and backward propagating modes, the total turbulent energy, and the normalized residual energy in the upstream and downstream region of a shock [27, 28, 30, 33]. The turbulent quantities are calculated from 64 second resolution plasma data sets. In this analysis we use a criterion that the ratio of fluctuations of the field in an interval and the mean field should be smaller than 1 i.e., $\langle \delta x^2 \rangle / X^2 << 1$. Here $\langle \delta x^2 \rangle$ represents the fluctuations of the field in each interval and X is the average field in the upstream and downstream region. The average value X is constant in the upstream and downstream region of the shock, and is determined from the Rankine Hugoniot values obtained from the ACE shock database (http://www.ssg.sr.unh.edu/mag/ace/ACElists/obs-list.html). The purpose of the $\langle \delta x^2 \rangle / X^2 << 1$ criterion is to exclude large quasi-periodic downstream (and upstream) fluctuations that may correspond to overshoots and undershoots associated with shock structure that are not accounted for in the R-H analysis. The turbulent quantities are calculated in each 5 minute interval in the case of the quasi-parallel shock of day 272 of year 2001, and in each 4 minute interval in the case of the quasi-perpendicular shock of day 250 of year 2000.

We also calculate the error corresponding to each of the observed values, and show them in the plots with an error bar. At first, we assume 1% uncertainty in the R, T, and N components of the solar wind velocity and the magnetic field. Then we calculate the error associated with each observed values using equations presented in Appendix A. We fit further the observed values by a least squares method [34]. Also, we fit the data using a Fourier series function,

$$y(t) = a_0 + \sum_{i=1}^{N} \Big(a_i \cos(i\omega t_i) + b_i \sin(i\omega t_i) \Big) \tag{14}$$

where the Fourier coefficients a_0, a_i and b_i are given by,

$$a_0 = \frac{\sum_{i=1}^{N} y_i}{N} \quad a_k = \frac{2}{N} \sum_{i=1}^{N} y_i \cos(k\omega t_i) \quad b_k = \frac{2}{N} \sum_{i=1}^{N} y_i \sin(k\omega t_i), \tag{15}$$

where t is time, $\omega = 2\pi/T$ is an angular frequency and T is a time period of the oscillation.

3.1. Comparison to the quasi-parallel shock of day 272 of 2001

Here we compare the numerical solution of equations (1)–(4) to the quasi-parallel shock of day 272 of 2001 from 9 hours upstream to 9 hours downstream of the shock. The coupled turbulence transport equations are solved for initial conditions: $f = 33.76$ km^{2}s^{-2}, $g = 116.39$ km^{2}s^{-2}, $E_D = -14.3$ km^{2}s^{-2}, and $\lambda = 2.1 \times 10^6$ km. The initial conditions for f, g, and σ_D are calculated by averaging the observed values in the upstream region of the shock, based on the mean fields derived from the Rankine-Hugoniot fitting procedure [35].

Figure 1 shows the comparison between the numerical solutions of Equations (1)–(4) and the observed values corresponding to the quasi-parallel shock of day 272 of 2001 as a function of X/X_{sh} or T/T_{sh}. The parameters X_{sh} and T_{sh} are the spatial and temporal positions of the shock. In Figure 1, the solid black curves represent the theoretical results, the blue "+" symbols the observed values, and the red "*" symbols have error bars associated with the observed values. Similarly, the solid straight red lines indicate a least square fitting to the observed values in the upstream and downstream region. The green sinosuidal curves are a fitting to the observed values using Fourier series function. To fit the observed values by a Fourier series function, we set $T = 40$ minute and i=1 in the Fourier series function (14) i.e., we consider only the lowest frequency of any potentially periodic behavior. In region immediately downstream of the shock, we exclude the very large amplitude fluctuations that are associated shock structure overshoots and undershoots.

Figure 1a illustrates the square root of the fluctuating magnetic energy $\langle b^2\rangle^{1/2}$ as a function of X/X_{sh} or T/T_{sh}. The theoretical $\langle b^2\rangle^{1/2}$ decreases slightly in the upstream region of the shock, increases at the shock, and then decreases gradually downstream of the shock. The least square and Fourier series fittings to the observed $\langle b^2\rangle^{1/2}$ values in the upstream and downstream region show that the theoretical and observed $\langle b^2\rangle^{1/2}$ are reasonably consistent with each other.

In Figure 1a, the theoretical $\langle b^2\rangle^{1/2}$ and the observed $\langle b^2\rangle^{1/2}$ both exhibit an enhancement at the shock i.e., turbulence is both amplified and generated at the shock in the following sense. From equations 1)–4), flow and Alfvén gradients are essentially zero and so the turbulence intensity $(f, g \equiv \langle \mathbf{z}^{\pm 2}\rangle)$ simply decays as spectral transfer leads to the dissipation of fluctuating energy. At the shock transition itself, the gradient terms are large and act as source terms in the abrupt deceleration region. This effectively leads to a compression of upstream turbulence as it transitions the shock larger. Notice too that the residual energy E_D changes in response to the flow, magnetic field, and density gradients leading to an amplification and redistribution of kinetic and magnetic fluctuating energy. This is discussed below.

Figures 1b–1f describe the fluctuating kinetic energy $\langle u^2\rangle$, the energy in backward propagating modes f, the energy in forward propagating modes g, the total turbulent energy E_T, and the normalized residual energy σ_D as a function of X/X_{sh} or T/T_{sh}, respectively. The theoretical results show that all the turbulent energies decrease slightly in the upstream region, are enhanced at the shock, and then gradually decrease downstream of the shock. The theoretical σ_D shows that the energy in the fluctuations is dominated by the magnetic energy in the upstream and downstream regions of the shock. However, the kinetic and magnetic energy become approximately equipartitioned at the shock. The least square and Fourier series function fitting indicate that the theoretical energies $\langle u^2\rangle$, f, g, E_T, and σ_D and the observed energies agree reasonably well to each other.

Figure 1g shows the normalized cross helicity σ_c as a function of X/X_{sh} or T/T_{sh}. The theoretical σ_c increases gradually in the upstream region, decreases at the shock, and again increases slowly in the downstream region. However, there appears to be a dip in the observed σ_c downstream of the shock followed by a recovery and slow increase. With the exception of the dip, the observed and theoretical values seem to be quite consistent. Figure 1h illustrates the normalized correlation length as a function of X/X_{sh}. It shows that the correlation length in the upstream and downstream regions increases, with a slight flattening at the shock.

3.2. Comparison to the quasi-perpendicular shock of day 250 of 2000

We compare the numerical solution of the turbulence transport equations (5)–(8) with the quasi-perpendicular shock of day 250 of 2000 for a period from 3 hours upstream to 3 hours downstream of the shock. The coupled equations (5)–(8) are solved using upstream initial conditions: $f = 76.77\ \mathrm{km^2s^{-2}}$, $g = 22.01\ \mathrm{km^2s^{-2}}$, $E_D = -16.82\ \mathrm{km^2s^{-2}}$, and $\lambda = 1.2\times 10^6$ km. Figure 2 illustrates the comparison between the theoretical results and observed values. In Figure 2, the solid black curves represent the theoretical results, the blue scattered "*" symbols with error bars identify the observed values and the associated error, the solid straight red lines and the green oscillatory curves correspond to a least square and Fourier series function fitting, respectively. To fit the observed values in the upstream and downstream region of the shock by the Fourier series function (14), we consider only the lowest frequency and set i=1 and T=14 minute.

Figure 2a shows the fluctuating magnetic energy $\langle b^2\rangle^{1/2}$ as a function of X/X_{sh} or T/T_{sh}. Evidently, both the theoretical and observed values in the downstream region are larger than in the upstream region. The least square and the Fourier series fitting to the observed values in the upstream and downstream region show that the theoretical $\langle b^2\rangle^{1/2}$ is consistent with the observed $\langle b^2\rangle^{1/2}$.

Figure 2b shows the fluctuating kinetic energy $\langle u^2\rangle$ as a function of X/X_{sh} or T/T_{sh}. The theoretical $\langle u^2\rangle$ exhibits similar trends to that observed in upstream and downstream of the shock. Figures 2c, 2d, and 2e illustrate the energy in backward propagating modes f, the energy in forward propagating modes g, and the total turbulent energy E_T as a function of X/X_{sh} or T/T_{sh}. We find that the turbulent energies f, g, and E_T are approximately constant in the upstream region, are enhanced at the shock, and decrease gradually in the downstream of the shock, consistent with the least square and Fourier series function fitting to the observed quantities.

Figure 2f plots the normalized residual energy σ_D as a function of X/X_{sh} or T/T_{sh}. The theoretical normalized residual energy is approximately constant in upstream and downstream of the shock, and is enhanced at the shock. Comparison of the theoretical σ_D and the observed σ_D shows reasonable consistency. The theoretical and observed σ_D are negative upstream and downstream of the shock, indicating that the magnetic energy dominates the fluctuations. Figure 2g is a plot of the normalized cross helicity σ_c and Figure 2h the normalized correlation length λ/L_C as a function of X/X_{sh} or T/T_{sh}. The theoretical σ_c is reasonably consistent with the observed σ_c. The normalized cross helicity σ_c is negative upstream and downstream of the shock, indicating that the intensity in backward modes is greater than in forward modes upstream and downstream of the shock. The normalized correlation length increases upstream and downstream of the shock, and flattens very slightly at the shock.

4. Conclusions

To study the interaction of turbulence with parallel and perpendicular shocks, we derived four coupled turbulence transport equations appropriate to parallel and perpendicular shocks, using the turbulence transport equations of Zank et al 2012. We solved the four coupled turbulence transport equations for parallel and perpendicular shocks, and compared the numerical solutions to the quasi-parallel shock of day 272 of year 2001, and the quasi-perpendicular shock of day 250 of year 2000. We assumed a 1% uncertainty in the R, T, and N components of the solar wind velocity and the magnetic field, and calculated uncertainties corresponding to various turbulent quantities using the results presented in Appendix A. We fitted the observed values in the upstream and downstream of the shock using a least square method and a low frequency Fourier series function fitting. The least square and Fourier series function fitting to the observed values, and the comparison of the theoretical results with the observed values indicates that the theoretical results show similar trends to the observed values. We can summarize the

findings of this work as,
1) all the turbulent energies are enhanced in transmission across parallel and perpendicular shocks, and decrease monotonically in the downstream region;
2) the correlation length increases upstream and downstream of a shock, with a slight flattening at the shock;
3) the cross helicity decreases at the shock in the parallel shock case, and increases at the shock in the perpendicular shock case, and
4) according to the theoretical model for the normalized residual energy, the energy in the fluctuations is dominated by the magnetic energy in the perpendicular shock case. For a parallel shock, the magnetic energy dominates the fluctuations in the upstream and downstream regions, whereas the two energies are approximately equipartitioned at the shock.

Acknowledgments
We acknowledge the partial support of the following grants: NASA NNX14AC08G, NASA/SWRI Subcontract A991132BT, NASA Subaward 37102-2, NNX14AJ53J GCR, NNX15AI65G.

Appendix A. Error associated with various turbulent quantities

Here we derive uncertainty relations for the energy in forward and backward propagating modes, the total turbulent energy, the residual energy, the normalized residual energy, the normalized cross helicity, the fluctuating kinetic and magnetic energy. To derive the uncertainty associated with the observations, we start from an error of propagation equation

$$\sigma_\xi^2 = \sum_{i=1}^{N} \left(\frac{\partial \xi}{\partial x_i} \right)^2 \sigma_i^2, \tag{A.1}$$

where σ_ξ is the error associated with the ξ variable and σ_i is the error associated with x_i. Following Equation A.1, if a function $\xi = x \pm y$ and σ_x and σ_y are the errors associated with x and y, then the error associated with ξ i.e., σ_ξ, is given by,

$$\sigma_\xi^2 = \sigma_x^2 + \sigma_y^2. \tag{A.2}$$

Similarly, if $\xi = x/y$, then

$$\sigma_\xi^2 = \xi^2 \left(\frac{\sigma_x^2}{x^2} + \frac{\sigma_y^2}{y^2} \right). \tag{A.3}$$

If $\xi = \sum_{i=1}^{N} x_i/N$ and the σ_i are the errors associated with x_i, then the error associated with ξ is given by

$$\sigma_\xi^2 = \frac{1}{N^2} \sum_{i=1}^{N} \sigma_i^2. \tag{A.4}$$

If $V = 1/(N-1)\sum_{i=1}^{N}(x_i - \bar{x})^2$ is a variance and σ_i are the errors associated with x_i, the error associated with the variance σ_V, following Equation (A.1), can be written in the form

$$\sigma_{\rm V}^2 = \left(\frac{2}{N-1}\right)^2 \sum_{i=1}^{N}(x_i - \bar{x})^2\sigma_i^2. \tag{A.5}$$

Using equations (A.1)–(A.5), we can derive the error for various turbulent quantities. For the residual energy $E_D = \langle u^2\rangle - \langle b^2/\mu_0\rho\rangle$, the error σ_{E_D} can be written in the form

$$\sigma_{E_D}^2 = \sigma_{\langle u^2\rangle}^2 + \frac{\sigma_{\langle b^2\rangle}^2}{\mu_0^2\langle\rho\rangle^2} + \frac{\langle b^2\rangle^2\sigma_{\langle\rho\rangle}^2}{\mu_0^2\langle\rho\rangle^4}, \tag{A.6}$$

where $\sigma_{\langle u^2\rangle}$, $\sigma_{\langle b^2\rangle}$, and $\sigma_{\langle\rho\rangle}$ are errors associated with the fluctuating kinetic energy $\langle u^2\rangle$, the fluctuating magnetic energy $\langle b^2\rangle$, and the mean density $\langle\rho\rangle$, respectively. Similarly, for the fluctuating solar wind velocity $\mathbf{u} = \mathbf{U} - \langle\mathbf{U}\rangle$ and the fluctuating magnetic field $\mathbf{b} = \mathbf{B} - \langle\mathbf{B}\rangle$, where $\mathbf{U}$ and $\mathbf{B}$ are the fields, and $\langle\mathbf{U}\rangle$ and $\langle\mathbf{B}\rangle$ are the mean fields, the errors $\sigma_{\mathbf{u}}$ for $\mathbf{u}$ and $\sigma_{\mathbf{b}}$ for $\mathbf{b}$ are given by

$$\sigma_{\mathbf{u}}^2 = \sigma_{\mathbf{U}}^2 + \sigma_{\langle\mathbf{U}\rangle}^2; \qquad \sigma_{\mathbf{b}}^2 = \sigma_{\mathbf{B}}^2 + \sigma_{\langle\mathbf{B}\rangle}^2, \tag{A.7}$$

where $\sigma_{\mathbf{U}}$ and $\sigma_{\mathbf{B}}$ are errors associated with background fields, and $\sigma_{\langle\mathbf{U}\rangle}$ and $\sigma_{\langle\mathbf{B}\rangle}$ are the errors associated with background mean fields.

The errors $\sigma_{\mathbf{z}^\pm}$ for the Elsässer variables $\mathbf{z}^\pm = \mathbf{u}\pm\mathbf{b}/\sqrt{\mu_0\rho}$ can be written in the form

$$\sigma_{\mathbf{z}_+}^2 = \sigma_{\mathbf{z}_-}^2 = \sigma_{\mathbf{u}}^2 + \frac{\sigma_{\mathbf{b}}^2}{\mu_0\rho} + \frac{b^2\sigma_\rho^2}{4\mu_0\rho^3}, \tag{A.8}$$

where σ_ρ is error associated with the density ρ. Using Equation (A.5), the errors associated with the variance of the Elsässer variables $\mathbf{z}^\pm$ can be written in the form

$$\sigma_{\langle z_+^2\rangle}^2 = \left(\frac{2}{N-1}\right)^2 \sum_{i=1}^{N}(z_i^+ - \bar{z}_+)^2\sigma_i^2; \qquad \sigma_{\langle z_-^2\rangle}^2 = \left(\frac{2}{N-1}\right)^2 \sum_{i=1}^{N}(z_i^- - \bar{z}_-)^2\sigma_i^2, \tag{A.9}$$

where σ_i is the errors associated with $\mathbf{z}_i^\pm$. The errors corresponding to the total energy E_T and the cross helicity E_C are the same, given by

$$\sigma_{E_T}^2 = \sigma_{E_C}^2 = \frac{1}{4}\left(\sigma_{\langle z_+^2\rangle}^2 + \sigma_{\langle z_-^2\rangle}^2\right). \tag{A.10}$$

Using Equation (A.3), the errors corresponding to the normalized cross helicity $\sigma_c = E_C/E_T$ and the normalized residual energy $\sigma_D = E_D/E_T$ can be written in the form

$$\sigma_{\sigma_c}^2 = \sigma_c^2\left(\frac{\sigma_{E_C}^2}{E_C^2} + \frac{\sigma_{E_T}^2}{E_T^2}\right); \quad \sigma_{\sigma_D}^2 = \sigma_D^2\left(\frac{\sigma_{E_D}^2}{E_D^2} + \frac{\sigma_{E_T}^2}{E_T^2}\right). \tag{A.11}$$

References

[1] Burlaga, L F, 1971 *Space Science Reviews* 12 600
[2] Richter A K, Hsieh K C, Luttrell A H, Marsch E, & Schwenn R, 1985 Washington DC American Geophysical Union Geophysical Monograph Series 35 33
[3] Gosling J T, 1983 *Space Science Reviews* 34 113
[4] Giacalone J, Jokipii J R, & Kota J, 1994 *J. Geophys. Res.* 99 19
[5] Zank G P, Rice W K M, & Wu C C, 2000 *J. Geophys. Res.* 105 25079
[6] Desai M I, Mason G M, Wiedenbeck M E, 2004 ApJ 611 1156
[7] Zank G P, Li G, & Verkhoglyadova O, 2007 *Space Science Reviews* 130 255
[8] Lee M A, Mewaldt R A, & Giacalone J, 2012 *Space Science Reviews* 173 247
[9] Reames D V, 2013 *Space Science Reviews* 175 53
[10] Verkhoglyadova O P, Zank G P, & Li G, 2015 *Physics Reports* 557 1
[11] Axford W I, Leer E, & Skadron G, 1977 International Cosmic Ray Conference 11 132
[12] Reames D V, 1999 *Space Science Reviews* 90 413
[13] Gonzalez W D, Tsurutani B T, & Clúa de Gonzalez A L, 1999 *Space Science Reviews* 88 529
[14] Carter B A, Yizengaw E, Pradipta R, 2015 *J. Geophys. Res.* 42 6554
[15] Jian L K, Russell C T, & Luhmann J G, 2011 *Solar Physics* 274 321
[16] Jian L K, Russell C T, Luhmann J G, Galvin A B, & Simunac K D C, 2013 American Institute of Physics Conference Series 1539 191
[17] Zank G P, Li G, Florinski V, Q Hu, D Lario, and C W Smith, 2006 *Journal of Geophysical Research (Space Physics)* 111 A06108
[18] Li G, Zank G P, & Rice W K M, 2003 *Journal of Geophysical Research (Space Physics)* 108 1082
[19] Rice W K M, Zank G P, & Li G, 2003 *Journal of Geophysical Research (Space Physics)* 108 1369
[20] Blanco-Cano X, Kajdič P, Aguilar-Rodríguez E, 2016 Journal of Geophysical Research (Space Physics) 121 992
[21] Kilpua E K J, Lumme E, Andreeova K, Isavnin A, & Koskinen H E J, 2015 Journal of Geophysical Research (Space Physics) 120 4112
[22] Jurac S, Kasper J C, Richardson J D, & Lazarus A J, 2002 *J. Geophys. Res.* 29 1463
[23] Oliveira D M, & Raeder J, 2015 Journal of Geophysical Research (Space Physics) 120 4313
[24] Oliveira D M, & Raeder J, 2014 Journal of Geophysical Research (Space Physics) 119 8188
[25] Hundhausen A J, 1972 Physics and Chemistry in Space 5
[26] Zank G P, Dosch A, Hunana P, 2012 ApJ 745 35
[27] Adhikari L, Zank G P, Bruno R, 2015 ApJ 805 63
[28] Adhikari L, Zank G P, Bruno R, 2015, Journal of Physics Conference Series 642 012001
[29] Elsässer W M, 1950 The hydromagnetic equations *Phys Rev* 79 183 183
[30] Zank G P, Matthaeus W H, & Smith C W, 1996 *J. Geophys. Res.* 101 17093
[31] Webb G M & McKenzie J F, 1984 Journal of Plasma Physics 31 337
[32] Vinas A F & Scudder J D, 1986 *J. Geophys. Res.* 91 39
[33] Adhikari L, Zank G P, Hu Q, & Dosch A, 2014 ApJ 793 52
[34] Press W H, Flannery B P, & Teukolsky S A, 1986 Cambridge: University Press 1986
[35] Hsieh K C, & Richter A K, 1986 *J. Geophys. Res.* 91 4157
[36] Dosch A, Adhikari L, & Zank G P, 2013 *American Institute of Physics Conference Series* 1539 155

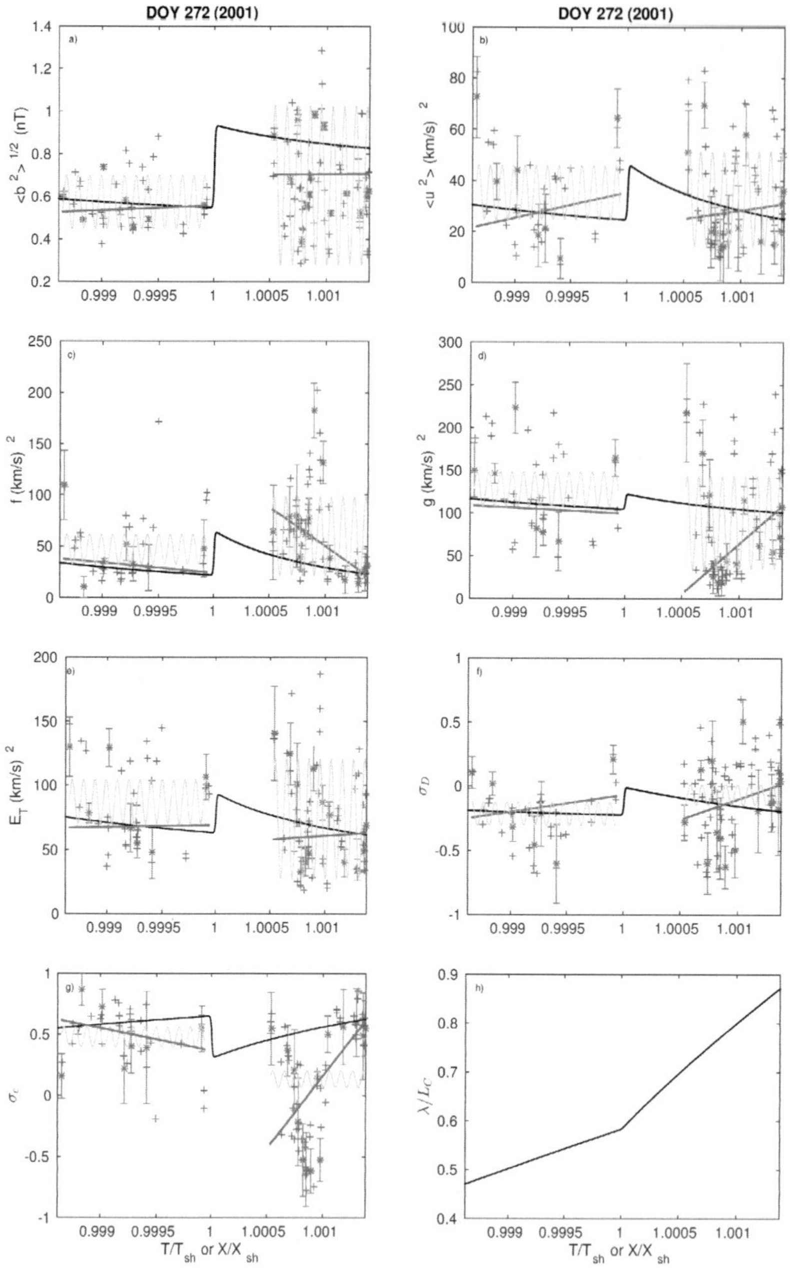

Figure 1. Comparison of solutions of turbulence transport equations (1)–(4) with the quasi-parallel shock of day 272 of 2001 as a function of T/T_{sh} or X/X_{sh}. Figure 1a shows the square-root of the fluctuating magnetic energy $\langle b^2\rangle^{1/2}$. Figure 1b illustrates the fluctuating kinetic energy $\langle u^2\rangle$. Figures 1c and 1d the energy in backward f and forward g propagating modes. Figures 1e and 1f the total turbulent energy E_T and the normalized residual energy σ_D. Figures 1g and 1h the normalized cross helicity σ_c and the correlation length λ. The parameter $L_C = 0.03$ AU is the correlation length at 1 AU [36].

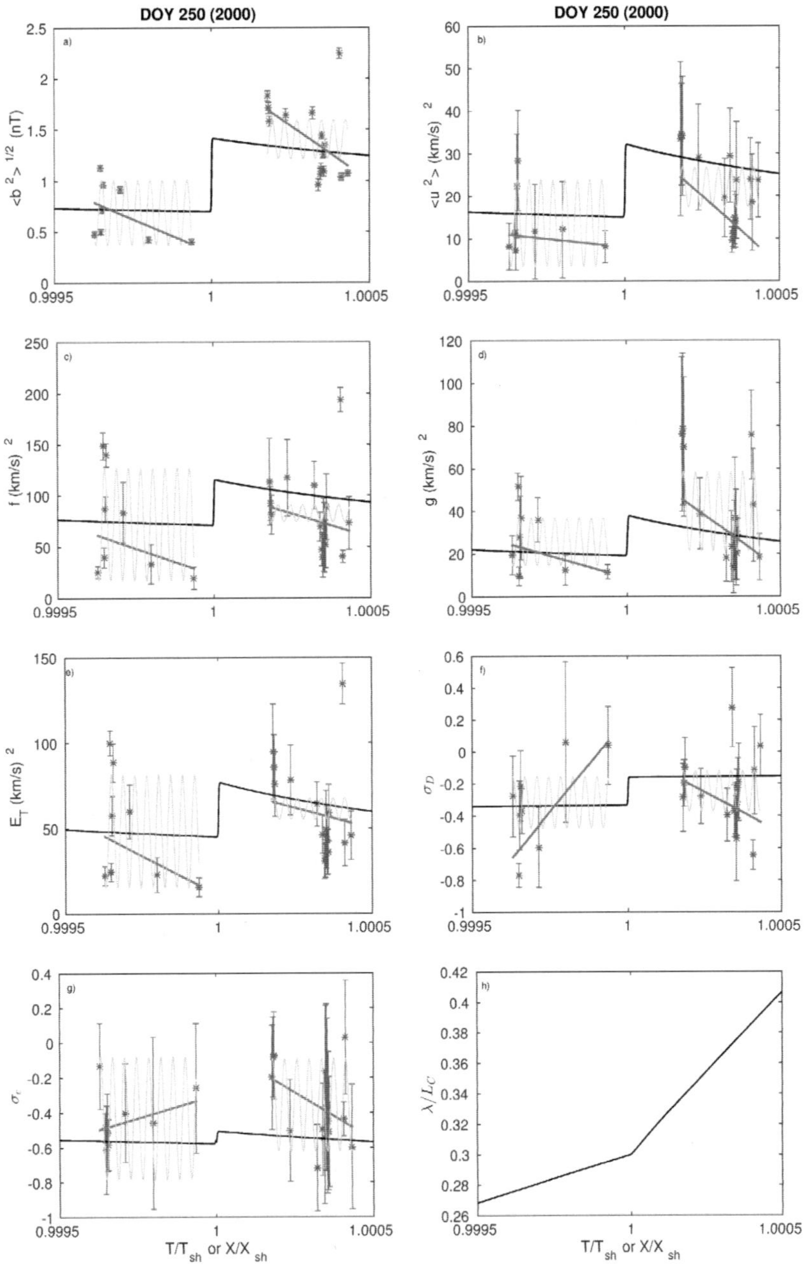

Figure 2. Comparison of solutions of the turbulence transport equations (5)–(8) with the quasi-perpendicular shock of day 250 of 2000 as a function of T/T_{sh} or X/X_{sh}. Figure 2a plots the square-root of the fluctuating magnetic energy $\langle b^2\rangle^{1/2}$; Figure 2b the fluctuating kinetic energy $\langle u^2\rangle$; Figures 2c and 2d the energy in backward f and forward g propagating modes; Figures 2e and 2f the total turbulent energy E_T and the normalized residual energy σ_D, and Figures 2g and 2h the normalized cross helicity σ_c and the correlation length λ. The parameter $L_C = 0.03$ AU is the correlation length at 1 AU.

Journal of Physics: Conference Series **767** (2016) 012002 doi:10.1088/1742-6596/767/1/012002

Energy spectra of ^{3}He-rich solar energetic particles associated with coronal waves

R Bučík[1,2], D E Innes[2,3], G M Mason[4] and M E Wiedenbeck[5]

[1] Institut für Astrophysik, Georg-August-Universität Göttingen, D-37077, Göttingen, Germany
[2] Max-Planck-Institut für Sonnensystemforschung, D-37077, Göttingen, Germany
[3] Max Planck/Princeton Center for Plasma Physics, Princeton, NJ 08540, USA
[4] Applied Physics Laboratory, Johns Hopkins University, Laurel, MD 20723, USA
[5] Jet Propulsion Laboratory, California Institute of Technology, Pasadena, CA 91109, USA

E-mail: bucik@mps.mpg.de

Abstract. In addition to their anomalous abundances, ^{3}He-rich solar energetic particles (SEPs) show puzzling energy spectral shapes varying from rounded forms to power laws where the later are characteristics of shock acceleration. Solar sources of these particles have been often associated with jets and narrow CMEs, which are the signatures of magnetic reconnection involving open field. Recent reports on new associations with large-scale EUV waves bring new insights on acceleration and transport of ^{3}He-rich SEPs in the corona. We examined energy spectra for 32 ^{3}He-rich SEP events observed by ACE at L1 near solar minimum in 2007-2010 and compared the spectral shapes with solar flare signatures obtained from STEREO EUV images. We found the events with jets or brightenings tend to be associated with rounded spectra and the events with coronal waves with power laws. This suggests that coronal waves may be related to the unknown second stage mechanism commonly used to interpret spectral forms of ^{3}He-rich SEPs.

1. Introduction

Small ^{3}He-rich solar energetic particle (SEP) events remain ~40 years after their discovery one of the less explored phenomenon in solar physics. The anomalous abundances of ^{3}He-rich SEPs have been explained by various mechanisms including gyro-resonant interactions with plasma wave turbulence [1, 2] or interactions with multiple magnetic islands [3].

The energy spectra of major ion species in some ^{3}He-rich SEP events are power laws or broken power laws but in other events ^{3}He and Fe exhibit rounded spectra toward the low energies with ^{3}He rollovers at ~100-600 keV nucleon^{-1} and Fe rollovers below ~100 keV nucleon^{-1} [4, 5]. It has been suggested that rounded spectra arise from the basic mechanism of ^{3}He enrichment and power laws involve a further stage of acceleration. It is well known that power law spectra are produced by shock acceleration models (see e.g. [6], for a review) but shock detection is not typically reported in ^{3}He-rich SEP events (e.g., via type II radio bursts). Notice that the spectral shape can deviate from power law when shock propagates into an inhomogeneous solar wind [7].

^{3}He-rich SEP sources have been associated with EUV jets and jet-like CMEs [8, 9, 10]. These are signatures of magnetic reconnection involving open field [11] which may create turbulence

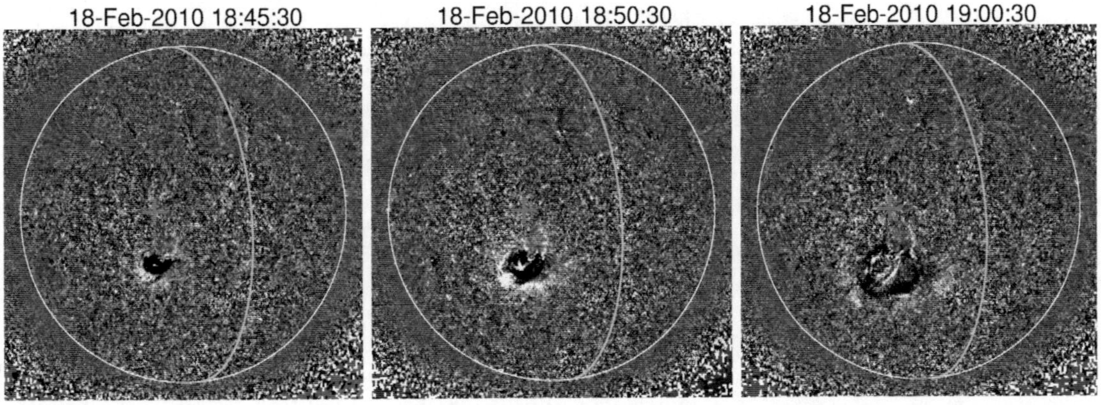

Figure 1. STEREO-A 195 Å EUV 2.5 min running differences images of the solar disk on 2010 February 18. Yellow arcs indicate the west limb from the Earth view and red pluses the L1 magnetic foot-point. At this time STEREO-A was separated from the Sun-Earth line by 65°.

and magnetic islands for a particle acceleration. With new observing capabilities on STEREO and SDO it has been recently shown that some ^{3}He-rich SEP events were associated with large-scale coronal EUV waves [12, 13]. These waves were found without CMEs or only with weak coronal outflows. A common view on EUV waves is that they present or contain magnetosonic waves which can even steepen to the shocks (see [14], for a review).

The aim of this paper is to examine relation between energy spectral characteristics of ^{3}He-rich SEPs and their flare signatures in the corona.

2. Observations and discussion

We examine energy spectra in 32 ^{3}He-rich SEP events observed at L1 during solar minimum conditions in 2007-2010. The events have been examined in detail in our recent survey [15]. Four events contain a high energy (>25 MeV) but relatively low intensity solar proton component [16], which might not be typical ^{3}He events but rather large gradual events or a mix of both. Three of these events show still very high ^{3}He enrichment and all have high Fe/O ratio, typical of ^{3}He-rich SEPs. Note that such cases if not identified may present one source of variability in ^{3}He-rich SEP events. Table 1 lists all the investigated events. Column 1 gives the event number, column 2 the event start day, column 3 the ^{3}He energy spectral form at ~0.1-2 MeV nucleon^{-1}, column 4 indicates whether the ion event had velocity dispersion, column 5 the EUV flare shape and column 6 the CME speed reported in SOHO/LASCO catalog. The EUV flares were examined using STEREO/EUVI [20]; the energetic ions were detected by ACE/ULEIS [21] and/or ACE/SIS [22]. Figure 1 shows an example of EUV wave associated with ^{3}He-rich SEP event on 2010 February 19. Shown are running differences images at three times to illustrate the wave expansion. Figure 2 presents the event integrated ^{4}He, ^{3}He, O, Fe fluence energy spectra for 30 events. The spectra for two remaining events (2007-May-23 and 2008-Feb-4) have been shown in [18]. The energetic ion fluences for events in 2009 were very weak. Specifically the ULEIS did not observe statistically significant ^{3}He fluences in 2009 November 3 event though at higher energies the SIS observed a ^{3}He-rich SEP event [19].

We are able to assign the ^{3}He spectral shape for 22 from 32 investigated ^{3}He-rich SEP events (see Table 1). In the remaining 10 events ^{3}He spectral points show complex variations (5 events) or we have insufficient number of spectral points (2 points as a maximum). The power law spectrum has been found in 12 events, where 3 events are with jet/brightening and 9 with a

Journal of Physics: Conference Series **767** (2016) 012002 doi:10.1088/1742-6596/767/1/012002

Table 1. ^{3}He-rich SEP events.

#	Start day	^{3}He spectrum[a]	Dispersion	EUV flare[b]	CME[c]
1	2007-Jan-14	PL	no	B	...
2	2007-Jan-24	R	yes	B	...
3	2007-May-23	PL	no	W	544
4	2008-Feb-4	?	no	B	...
5	2008-Jun-16	?(PL)	?	J	...
6	2008-Nov-4	?(PL)	yes	W	732
7	2009-Apr-29	...	?	W	239
8	2009-May-1	...(R)	?	W	...
9	2009-Jul-5	...	?	W	yes
10	2009-Oct-6	PL	?	B	...
11	2009-Nov-3	...	?	W	yes
12	2009-Dec-22	PL	no	W	318
13	2010-Jan-16	PL	yes	J	...
14	2010-Jan-27	?(PL)	yes	W	...
15	2010-Jan-31	...(PL)	?	W	219
16	2010-Feb-8	PL	no	W	yes
17	2010-Feb-12	PL	no	W	509
18	2010-Feb-19	R	yes	W/B	223
19	2010-Mar-4	PL	yes	W	374
20	2010-Mar-19	PL	yes	W	...
21	2010-Jun-12	PL	no	W	486
22	2010-Sep-1	R	no	W	1304
23	2010-Sep-2	R	no	J	...
24	2010-Sep-4	R	yes	B	...
25	2010-Sep-17	?	no	J	yes
26	2010-Oct-17	PL	yes	W	304
27	2010-Oct-19	R	yes	B	385
28	2010-Nov-2	R	yes	W	253
29	2010-Nov-14	R	no	B	442
30	2010-Nov-17	R	yes	J	639
31	2010-Nov-29	R	?	W	505
32	2010-Dec-10	PL	?	W	299

[a] at 0.1-2 MeV nucleon^{-1}; PL-power law, R-rounded, ?-unclear; the spectral form in parentheses is for Fe and extends down to 30 keV nucleon^{-1}; spectra in events #2-6, 14, 26-30 reported in [17, 18, 13, 12]

[b] B-brighthening, J-jet, W-wave

[c] CME speed in km s^{-1} from LASCO catalog (cdaw.gsfc.nasa.gov/CME_list); yes-CME observed with STEREO/COR-1

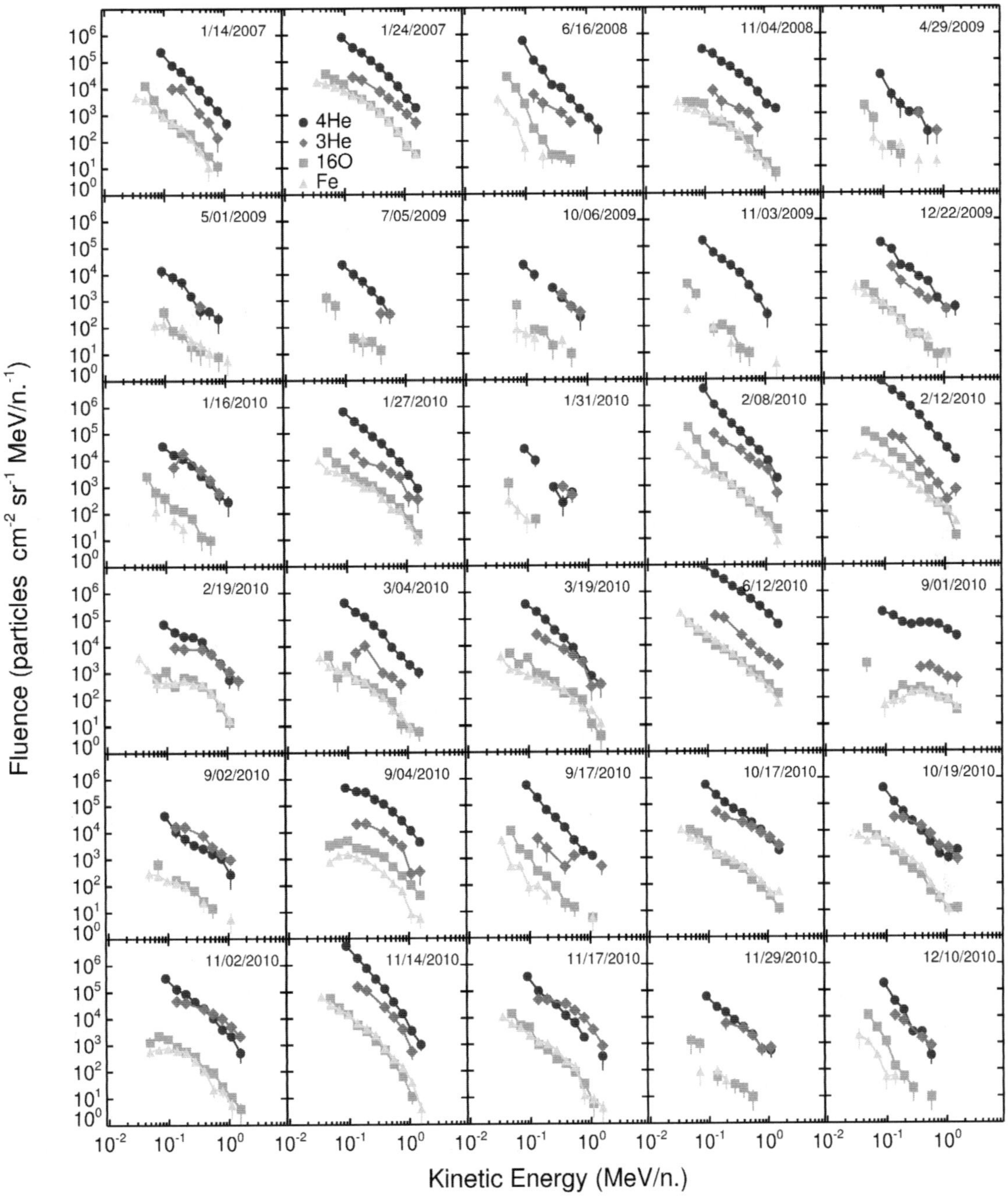

Figure 2. Energy spectra of ^{4}He, ^{3}He, O, Fe at ~0.03-2 MeV nucleon^{-1} observed with ACE/ULEIS for 30 ^{3}He-rich SEP events in 2007-2010.

Journal of Physics: Conference Series **767** (2016) 012002 doi:10.1088/1742-6596/767/1/012002

coronal wave. The rounded spectral shape is found in 10 events, where 6 are with jet/brightening and 4 with a wave. Since in the two class scheme the Fe spectrum behaves as ^{3}He, we examined the Fe spectrum in the cases where ^{3}He is unclear. This adds 5 more events. We found 16 events with ^{3}He or Fe having power laws where 4 events were with jet/brightening and 12 with a coronal wave. This confirms the previous finding depending solely on ^{3}He spectrum. Including Fe we now have 11 events with rounded spectra where 6 are with jet/brightening and 5 with a wave, again consistent with the previous result. This shows that events with jet/brightening have a slight tendency to have rounded ^{3}He spectra (6 of 9) and events with a wave to have power law ^{3}He spectra (9 of 13). This is consistent with a recent study reporting similar tendency for 13 ^{3}He-rich SEP events with jets and 4 events with EUV waves [12].

Excluding 4 events with high energy solar proton components (#12, 17, 21, 22), which are associated with EUV waves and all but one have power law ^{3}He spectra, the results will not change dramatically. The ^{3}He-rich SEP events with EUV waves have still a tendency to have power law ^{3}He spectra (6 of 9) or power law ^{3}He or Fe spectra (9 of 13). Exclusion of all events with a high energy proton component may be too conservative as for example one such event, 2009-Dec-22, have very high ^{3}He and Fe enrichments (386 keV nucleon^{-1} ^{3}He/^{4}He$\sim$0.23, Fe/O$\sim$1.16). The 2 of 3 above events with the proton component which do not have an occulted source active region (AR) have reported type II radio bursts (ftp.ngdc.noaa.gov). The CMEs which accompany these solar proton events were generally not too fast (318, 486, 509, 544, and 1304 km s^{-1} as reported by LASCO catalog). The slowest one in 2009-Dec-22 event was also associated with a type II radio burst. We get similar results if we consider coronal wave events only with slow CMEs ($<$350 km s^{-1}) or with no CME: ^{3}He power law spectrum in 5 from 7 events and ^{3}He (or Fe) power law in 7 from 10 events.

In our earlier study [13] we report rounded ^{3}He, Fe spectra in the event with a weak EUV wave, probably without a significant influence on energetic ions. Therefore we look at the wave associated events with ^{3}He rounded spectra (2010-Feb-19, 2010-Sep-1, 2010-Nov-2, 2010-Nov-29). The coronal wave in February 19 event was ejected about 30 min earlier, not at the time of the type III radio burst. Note this is the only case in this survey when the wave was not co-temporal with the flare and/or type III burst. Though overall the wave looks quite significant in the February 19 event (see Figure 1), maybe it was ejected too early to have an effect on the later particle injection. The wave in the September 1 event had a large spatial scale but faded before reaching the L1 foot-point. The February 19 and September 1 events have all ^{4}He, ^{3}He, O, Fe spectra rounded, the feature has not been previously reported (see later). The November 2 and November 29 events are associated with quite narrow waves with less bright fronts. Furthermore the November 2 wave has the narrowest front in this study and the source AR in the November 29 event was extraordinary separated from the nominal magnetic connection. Thus the influence of the waves on energetic ion spectra in these two events would be minimal.

We have also 3 events with power law ^{3}He spectra (2007-Jan-14, 2009-Oct-6, 2010-Jan-16) associated with a brightening or a jet. The October 6 event spectrum could be still rounded as we do not have spectral points below 400 keV nucleon^{-1}. Other two events have one, the lowest energy (100 keV nucleon^{-1}) spectral point, deviated from the power law dependence but this can be an instrumental issue. It has been discussed that propagation through interplanetary space is probably not a cause of a common spectral shapes observed in ^{3}He-rich SEP events [5]. Indeed, the 2007 January 14 event is a short duration event with a non-dispersive onset while the 2010 January 16 event is a dispersive one, though both have similar, power law, spectral shapes. The 2009 October 6 event is too weak to observe a velocity dispersion. In this study about half of the events with rounded ^{3}He spectra have a dispersive onset (6 of 10; 1 event with unclear dispersion) indicating nearly scatter-free propagation and half of the events with power law ^{3}He spectra have non-dispersive onsets (6 of 12; 2 events are unclear).

We found several ^{3}He-rich SEP events which differ from the two class scheme introduced in

[4]. For example the following two events have only the ^{3}He spectrum rounded, while the Fe is power law: the 2010 November 17 event and maybe also the 2010 November 14 event. Neither of these is associated with a coronal wave and both have the same solar source. Since we could not see this feature in other events (and no other reports are known) this may be due to specific conditions in a particular solar source region. However note on some other events where Fe shows a clear power law shape, ^{3}He looks more complex (e.g., 2008-Nov-4, 2010-Jan-27) which might be of the previous type. Furthermore we found other events where all ^{4}He, ^{3}He, O, Fe spectra are rounded. Typical examples are the events of 2010 September 1 (already mentioned) and 2010 September 4. The September 1 event is particularly unusual, with the ^{4}He spectrum continuing to rise at the lowest energies after the turnover. These two events are associated with different ARs: September 1 with a coronal wave and a high speed CME (with >25 MeV proton event) and September 4 with a brightening. There is possibly another ^{3}He-rich SEP event with all spectra rounded, the above mentioned 2010 February 19 event. Here after a turnover the Fe spectrum continues to rise at the lowest energies. This event was associated with a brightening but the wave ejected earlier was still present during the type III burst. Rounded ^{4}He spectra have been predicted by a MHD turbulence model at lower (<20 keV nucleon^{-1}) energies [23] not covered by available instruments.

3. Summary

We examined energy spectra for 32 ^{3}He-rich SEP events and identified ^{3}He spectral shapes for 22 events. We found a slight tendency for events with jets/brightenings in their solar source to have a rounded ^{3}He spectrum and for events with coronal waves in their source to have a power law ^{3}He spectrum. Though more events need to be examined to confirm this tendency the second stage mechanism behind the ^{3}He-rich SEP acceleration may be related to coronal EUV waves even when CMEs are not seen or are represented only by slow white-light outflows.

Half of the events with rounded ^{3}He spectra (5 of 10) show peculiar behaviour not previously reported. We noticed events where clearly only the ^{3}He spectrum is rounded and also events where all major species show rounded spectra. This may present further constraints on modelling of ion acceleration/propagation in ^{3}He-rich SEP events.

Acknowledgments

The work of RB is supported by the Deutsche Forschungsgemeinschaft (DFG) under grant BU 3115/2-1. RB thanks Gary Zank for the invitation to the 15th AIAC that led to this paper.

References

[1] Fisk L A 1978 *Astrophys. J.* **224** 1048
[2] Miller J A 1998 *Space. Sci. Rev.* **86** 79
[3] Drake J F and Swisdak M 2012 *Space. Sci. Rev.* **172** 227
[4] Mason G M, Dwyer J R and Mazur J E 2000 *Astrophys. J. Lett.* **545** L157
[5] Mason G M *et al* 2002 *Astrophys. J.* **574** 1039
[6] Verkhoglyadova O P, Zank G P and Li G 2015 *Phys. Rep.* **557** 1
[7] Zank G P, Rice W K M and Wu C C 2000 *J. Geophys. Res.* **105** 25079
[8] Kahler S W, Reames D V and Sheeley Jr N R 2001 *Astrophys. J.* **562** 558
[9] Wang Y-M, Pick M and Mason G M 2006 *Astrophys. J.* **639** 495
[10] Nitta N V, Reames D V, DeRosa M L, Liu Y, Yashiro S and Gopalswamy N 2006 *Astrophys. J.* **650** 438
[11] Shibata K *et al* 1992 *Publ. Astron. Soc. Japan* **44** L173
[12] Nitta N V, Mason G M, Wang L, Cohen C M S and Wiedenbeck M E 2015 *Astrophys. J.* **806** 235
[13] Bučík R, Innes D E, Guo L, Mason G M and Wiedenbeck M E 2015 *Astrophys. J.* **812** 53
[14] Warmuth A 2015 *Living Rev. Solar Phys.* **12** 3
[15] Bučík R, Innes D E, Mason G M and Wiedenbeck M E 2016 *Astrophys. J.* submitted
[16] Richardson I G, von Rosenvinge T T, Cane H V, Christian E R, Cohen C M S, Labrador A W, Leske R A, Mewaldt R A, Wiedenbeck M E and Stone E C 2014 *Solar Phys.* **289** 3059

[17] Wiedenbeck M E *et al* 2010 *AIP Conf. Proc.* **1216** 621
[18] Mason G M, Nitta N V, Cohen C M S and Wiedenbeck M E 2009 *Astrophys. J. Lett.* **700** L56
[19] Wiedenbeck M E, Mason G M, Cohen C M S, Nitta N V, Gómez-Herero R and Haggerty D K 2013 *Astrophys. J.* **762** 54
[20] Howard R A *et al* 2008 *Space. Sci. Rev.* **136** 67
[21] Mason G M *et al* 1998 *Space. Sci. Rev.* **86** 409
[22] Stone E C *et al* 1998 *Space. Sci. Rev.* **86** 357
[23] Liu S, Petrosian V and Mason G M 2006 *Astrophys. J.* **636** 462

On the scaling features of magnetic field fluctuations at non-MHD scales in turbulent space plasmas

G Consolini[1], F Giannattasio[1], E Yordanova[2], Z. Vörös[3], M F Marcucci[1], M Echim[4,5] and T Chang[6]

[1] INAF-Istituto di Astrofisica e Planetologia Spaziali, Roma, Italy
[2] Swedish Institute for Space Physics, Uppsala, Sweden
[3] Space Research Institute, Austrian Academy of Sciences, Graz, Austria
[4] Belgian Institute for Aeronomy, Brussels, Belgium
[5] Institute for Space Sciences, Magurele, Romania
[6] Kavli Institute for Astrophysics and Space Research, MIT, Cambridge, MA, USA

E-mail: giuseppe.consolini@iaps.inaf.it

Abstract. In several different contexts space plasmas display intermittent turbulence at magneto-hydro-dynamic (MHD) scales, which manifests in anomalous scaling features of the structure functions of the magnetic field increments. Moving to smaller scales, i.e. below the ion-cyclotron and/or ion inertial length, these scaling features are still observed, even though its is not clear if these scaling features are still anomalous or not. Here, we investigate the nature of scaling properties of magnetic field increments at non-MHD scales for a period of fast solar wind to investigate the occurrence or not of multifractal features and collapsing of probability distribution functions (PDFs) using the novel Rank-Ordered Multifractal Analysis (ROMA) method, which is more sensitive than the traditional structure function approach. We find a strong evidence for the occurrence of a near mono-scaling behavior, which suggests that the observed turbulent regime at non-MHD scales mainly displays a mono-fractal nature of magnetic field increments. The results are discussed in terms of a non-compact fractal structure of the dissipation field.

1. Introduction

Turbulence is very common in the framework of space plasmas [1]. For instance, solar wind clearly displays the occurrence of turbulent fluctuations in several quantities, such as magnetic field and plasma parameters (velocity, plasma density, etc.). The turbulent nature of magnetized plasmas mainly manifests in power law spectral densities with power-law exponents near the one predicted by the Kolmogorov's (K41) and/or Iroshnikov-Kraichnan (IK) theory of turbulence. Furthermore, the observed turbulence at MHD scales is accompanied by anomalous scaling features, which are the counterpart for the occurrence of *intermittency*.

Intermittency, i.e. a non-homogeneous cascading process, is, indeed, a very well documented property of magnetic field fluctuations (increments) in turbulent space plasmas at MHD scales, and manifests in an anomalous scaling of the structure functions $S_p(\delta r)$ of magnetic field increments,

$$S_p(\delta r) = S_q(\delta r)^{\frac{p}{q}+\beta(p)} \tag{1}$$

where δr is a scale separation and $\beta(p)$ is a correction to scaling, which depends on the moment order p. This anomalous scaling is a manifestation of the multifractal nature of the turbulent cascading mechanism, which manifests in a dependence of the shape of the PDFs of magnetic field increments on the scale δr.

While the occurrence of anomalous scaling features at MHD scales in turbulent plasmas is a well documented phenomenon, the situation is less clear when one moves to non-MHD scales, i.e., typically at scales below the ion-gyroradius ρ_L and/or the ion inertial length η_i (typically, proton in solar wind). Indeed, at these scales there are controversial results indicating the occurrence of both anomalous and global scaling features, so that it is still not clear if the nature of fluctuations/increments at non-MHD scales is mono-fractal or multifractal [2, 3, 4, 5].

The multi- and/or mono-fractal character of fluctuations in the non-MHD/kinetic domain is very relevant to model the nature of the fluctuations at these scales. Indeed, because stochasticity plays a relevant role in the cascading mechanism (Landau's remark to K41 theory of turbulence), scale invariance should be not global but, only, local [6]. As a consequence, intermittency and inhomogeneity of the dissipation field is expected to grow indefinitely going to smaller and smaller scales [7]. Conversely, a mono-fractal character of magnetic field fluctuations/increments at non-MHD scales could be the consequence of a simple non-compact fractal structure of the dissipation field at these scales, as, for instance, it has been suggested in the case of fluid turbulence [8]. This would manifest in a saturation of the kurtosis/flattness of the magnetic field increments at these non-MHD scales.

Here, we show preliminary results on the scaling features of magnetic field increments at non-MHD scales for a period of fast solar wind conditions as observed by ESA-Cluster mission. The analysis is done using the novel method of Rank-Ordering-Multifractal-Analysis (ROMA) introduced by Chang and Wu [9].

2. Data Description and Methods

To investigate the scaling features of magnetic field increments at non-MHD scales, we have considered a period of fast solar wind observed by Cluster satellites and characterized by quasi-stationary conditions. In detail, the observations refers to January, 20th, 2007 from 12:00 UT to 14:00 UT when the Cluster satellites encountered a stream of fast solar wind characterized by a mean velocity $\bar{v} \sim 600$ km/s, a mean magnetic field $\bar{B} \sim 4$ nT, a proton density $n_p \sim 2$ cm^{-3}, a temperature $T_p \sim 30$ eV, and a plasma beta $\beta \sim 1.6$.

Figure 1 shows the three magnetic field components for the selected time interval as observed by the FGM and STAFF experiments on-board of Cluster mission.

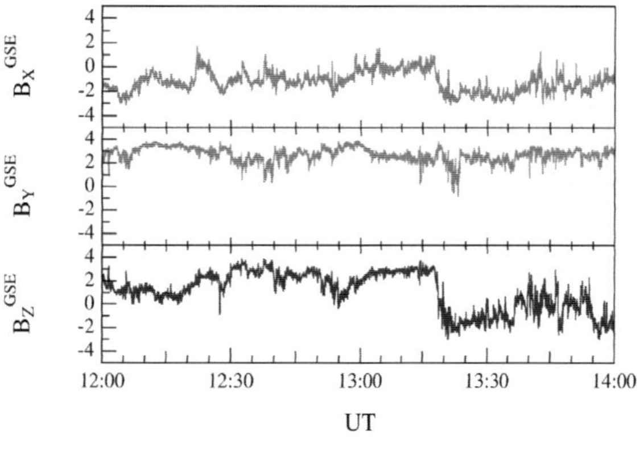

Figure 1. The magnetic field measurements relative to the period of fast solar wind observed by Cluster satellite.

The data used to investigate the scaling features refer to FGM and STAFF experiments on-board of Cluster. The merging is done by low- and high-pass filtering of respectively FGM and

STAFF data, with a cut-off frequency of 1 Hz using a finite response filter. In particular, data have been merged so to get the very high resolution of 450 samples per second. Magnetic field is measured in the GSE coordinate system.

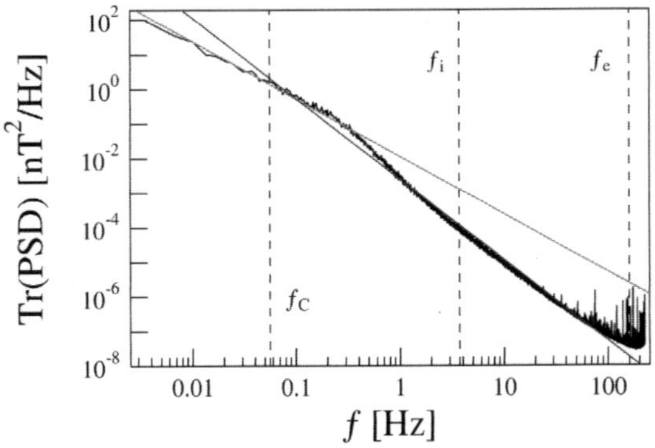

Figure 2. The trace of the PSD. The dashed vertical lines refer to the proton gyrofrequency f_C, and to the frequency corresponding to the proton and electron inertial lengths $f_k \sim \bar{v}/\rho_k$. Red and blue lines are power-law best fits.

Figure 2 shows the trace of the power spectral density (PSD) matrix of the magnetic field, which is defined as

$$\mathrm{Tr}(PSD) = \sum_i S_i(f) \tag{2}$$

where $S_i(f)$ is the power spectral density of the i-component of the magnetic field.

The PSD is characterized by two main spectral domains; a nearly $f^{-5/3}$ domain below the proton gyrofrequency f_C and a nearly $f^{-7/3}$ power-law domain at frequencies above f_C, which corresponds to a brown-to-dark spectrum in the non-MHD domain.

To analyze the scaling features of the magnetic field increments, $\delta b_i^\tau(t) = b_i(t+\tau) - b_i(t)$, we have applied the novel method of Rank-Ordered Multifractal Analysis (ROMA), introduced by Chang and Wu [9] to characterize the multifractal properties of a signal.

ROMA is based on the construction of invariants in 2nd order critical points displaying multi-scaling properties as a results of competing renormalization group (RG) fixed points. Its central idea is to analyze the singularity features of signal increments by grouping them according to their re-scaled sizes. In particular, the ROMA method consists in ranking the domain of scaled increments into separate intervals and in determining the associate scaling exponent s that satisfies the equation

$$\tau^s P(\delta x; \tau) = P_s(\tau^{-s}\delta x), \tag{3}$$

where δx are the signal increments at the timescale τ, and $P(\delta x;\tau)$ is the corresponding probability density function (PDF) of the signal increments. As a result, it is possible to construct an invariant shape for the PDFs and, simultaneously to get the scaling features of fluctuations/increments characterized by a different amplitude in terms of a rank-ordered multifractal (ROMA) spectrum $s(Y)$, where

$$Y = \tau^{-s(Y)}\delta x(\tau). \tag{4}$$

The ROMA spectrum $s(Y)$ can be found by solving the following functional equation for a range-limited structure function,

$$S'(\mid \delta x \mid; \tau) = \int_{\tau^s Y_i}^{\tau^s Y_{i+1}} \mid \delta x(\tau) \mid^q P(\mid \delta x \mid; \tau) d \mid \delta x \mid \simeq \tau^{qs}, \tag{5}$$

Journal of Physics: Conference Series **767** (2016) 012003 doi:10.1088/1742-6596/767/1/012003

and its knowledge allows the construction of a scale invariant PDF $\mathcal{I}(Y)$ defined according to the following transformation:

$$P(\delta x;\tau) \;\rightarrow\; \mathcal{I}(Y) = \tau^{s(Y)} P(\delta x;\tau). \tag{6}$$

The ROMA spectrum $s(Y)$ is, thus, the real quantity able to characterize the different scaling features of fluctuations/increments of different amplitude Y. This is a peculiar feature of ROMA method, which differently from other methods is not based on average behavior. In other words, ROMA is less affected by scaling exponents of the most probable fluctuation amplitude than the partition function based methods and, thus, is much more efficient in characterizing multifractal signals. In this framework, ROMA method is able to characterize the different universality classes of rank-ordered amplitude fluctuations/increments implying thus the possibility to envisioning the different physical processes.

3. Analysis and Results

To investigate the scaling features of magnetic field fluctuations/increments in the non-MHD domain (i.e. at timescales τ shorter than the proton gyroperiod T_C) we have confined our analysis to temporal scales $\tau < 1$ s. Indeed, this timescale of 1s separates two scaling domains as clearly shown in Figure 3 where we report the 2nd-order structure function, $S_2(\tau)$, of the B_Z component.

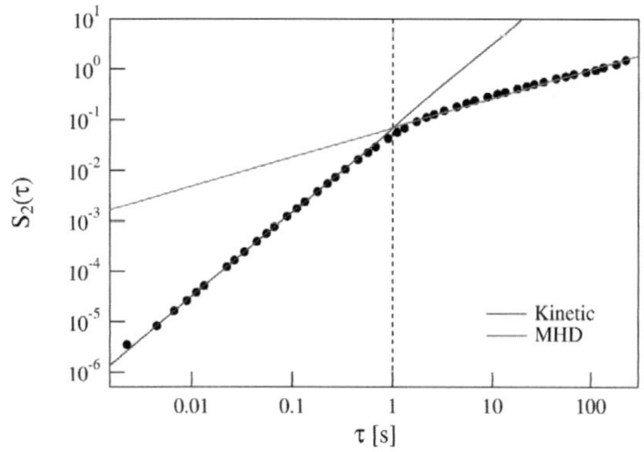

Figure 3. The 2nd-order structure function, $S_2(\tau)$, of the B_Z component. The dashed vertical line indicates the typical scale separation between the MHD domain and the kinetic one. Red and blue lines are power-law best fits.

The evaluation of the ROMA spectrum $s(Y)$ has been done solving Eq. 4 for different values of the rank-ordered scaled variable Y at different moment order q as described in Ref. [9, 10, 11].

Figures 4 and 5 show the obtained ROMA spectra, $s(Y)$, for the kinetic ($\tau < 1$ s) domain and MHD ($\tau > 1$ s) one, respectively.

While the ROMA spectrum $s(Y)$ in the MHD inertial domain shows a large variability ($\Delta s(Y)/\langle s(Y)\rangle \sim 70$ %), the situation is less pronounced at non-MHD scales (kinetic domain) where the variability is very small ($\Delta s(Y)/\langle s(Y)\rangle \sim 15$ %). The results from ROMA confirm the multifractal nature of the magnetic field increments in the MHD domain and simultaneously indicate that at non-MHD domain the scaling features of the magnetic field fluctuations are compatible with the quasi mono-fractal nature, being the spectrum nearly flat. Furthermore, we observe that the scaling features of the different magnetic field components do not differ from each other in a significative manner, so that we can assume that scaling features at non-MHD kinetic scales are essentially isotropic.

The observed average scaling exponent $\langle s(Y)\rangle_{Y\leq 20}$ for the kinetic domain is $[0.84 \pm 0.05]$, a value which is quite well in agreement with other observations in different contexts at similar scales [4, 5].

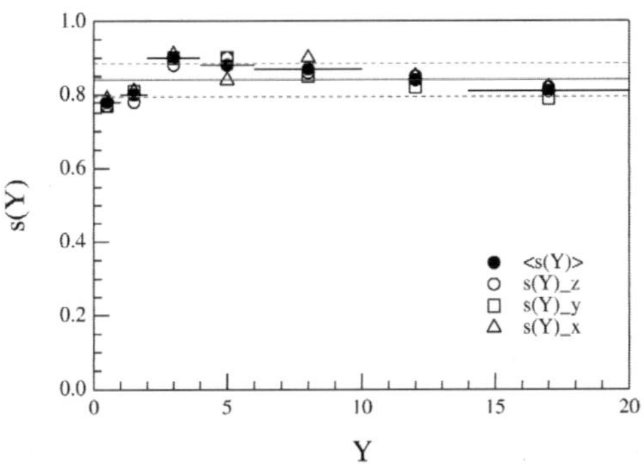

Figure 4. ROMA spectrum $s(Y)$ associated with the non-MHD/kinetic domain ($\tau < 1$ s) for the three magnetic field components. Horizontal lines indicates the mean value (blue line) and the values corresponding to $\pm 1\sigma$ (dashed red lines).

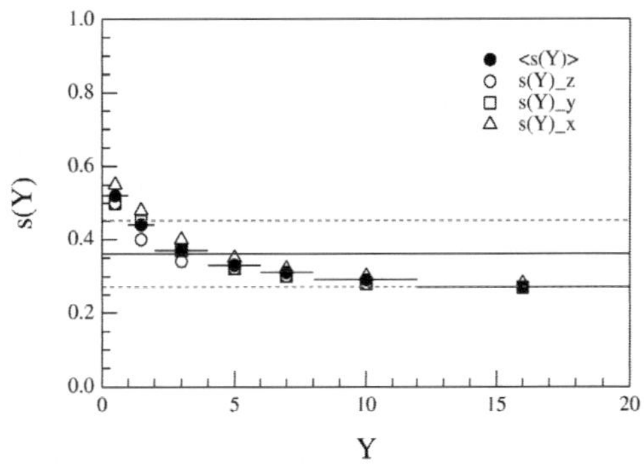

Figure 5. ROMA spectrum $s(Y)$ associated with the MHD domain ($\tau > 1$ s) for the three magnetic field components. Horizontal lines indicates the mean value (blue line) and the values corresponding to $\pm 1\sigma$ (dashed red lines).

As a result of ROMA method we have found a strong indication that the magnetic field fluctuations/increments at non-MHD kinetic scales are essentially mono-fractal. As a consequence of that, we should be able to construct an invariant PDF by simply collapsing all the PDFs at the different timescales τ onto a single invariant shape applying the following one-exponent scale transformation reported in Eqs. 4 and 6.

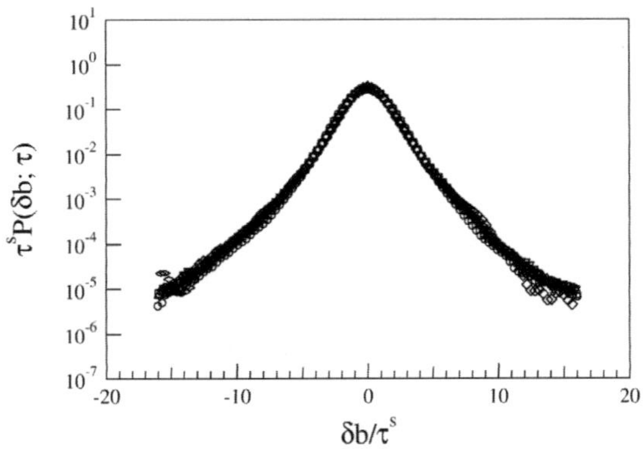

Figure 6. One-exponent collapsing of PDFs of the increments of the magnetic field B_Z component relative to the non-MHD domain ($\tau \in [4, 141]$ ms).

Figure 6 shows the PDFs' collapsing for the B_Z-component magnetic field increments in the range $\tau \in [4, 141]$ ms, using the average scaling exponent $s \sim 0.84$. PDFs' collapsing is very good supporting the mono-fractal (mono-scaling) nature of the observed fluctuations/increments. Similar results are found for the other two components X and Y.

4. Discussion and Conclusions

The results obtained by the application of ROMA method to a case study of solar wind magnetic field fluctuations support the existence of a different character of the scaling features of fluctuations/increments at MHD and non-MHD (kinetic) scales. In particular, while in the MHD domain, and more specifically in the inertial range, these fluctuations are multifractal as expected in the case of a non-homogeneous dissipation pattern in turbulent media, in the kinetic domain we have found a strong indication of a nearly global scale invariance, i.e. a mono-fractal nature. This result on the mono-fractal character of magnetic field fluctuations/increments at non-MHD scales agrees with the previous findings in solar wind by Kiyani et al. [4], and also with other results from different space plasma context, such as, for instance, magnetic field fluctuations during magnetic reconnection [5] and geomagnetic tail current disruption events [3]. Furthermore, a comparison of the observed scaling exponents in the non-MHD domain indicates that there is a certain degree of universality in the scaling features, being $s \sim [0.8 \pm 0.1]$.

A possible physical scenario of the mono-fractal nature of magnetic field fluctuations/increments at non-MHD scales may be that the observed global scale-invariance is the counterpart of the fractal nature the dissipative structures at these scales. In this framework, the multifractal character of fluctuations at MHD scales is lost implying a saturation of the intermittent character. A similar scenario seems to occur in the case of fluido-dynamic turbulence (see e.g., [8]).

In turbulent space plasmas dissipation is generally expected to occur where currents flow. Thus, a reasonable candidate of the dissipative structure is a filamentary, not-space filling, current web at scales below the ion (proton) inertial length. In other words, the fluctuation zoo associated with the turbulent magnetic field may generate multiscale coherent structures, which at these scales would imply a filamentation of the current, i.e. a fractal structure of the the dissipative field at the finest scales [12, 13].

Following Milovanov et al. [14] we can try to infer the effective fractal dimension d_s of this filamentary current web from the spectral exponent and/or the Hurst exponent, $\mathcal{H}$, of the fluctuations at non-MHD scales. In our specific case we get for the effective dimension d_s,

$$d_s = \langle s(Y) \rangle + 1 \sim 1.84, \qquad (7)$$

where we have assumed $\langle s(Y) \rangle \simeq \mathcal{H}$. The effective dimension d_s is below the embedding one $E = 3$, confirming the non compact nature of the current structure. This resultt is also confirmed by the well-known *Alexander-Orbach conjecture* [15],

$$D = \frac{2d_s}{2+\delta}. \qquad (8)$$

where d_s is the fractal dimension of a percolating network, *connectivity index* $\tilde{\delta}$ is the connectivity index and *fracton dimension* D (where $D = 4/3$ for percolation clusters). Using the above relation we obtain for the connectivity index $\tilde{\delta} \sim 3/4$, which again confirms that we are dealing with a non compact structure of the dissipation field (indeed, for compact sets the connectivity index $\tilde{\delta} = 0$) .

In conclusion, in this preliminary work we have clearly found a further confirmation of the occurrence of the quasi mono-fractal nature of the magnetic field fluctuations/increments at non-MHD scales, using the ROMA technique. A possible physical scenario for this mono-scaling nature and the related global scale-invariance was also suggested. In detail, the proposed scenario relates the observed mono-fractality to a complex topology (filamentary fractal web) of dissipation structure (perhaps the current web) instead of a special kind of wave-turbulence, not displaying multi-fractal properties in the cascading process. Recent, non-MHD simulations and observations support the existence of a complex topology of structures at these scale and the formation of a filamentary current web [16, 17, 18, 19].

The next step to confirm this scenario will involve a comparison between ad-hoc simulations on this domain and the analysis of high resolution data coming from multi-satellite missions (ESA-CLUSTER, NASA-MMS) and/or proposed mission (e.g., ESA-THOR).

Acknowledgments
We acknowledge the CLUSTER FGM and STAFF P.I.s and teams and the ESA-Cluster Active Archive team for making available the data used in this work. G. Consolini thanks A. Milovanov (ENEA, Italy) for fruitful discussions. This research has received funding from the European Community's Seventh Framework Programme [FP7/2007-2013] under Grant agreement no. 313038/STORM. Z. Vörös was supported by the Austrian Funds FWF under projects P24740-N27. M. Echim thanks also project PN-II-ID-PCE-2012-4-0418.

References

[1] Bruno R and Carbone V 2013 *Liv. Rev. Solar Phys.* **10** 2
[2] Alexandrova O, Carbone V, Veltri P and Sorriso-Valvo L 2008 *Astrophys. J.* **674** 1153
[3] Consolini G, Kretzschmar M, Lui A T Y, Zimbardo G and Macek W E 2005 *J. Geophys. Res.* **110** A07202
[4] Kiyani K H, Chapman S C, Khotyaintev Yu V, Dunlop M W and Sahraoui F 2009 *Phys. Rev. Lett.* **103** 075006
[5] Consolini G, Grandioso S, Yordanova E, Marcucci M F and Pallocchia G 2015 *Astrophys. J.* **804** 19
[6] Frisch U 1995 Turbulence A Kolmogorov Legacy (Cambridge: cambridge Uni. Press)
[7] Kraichnan R H 1967 *Phys. Fluids* **10** 2080
[8] Chevillard L, Castaing B and Lévêque E 2005 *Eur. Phys. J.* B **45** 561
[9] Chang T S and Wu C C 2008 *Phys. Rev.* E **77** 045401
[10] Consolini G and De Michelis P 2011 *Nonlin. Processes Geophys.* **18** 277
[11] Chang T S, Wu C C, Podesta J, Echim M, Lamy H and Tam S W Y 2010 *Nonlin. Processes Geophys.* **17** 545
[12] Chang T S 1999 *Phys. Plasmas* **6** 4163
[13] Zelenyi L M and Milovanov A V 2004 *Phys. Uspeki* **47** 749.
[14] Milovanov A V, Zelenyi L M and Zimbardo G 1996 *J. Geophys. Res.* **101** 19903
[15] Alexander S and Orbach R 1982 *J. Physique Lett.* **43** 625
[16] Perrone D, Valentini F, Servidio S, Dalena S and Veltri P 2013 *Astrophys. J.* **762** 99
[17] Wan M, Matthaeus W H, Roytershteyn V, Karimabadi H, Parashar T, Wu P and Shay M 2015 *Phys. Rev. Lett.* **114** 175002
[18] Vörös Z, Yordanova E, Echim M M, Consolini G and Narita Y 2016 *Astrophys. J. Lett.* **819** L115
[19] Yordanova E, Vörös Z, Varsani A et al. 2016 *Geophys. Res. Lett.* **43** 5969

Charge-to-mass dependence of heavy ion spectral breaks in large gradual solar energetic particle events

M. I. Desai[1,2], G. M. Mason[3], M. A. Dayeh[1], R. W. Ebert[1], D. J. McComas[4], G. Li[5], C. M. S. Cohen[6], R. A. Mewaldt[6], N. A. Schwadron[7], and C. W. Smith[7]

[1]Southwest Research Institute, 6220 Culebra Road, San Antonio, TX 78023, USA
[2]University of Texas at San Antonio, One UTSA Circle, San Antonio, TX 78249, USA
[3]Johns Hopkins University/Applied Physics Laboratory, Laurel, MD 20723, USA
[4]Department of Astrophysical Sciences, Princeton University, NJ 08544, USA
[5]CSPAR, University of Alabama in Huntsville, Huntsville, AL 35756, USA
[6]California Institute of Technology, Pasadena, CA 91125, USA
[7]University of New Hampshire, 8 College Road, Durham, NH 03824, USA

E-mail: mdesai@swri.edu

Abstract. We fit the ~0.1-500 MeV/nucleon H-Fe spectra in 46 large SEP events surveyed by [1] with the double power-law Band function to obtain a normalization constant, low- and high-energy parameters γ_a and γ_b; and break energy E_B. We also calculate the low-energy power-law spectral slope γ_1. We find that: 1) γ_a, γ_1, and γ_b are species-independent within a given SEP event, and the spectra steepen with increasing energy; 2) E_B's are well ordered by Q/M ratio, and decrease systematically with decreasing Q/M, scaling as $(Q/M)^\alpha$ with α varying between ~0.2-3; 3) α is well correlated with Fe/O at ~0.16-0.23 MeV/nucleon and CME speed; 4) In most events: α <1.4 and the spectra steepen significantly at higher energy with γ_b-γ_a >3; and 5) Seven out of nine extreme SEP events (associated with faster CMEs and GLEs) are Fe-rich and have α >1.4 with flat spectra at low and high energies yielding γ_b-γ_a <3. The species-independence of γ_a, γ_1, and γ_b and the systematic Q/M dependence of E_B within an event, as well as the range of values for α suggest that the formation of double power-laws in SEP events occurs primarily due to diffusive acceleration at near-Sun CME shocks and not due to scattering in the interplanetary turbulence. In most events, the Q/M-dependence of E_B is consistent with the equal diffusion coefficient condition while the event-to-event variations in α are probably driven by differences in the near-shock wave intensity spectra, which are flatter than the Kolmogorov turbulence spectrum but still weaker compared to that inferred for the extreme events. The weaker turbulence allows SEPs to escape more easily, resulting in weaker Q/M-dependence of E_B, (lower α values) and spectral steepening at higher energies. In extreme events, the flatter spectra at high- and low-energy and stronger Q/M-dependence of E_B (larger α values) occur due to enhanced wave power, which also enables the faster CME shocks to accelerate flare suprathermals more efficiently than ambient coronal ions.

1. Introduction

Large gradual solar energetic particle (SEP) events are believed to be accelerated via diffusive shock acceleration (DSA) mechanisms at shock waves driven by fast coronal mass ejections (CMEs) that plough through the solar corona and interplanetary (IP) medium (e.g., [2–4]). Such large SEP events, if sufficiently intense, can significantly increase radiation levels in the near-Earth environment, thus damaging technological assets and adversely affecting the health and safety of humans in space (e.g., [5]). Previous studies have shown that the differential energy spectra of H-Fe nuclei in large SEP events exhibit two distinct (or broken) power-laws above and below a characteristic roll-over or break energy, with the break energy typically decreasing for the heavier ion species, or more precisely, with the ion's charge-to-mass or Q/M ratio (e.g., [6–9]). [10] suggested that this systematic Q/M-dependence occurs because the energy spectra, usually plotted in MeV/nucleon, steepen or roll over at the same value of the diffusion coefficient for different species, which depends on ion rigidity or the Q/M ratio (see [10–12]). [10] also showed that this behavior can be characterized by $\frac{E_X}{E_H} \propto (\frac{Q_X}{M_X})^{\alpha}$, where E_X is the spectral break energy of species X; E_H is the spectral break energy of H; and Q_X and M_X are the ionic charge and atomic mass of species X.

2. Instrumentation, Event Selection, and Data Analyses

We use energetic ion data from (1) the Ultra-Low-Energy Isotope Spectrometer (ULEIS: [13]); (2) the Solar Isotope Spectrometer (SIS: [14]); and the Electron, Proton, and Alpha monitor (EPAM: [15]) on board NASA's Advanced Composition Explorer (ACE: [16]) launched in 1997 August. We also use proton data from the Proton and Electron Telescope (PET) on board the Solar, Anomalous, and Magnetospheric Particle Explorer (SAMPEX/PET: [17]); the Energetic and Relativistic Nuclei and Electron experiment (ERNE: [18]) on board the joint ESA/NASA Solar and Heliospheric Observatory (SoHO); and the Energetic Particle Sensor (EPS) on NOAA's Geostationary Operational Environmental Satellites (GOES, series 8-15).

[1] describes our event selection criteria and method for identifying sampling intervals for 46 isolated, large SEP events observed at 1 AU from November 1997 through April 2014 that were not accompanied by ≥0.1 MeV/nucleon intensity enhancements associated with the passage of in-situ CME-driven IP shocks (e.g., [19]). Such enhancements are known as Energetic Storm Particles or ESPs. For each of these 46 SEP events, we followed the [20] procedure and used ULEIS and SIS oxygen time-intensity profiles and 1/ion speed spectrograms (see Figure 1) to select energy-dependent sampling intervals to account for velocity dispersion caused by the later arrival of slower ions compared to the faster ions. An example of the application of this technique is shown as the red trapezoid in Figure 1b. This interval also includes an IP shock that arrived at ACE at 2116 UT on 2011 November 28 (`http://www.srl.caltech.edu/ACE/ASC/DATA/Shocks/shocks_good.html`). This SEP event was included in our survey because the in-situ CME shock was not accompanied by a significant local ESP event above ~0.1 MeV/nucleon.

Tables 1 and 2 of [1] provide the solar source properties, fluence sampling intervals, the ~0.5-2.0 MeV/nucleon ^{3}He/^{4}He ratio, and the Fe/O ratios at ~0.16-0.23 MeV/nucleon and ~15-21 MeV/nucleon associated with these 46 events. In this study for each SEP event, we used ACE/ULEIS, ACE/SIS, GOES/EPS, SoHO/ERNE, and when available, SAMPEX/PET, to obtain the event-integrated ~0.1-500 MeV/nucleon fluence spectra for 11 species in the range H-Fe, as shown in the three examples in Figure 2. The proton spectra from various instruments in each SEP event at overlapping energies are in excellent agreement within the stated ~20% uncertainties that account for differences between various instruments (see [21]).

We fit each spectrum using the non-linear least-squares Levenberg-Marquardt technique and minimized the χ^2 to fit the 4 parameter Band function (see [22]; Eq. 1) given by:

$$dJ/dE = CE^{-\gamma_a} \exp\left(-\frac{E}{E_B}\right) \textbf{ for } E \leq (\gamma_b - \gamma_a) E_B$$

$$dJ/dE = CE^{-\gamma_b} \left[(\gamma_b - \gamma_a) E_B\right]^{\gamma_b - \gamma_a} \exp(\gamma_b - \gamma_a) \textbf{ for } E \geq (\gamma_b - \gamma_a) E_B \quad (1)$$

Here C is the normalization constant, γ_a and γ_b are the low-energy and high-energy Band-parameters, and E and E_B are the kinetic and spectral break energy, respectively. Units of E

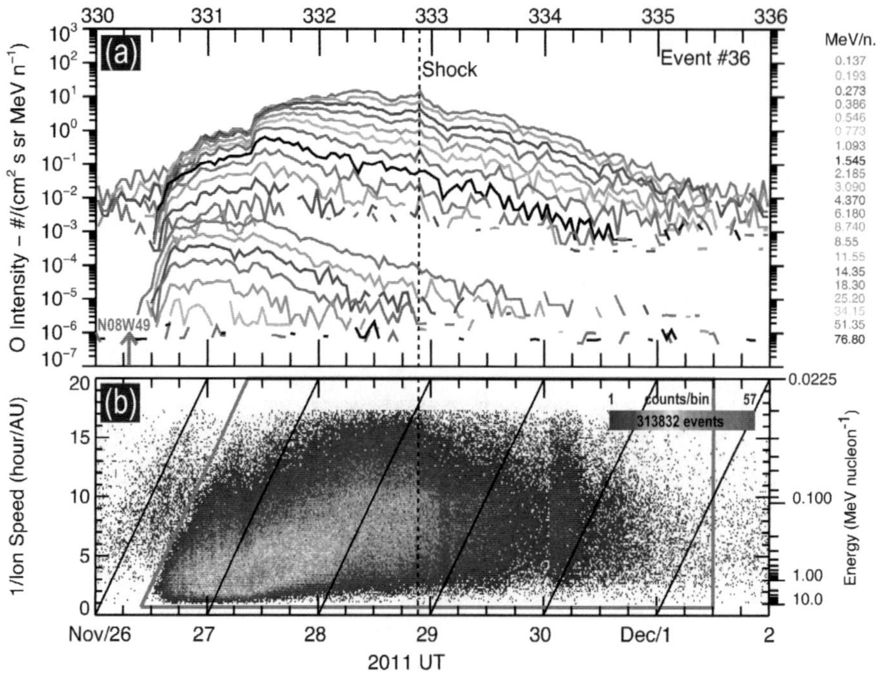

Figure 1. (a) Hourly-averaged time-intensity profiles of ~0.1-100 MeV/nucleon oxygen nuclei measured by ACE/ULEIS and ACE/SIS during a large gradual solar energetic particle (SEP) event that occurred on 2011 November 26-December 1. The pink arrow indicates the flare occurrence time. (b) 1/ion speed spectrogram of ~0.1-10 MeV/nucleon C-Fe nuclei measured by ULEIS. The red region represents the SEP sampling interval that accounts for velocity dispersion of ions traveling along a 1.2 AU path length (slanted black lines). The arrival time of an IP shock at ACE on 2011 November 28, 2116 UT is taken from the ACE IP shock list; see [1] for more details.

and E_B are MeV/nucleon. For each Band-fit parameter, we obtain the formal 1σ uncertainty from the off-diagonal terms of the covariance matrix [23]. As discussed in [1], we found that for most species in most SEP events (see Figures 2a, 2c, and 2e), the fits are visually and statistically reasonable, with reduced χ^2 values having ~50% probabilities (also see [9]). Note that the Band function has no physical basis, and we use it here because it helps us to characterize the behavior of SEP spectra over a broad energy range using only 4 free parameters.

Table 2 in [21] provides the H, O, and Fe Band function fit parameters for the 46 SEP events. For each SEP event, we also fitted the roll-over or break energy E_X of each species X normalized to the proton spectral break energy, E_H with a linear function of the form $\log(\frac{E_X}{E_H}) = \log(n_0) + \alpha \log(\frac{Q_X}{M_X})$, whose slope α is the power-law exponent discussed in Section 1. Examples of three different types of Q/M-dependence of E_X/E_H, i.e., three different values for α, are shown in Figures 2b, 2d, and 2f.

3. Properties of Spectral Slopes

[1] shows that the non-orthognality of the Band function results in strong coupling between the O Band-parameters γ_a and E_B, and that γ_a can be significantly different from what is commonly described as the low-energy power-law spectral slope γ_1. In order to obtain a physically meaningful quantity that represents the low-energy portion of the SEP spectra below the break energies more accurately, we calculate the low-energy spectral slope γ_1 between ~0.1-

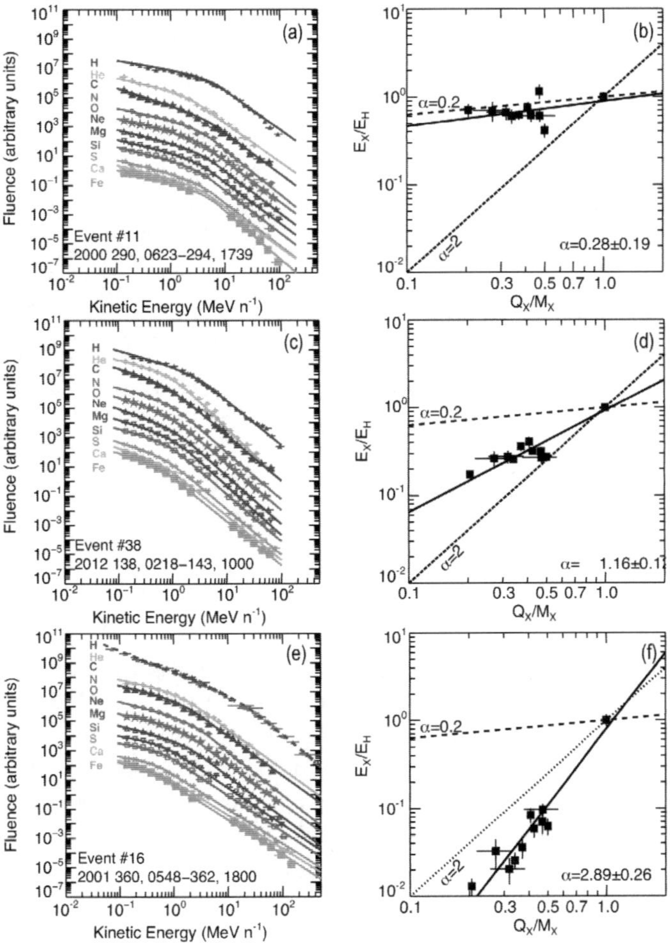

Figure 2. (a, c, e) Event-integrated differential fluences versus energy of ~0.1-500 MeV/nucleon H-Fe nuclei during 3 large SEP events. The energy spectra for different species are offset for clarity. Solid lines are fits to the spectra using the Band function (see Eq. 1; [22]). In all panels, the H spectra (top-most blue symbols) are from ACE/ULEIS, SoHO/ERNE, and GOES-8/EPS, and the He-Fe spectra are from ACE/ULEIS and ACE/SIS. In (e) red data points superposed on the top-most blue symbols are proton data from ACE/EPAM, GOES-8/EPS, and SAMPEX/PET, and the top-most dotted red-curve shows the corresponding Band-function fit from [9]. (b, d, f) Spectral break energy E_X of species X normalized to E_H – break energy of H vs. the ion's charge-to-mass (Q/M) ratio. The solid line is the fit to the data $\log(\frac{E_X}{E_H}) = \log(n_0) + \alpha \log(\frac{Q_X}{M_X})$. Dashed lines are the same equation with $\alpha = 2$, and dotted lines are the same equation with $\alpha = 0.2$.

1 MeV/nucleon for each species in all SEP events using Eq. 1 (top) and the corresponding Band-parameters γ_a and E_B.

Figure 3a shows a scatter plot of γ_b vs. γ_a obtained for each individual species in all SEP events. As seen for the O spectral slopes in [1], we note that most SEP spectra are flatter at lower energies and steepen above the break energy. Figures 3b and 3c investigate species-associated variations in (a) γ_1 and γ_a, and (b) γ_b within an event by plotting the distributions of their corresponding mean deviations. The mean deviation for each parameter in each event is calculated with respect to the corresponding species-averaged value of that parameter within that event. We note the following: (1) The mean deviations of γ_1, γ_a, and γ_b have narrow distributions that result in well-behaved Gaussian-like distributions (black curves) with small 1σ values. These results indicate that, within a given SEP event, the three parameters γ_1, γ_a, and γ_b, have remarkably similar values for all species, for each event, both the low-energy and high-energy spectral slopes are similar within ~10-15%. Thus, in a given event, species-associated spectral variations are driven primarily by differences in the break energy at which the spectra steepen. Three examples of such species-associated differences in E_B are shown in Figures 2b, 2d, and 2f (see [21]).

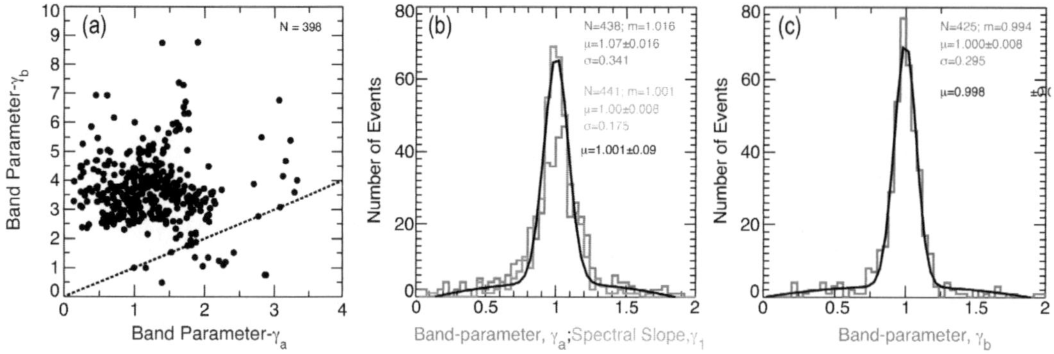

Figure 3. (a) Scatter-plots of γ_a vs. γ_b. Dotted line shows spectra for which $\gamma_a = \gamma_b$. All parameters with relative uncertainties >100% are excluded. Histograms of mean deviations of the (b) Band-parameter γ_a (red), and low-energy spectral slope γ_1 (blue); and (c) Band-parameter, γ_b from the corresponding species-averaged value in each event. N=number of data points; m=median; μ=mean and standard error of the mean; and σ=1 standard deviation of the distribution. The solid black curves show Gaussian fits, with mean and 1σ standard deviation, for the distributions of γ_1 in (a) and γ_b.

4. Properties of Power-Law Exponent α

Figure 4 investigates (a) the statistical properties of α and its relationship with (b) CME speed, (c) the ~0.16-0.23 MeV/nucleon Fe/O ratio, and (d) the difference between the high- and low-energy Band parameters $\gamma_b - \gamma_a$. Figure 4a shows the histogram of α along with the mean, standard deviation, and median value of the distribution. The CME speeds are obtained from `http://umbra.nascom.nasa.gov/SEP/`, `ftp://ftp.ngdc.noaa.gov/STP/swpc_products/daily_reports/solar_event_reports/`, `cdaw.gsfc.nasa.gov/CME_list/` ([24–27]). For the 33 events shown in Figure 4 we note the following.

(i) α has a mean value of 1.27, median value of 1.16, and lies between ~0.2-3, with values for 3 SEP events greater than 2; two of these events were also accompanied by Ground Level Enhancements or GLEs (e.g., see Figure 1f).

(ii) α exhibits statistically significant, positive trends with the CME speed and Fe/O ratio, with values for correlation coefficients of r ~0.414 and r ~0.39, respectively, which have probabilities of <2.5% of being exceeded by uncorrelated pairs of parameters. It is also evident that the correlations with both the CME speed and Fe/O are largely due to the presence of events associated with high CME speeds and GLEs.

(iii) α is not correlated with $\gamma_b - \gamma_a$, but SEP events with α ~0.2-1.4 exhibit a larger range of values for $\gamma_b - \gamma_a$, i.e., events for which the spectra steepen significantly at higher energies have α values between ~0.6-1.4. Figure 4d also shows that extreme SEP events have $\alpha \geq 1.4$ and $\gamma_b - \gamma_a < 3.$, i.e., extreme SEP events have stronger Q/M-dependence in E_B's and flatter spectra at high and low energies.

5. Summary and Discussion

We fit the event-integrated fluence spectra of ~0.1-500 MeV/nucleon H-Fe in the 46 SEP events with the 4 parameter Band function and investigate properties of the SEP Band-parameters γ_a, γ_b, and E_B (also see [21]). We also calculate the low-energy power-law spectral slope γ_1. Our results are:

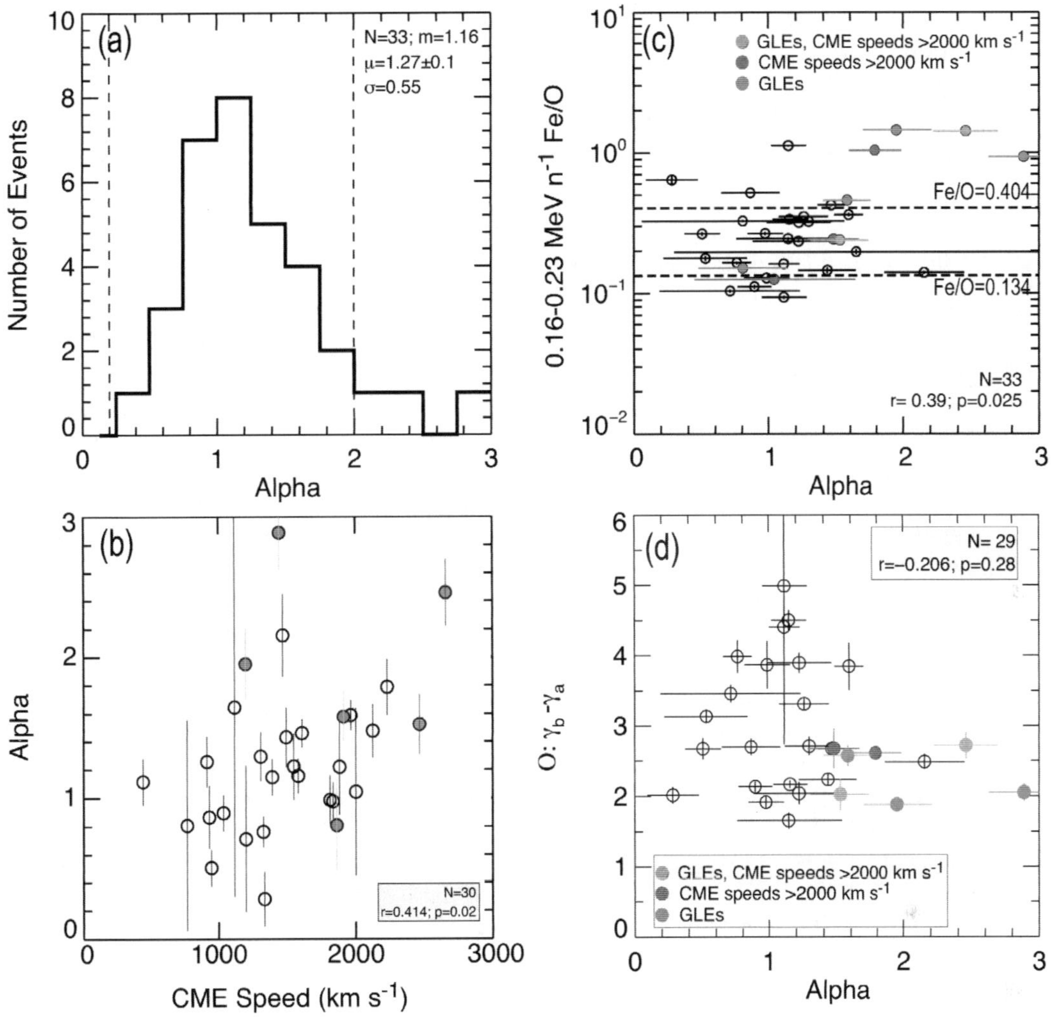

Figure 4. (a) Histogram of α. (b) α vs. CME speed (km/s). (c) α vs. ~0.16-0.23 MeV/nucleon Fe/O. (d) α vs. the difference $\gamma_b - \gamma_a$. N=number of events plotted; μ, σ, and m as defined in Figure 3; r=correlation coefficient and p=probability that the absolute value of r can be exceeded by an uncorrelated pair of parameters. Green symbols show SEP events accompanied by Ground Level Enhancements or GLEs, red symbols denote SEP events associated with CMEs that have speeds >2000 km/s, blue symbols are SEPs that were accompanied by GLEs and >2000 km/s speed CMEs. Dashed lines show Fe/O ratios at 0.404 and 0.134, which are average values in several large SEP events at ~0.32-0.5 MeV/nucleon [20] and at ~5-12 MeV/nucleon [28], respectively.

(i) Figures 2b, 2d, and 2f show that the E_B's in 3 SEP events vary systematically according to the ion's Q/M ratio. We also find that this dependence can be characterized by a single parameter α – given by fitting a linear function of the form $\log(\frac{E_X}{E_H}) = \log(n_0) + \alpha \log(\frac{Q_X}{M_X})$ – in 33 of the 46 events surveyed.

(ii) Figure 3a shows that γ_a lies between ~0.1-3 and γ_b lies between ~0.5-9. γ_a is also typically smaller than γ_b, implying that the energy spectrum of each species in a given event steepens with increasing energy.

Journal of Physics: Conference Series **767** (2016) 012004 doi:10.1088/1742-6596/767/1/012004

(iii) Figures 3b and 3c show that in a given SEP event, γ_a, γ_1, and γ_b are species-independent, since the mean deviation of each parameter exhibits a Gaussian-like distribution with a small 1σ of $\leq$0.08.

(iv) Figure 4a shows that for 33 SEP events, α varies between ~0.2-3. SEPs associated with GLEs and faster CMEs with speeds >2000 km/s have $\alpha \geq 1.4$ vs. $\alpha < 1.4$ for the rest of the events.

(v) Figures 4b and 4c show that α exhibits statistically significant positive trends with both, the CME speed and the ~0.16-0.23 MeV/nucleon Fe/O; extreme SEPs are associated with the fastest CMEs and highest Fe/O ratios.

(vi) Figure 4d shows that α is not correlated with the difference in the spectral slopes $\gamma_b - \gamma_a$; $\gamma_b - \gamma_a$ varies between ~2-5 in events with $\alpha \leq 1.4$, while extreme SEPs have $\alpha \geq 1.4$ and $\gamma_b - \gamma_a < 3$.

5.1. Properties of CME shocks and associated turbulence

Comparing our survey to prior studies, we note that some of the 5 events studied by [10] and [12] also included the local shock-accelerated ESP component that accompanied the larger SEP event. In contrast, we eliminate all events with possible contributions from local IP shock-associated populations (see [1]). Further, we use event-integrated fluences, rather than time-intensity profiles (see [29]), to study the SEP spectral properties. In particular, [29] used a detailed model of IP propagation and showed that transport from the inner solar system can lower the break energy systematically for all species, as well as lower the slopes by ~10-20%, but that the basic spectral form remains intact (see Fig. 14 in [29]). Alternatively, we note that [30] fitted the double power-law proton spectra in 9 of the 16 GLE events studied by [10] with an analytical model that included interplanetary transport effects, and found that single power-law spectra injected by CME shocks near the Sun can exhibit spectral breaks at 1 AU due to scatter-dominated transport through the interplanetary medium. However, the [30] model predicts that α in GLE-associated SEP events should lie in the range ~0.18-0.75, which is clearly inconsistent with the $\alpha > 1.58$ observed in 5 of the 7 GLEs in our survey (see Figures 2f and 4). On this basis, we contend that the formation of the double power-law SEP spectra, their associated properties, and the observed Q/M-dependence of E_B primarily reflect conditions near the distant CME-driven shocks where the acceleration takes place, and are not significantly affected by contributions from local interplanetary shock-accelerated populations nor by Q/M-dependent transport and scattering in the interplanetary turbulence en route to 1 AU (e.g., see [12; 29; 31]).

A fundamental prediction of early 1-dimensional (1D) steady-state (as well as the more recent time-dependent DSA-based SEP models) is that, in a given event, the differential energy spectrum of the accelerated particles below the break energy is characterized by a low-energy power-law spectral slope γ given by $\frac{dj}{dE} \propto E^{-\gamma}$ (e.g., [4; 32–34]). These models also predict that γ is independent of ion species, and is determined solely by $\gamma \approx (H+2)/(2H-2)$, where H is the strength or compression ratio of the CME-driven shock. Our results show that both, the SEP Band-parameter γ_a and the low-energy spectral slope γ_1 in a given SEP event are remarkably similar for all species, and that such species-independent spectral slopes are observed at both low and high energies for most of the events in our survey. We therefore suggest that, to first order, the formation of double power-law spectra in large SEP events is consistent with DSA at near-Sun CME shocks (e.g., [34]).

We now use the DSA-predicted relationship between γ (here we use the species-averaged γ_1 for each event) and H to infer compression ratios of near-Sun CME shocks. Figure 5a compares the relationship between the inferred compression ratio and CME speed. The figure shows that the inferred shock compression ratios in 37 events lie between ~1-5.5, with $H > 4$ for 4 events.

These values are consistent with the predicted range of values for CME shock compression ratios in large SEP events (see [34]), and are also well within the constraints of the Rankine-Hugoniot discontinuity conditions for the allowable range of ~1-4 and upper limit of < 4 for shocks in non-relativistic plasmas (e.g., [35]). We note that all cases of $H > 4$ have sizeable uncertainties. H also exhibits a weak but positive correlation with CME speed; for 37 events, a value of r ~0.38 has <2% probability, and excluding the 4 events with $H > 4$, yields r ~0.32 which has <7% chance of being exceeded by uncorrelated pairs of parameters.

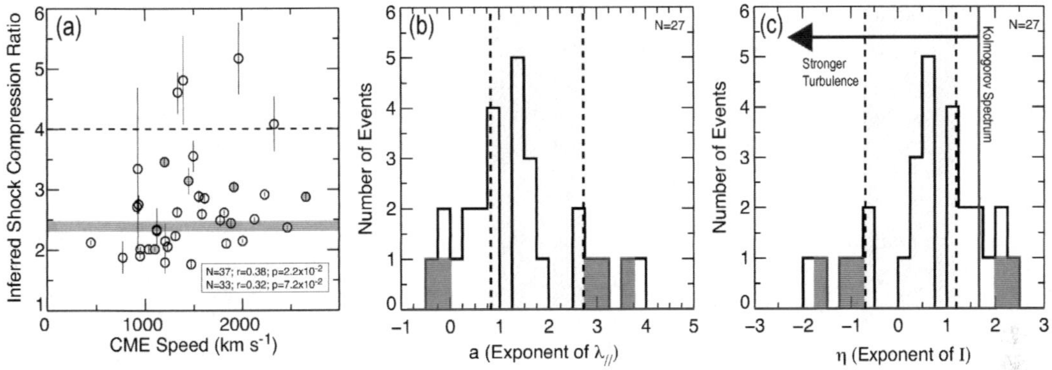

Figure 5. (a) Scatterplot of the inferred shock compression ratio vs. CME speed. Histograms of (b) a, the inferred power-law exponent of the M/Q-dependence of the scattering mean free path $\lambda_{\parallel} \propto (M/Q)^a$, and (c) η, the inferred power-law exponent of the wave intensity spectrum I near the CME-driven shock given by $I \propto k^{-\eta}$. Green data points in (a) denote GLEs. Orange shaded region in (a) encompasses the mean value and the standard error of the mean, calculated by excluding the 4 outliers with values >4 and the six GLE events. Shaded yellow regions in (b) and (c) depict the range of values inferred by [12] for 5 large SEP events. Blue vertical line in (c) represents $\eta = 5/3$ – the typical interplanetary Kolmogorov spectrum; red histograms represent 5 GLE events.

The heavy ion fluence spectra in most SEP events are flat at energies below ~1 MeV/nucleon and steepen above a roll-over or break energy, which decreases systematically with the ion's Q/M ratio. The Q/M-dependence of E_B's in a given SEP event is well represented by the function $\log(\frac{E_X}{E_H}) = \log(n_0) + \alpha \log(\frac{Q_X}{M_X})$ and characterized by the power-law exponent α. The values of α in 33 SEP events lie between ~0.2-3, which encompasses the range of α values found by [10; 12]. Thus, with the exception of 3 events with $\alpha > 2$ (see Figure 4a), the range of values for α in our survey is consistent with the corresponding range of ~0.2-2 predicted by [36]. In this model, the Q/M-dependence of the spectral break energies in a given SEP event occurs due to the "equal diffusion coefficient" or the "equal acceleration time" condition, and the event-to-event variations in the power-law exponent α are driven by the differences in the slopes of the turbulence spectra expected near shocks with different obliquity.

Assuming that the spectral break energies for different species in a given SEP event occur at the same value of the diffusion coefficient $\kappa_{\parallel}$, which scales as $(M/Q)^a$ with observed values of $a \approx 0.8 - 2.7$, [12] followed [37] and inferred that the power-law index η of the turbulence or wave intensity spectrum, given by $I(k) \propto k^{-\eta}$, near the CME shock acceleration region, ranged between 1.2 to -0.7. Here $\eta = 2 - a$, and a is related to the exponent α in our survey by $a = \alpha(2 - \alpha)$. We now follow the approach of [10; 12] to infer the power-law exponent a, which determines the scaling between the particle diffusion coefficient and the ion's M/Q ratio, and the power-law index η of the wave intensity spectrum for 27 SEP events in our survey. In this analysis, we only include events that satisfied the following: 1) fitted values of α and the

Journal of Physics: Conference Series **767** (2016) 012004 doi:10.1088/1742-6596/767/1/012004

inferred values of a and η have relative uncertainties <100%, and 2) $-4 < \eta < +4$.

Figure 5 plots histograms of (b) a and (c) η, calculated using the values of α in 27 SEP events. Note that within the estimated uncertainties, a in 15 out of 27 SEP events lies in the range ~0.75-2.75 and is comparable to that obtained by [12], as shown by the yellow shaded region. Also, a varies between ~-0.33-3.9, which is roughly consistent with the typical range of ~0.5-7 proposed recently by [34]; in this model $a < 1$ implies weak dependence of $\lambda_{\parallel}$ on the ion's Q/M ratio, while $a > 1$ indicates that $\lambda_{\parallel}$ depends strongly on Q/M (also see [36; 38–40]).

We remark that η in 27 events lies between -1.87 and 2.33; η in 15 events lies between ~-0.7–1.2, as reported by [12]. In 9 events, $\eta > 1.2$ — the largest value reported by [12]. Overall, in 4 events $\eta > 5/3$, which represents cases where the turbulence intensity spectra near the distant CME shocks may be significantly steeper than the typical interplanetary Kolmogorov $k^{-5/3}$ turbulence spectrum. In contrast, $\eta \leq 5/3$ in 23 events, which implies that the turbulence spectra near the corresponding CME shocks are probably significantly flatter than the Kolmogorov spectrum.

5.2. Extreme SEP events and their implications for theoretical models

All 9 extreme SEP events have $\alpha \geq 1.4$, low values for γ_a, γ_b, and $\gamma_b - \gamma_a < 3$. This indicates that the corresponding spectra are relatively flat with similar spectral slopes at low and high energies. The fact that $\alpha > 2$ in 2 of the 7 GLE-associated SEP events in our study, taken together with the general result that higher values of $\alpha \geq 1.4$ are typically observed in SEP events that are also associated with faster CMEs and GLEs, indicates that spectral properties in these extreme SEP events are most likely governed by highly efficient trapping and stronger Q/M-dependent scattering due to substantially enhanced wave power near the distant CME-driven shocks (see also [12; 36]).

The above scenario is also consistent with the following inferred results shown in Figure 5:

(i) Figure 5a shows that the inferred values for the shock compression ratio H in extreme events tend to be somewhat larger than the event average, and that the correlations between H and α are likely to be driven by these events.

(ii) Figure 5b (red histograms) shows that $\lambda_{\parallel}$ has a strong Q/M-dependence with $a \geq 2.8$ in 3 events and $a < 0.2$ in 2 events.

(iii) Figure 5c (red histograms) shows that $\eta > 2$ or $\eta < -1.1$ in these 5 events, which corresponds to substantially enhanced wave power.

(iv) The 4 remaining extreme SEP events (not shown in Figures 5b and 5c) had amongst the strongest observed Q/M-dependence in E_B's, i.e., $\alpha > 1.6$, which results in $|\eta| > 4$.

The above results also imply that the scattering and trapping of particles near the distant CME shocks during at least 2 of the extreme SEP events is so strong that the Q/M-dependence of the spectral break energies exceeds the limit of the equal diffusion or equal resonance condition described by [36], thus indicating that the underlying mechanisms have not yet been fully incorporated in current theoretical models.

[41] reported that observations at several interplanetary shocks near 1 AU were consistent with their nonlinear guiding center theory-based model that described DSA at parallel and perpendicular shocks in the presence of wave excitation and ambient upstream turbulence, respectively. In this model, ambient turbulence near perpendicular shocks allows lower-energy particles to be accelerated as efficiently as parallel shocks, producing single power-law like spectra. In contrast, the equal resonance condition and occasional wave excitation near parallel shocks results in more efficient trapping and rapid acceleration of particles to higher energies, creating spectra that exhibit double power-laws or exponential roll-overs with the maximum energy scaling as $(Q/M)^{\alpha}$.

While our results are qualitatively consistent with the [41] model and other models in which the strongest Q/M dependence of the heavy ion spectral breaks occur at quasi-parallel shocks (e.g., [36]), we note that such shocks also have lower injection thresholds, and are therefore expected to primarily inject and accelerate the low-energy solar wind or ambient coronal ions (e.g., [42]; however, see [43] for an alternative viewpoint). In contrast, and consistent with the results reported in [1], we find that many of the extreme events that exhibit strong Q/M-dependent spectral break energies are also Fe-rich and ^{3}He-rich (see [21]). Specifically, we note that the ~0.16-23 MeV/nucleon Fe/O ratio is enhanced between factors of ~2-10 compared to the average SEP value of ~0.134 in 7 of the 9 extreme SEP events, while the ~0.5-2.0 MeV/nucleon ^{3}He/^{4}He ratio is enhanced by more than an order of magnitude over the corresponding solar wind value in 5 extreme SEPs (see [21]). This points to the importance of contributions of suprathermal flare-origin material to the seed populations for fast CME shocks, even in those cases where the turbulence levels are significantly enhanced and the shocks may be quasi-parallel. We suggest that in such events, the enhanced turbulence traps, injects, and accelerates the higher-energy suprathermals much more efficiently than the co-existing lower-energy solar wind or coronal suprathermal ions. Simultaneously, the equal diffusion coefficient condition causes the spectral break energies to exhibit stronger Q/M-dependence, occasionally exceeding the equal resonance condition limit, as in the case of 2 SEP events that produced GLEs. We suggest that our results can be reconciled with SEP models provided that they include suprathermal flare-origin material as an important component of the seed population that is available for acceleration at near-Sun CME shocks.

Acknowledgments
We are grateful to the members of the Space Physics Group at the University of Maryland and the Johns Hopkins Applied Physics Laboratory (JHU/APL) for the construction of the ULEIS instrument and to members of the Space Radiation Laboratory at the California Institute of Technology for the construction of the SIS instrument. We acknowledge use of the NOAA GOES and SoHO/ERNE proton data. Work at SwRI is partially supported by NASA grants NNX13AE07G and NNX13AI75G, NASA contracts NNX10AT75G and NNN06AA01C, and NSF Grants AGS-1135432 and AGS-1460118. Work at APL was supported by NASA grants NNX13AR20G/115828 and 44A- 1091698.

References

[1] Desai M I, Mason G M, Dayeh M A, Ebert R W, McComas D J, Li G, Cohen C M S, Mewaldt R A, Schwadron N A and Smith C W 2016 *The Astrophysical Journal* **816** 68

[2] Reames D V 1999 *Space Sci. Rev.* **90** 413–491

[3] Reames D V 2013 *Space Science Reviews,* 6

[4] Lee M A 2005 *Astrophys. J. Supplement Series* **158** 38–67

[5] Desai M I and Giacalone J 2016 *Living Reviews in Solar Physics* **5** 1

[6] McGuire R E, von Rosenvinge T T and McDonald F B 1984 *Solar/Interplanetary Intervals* ed Shea M A, Smart D F and McKenna-Lawlor S M P p 157

[7] Ellison D C and Ramaty R 1985 *Astrophys. J.* **298** 400–408

[8] Mazur J E, Mason G M, Klecker B and McGuire R E 1992 *Astrophysical Journal* **401** 398–410

[9] Mewaldt R A, Looper M D, Cohen C M S, Haggerty D K, Labrador A W, Leske R A, Mason G M, Mazur J E and von Rosenvinge T T 2012 *Space Sci. Rev.* **171** 97–120

[10] Mewaldt R A, Cohen C M S, Mason G M, Labradora A W, Looper M L, Haggerty D E, Maclennan C G, Cummings A C, Desai M I, Leske R A, Li G, Mazur J E, Stone E C and

Wiedenbeck M E 2005 *The Physics of Collisionless Shocks: 4th Annual IGPP. International Astrophysics Conference* (*ASP Conference Series* vol 781) ed Li G, Zank G P and Russell C T pp 227–232

[11] Tylka A J, Boberg P R, McGuire R E, Ng C K and Reames D V 2000 *Acceleration and Transport of Energetic Particles Observed in the Heliosphere* (*ASP Conference Series* vol 528) ed Mewaldt R A, Jokipii J R, Lee M A, Möbius E and Zurbuchen T H pp 147–152

[12] Cohen C M S, Stone E C, Mewaldt R A, Leske R A, Cummings A C, Mason G M, Desai M I, von Rosenvinge T T and Wiedenbeck M E 2005 *J. Geophys. Res.* **110** A09S16

[13] Mason G M, Gold R E, Krimigis S M, Mazur J E, Andrews G B, Daley K A, Dwyer J R, Heuerman K F, James T L, Kennedy M J, Lefevere T, Malcolm H, Tossman B and Walpole P H 1998 *Space Sci. Rev.* **86** 409–448

[14] Stone E C, Cohen C M S, Cook W R, Cummings A C, Gauld B, Kecman B, Leske R A, Mewaldt R A, Thayer M R, Dougherty B L, Grumm R L, Milliken B D, Radocinski R G, Wiedenbeck M E, Christian E R, Shuman S and von Rosenvinge T T 1998 *Space Science Reviews* **86** 357–408

[15] Gold R E, Krimigis S M, Hawkins S E I, Haggerty D K, Lohr D A, Fiore E, Armstrong T P, Holland G and Lanzerotti L J 1998 *Space Science Reviews* **86** 541–562

[16] Stone E C, Frandsen A M, Mewaldt R A, Christian E R, Margolies D, Ormes J F and Snow F 1998 *Space Science Reviews* **86** 1–22

[17] Cook W R, Cummings A C, Cummings J R, Garrard T L, Kecman B, Mewaldt R A, Selesnick R S, Stone E C, Baker D N and von Rosenvinge T T 1993 *IEEE Transactions on Geoscience and Remote Sensing (ISSN 0196-2892)* **31** 565–571

[18] Torsti J, Valtonen E, Lumme M, Peltonen P, Eronen T, Louhola M, Riihonen E, Schultz G, Teittinen M, Ahola K, Holmlund C, Kelha V, Leppälä K, Ruuska P and Strömmer E 1995 *Solar Physics* **162** 505–531

[19] Desai M I, Mason G M, Dwyer J R, Mazur J E, Gold R E, Krimigis S M, Smith C W and Skoug R M 2003 *The Astrophysical Journal* **588** 1149–1162

[20] Desai M I, Mason G M, Gold R E, Krimigis S M, Cohen C M S, Mewaldt R A, Mazur J E and Dwyer J R 2006 *Astrophys. J.* **649** 470–489

[21] Desai M I, Mason G M, Dayeh M A, Ebert R W, McComas D J, Li G, Cohen C M S, Mewaldt R A, Schwadron N A and Smith C W 2016 *submitted to The Astrophysical Journal*

[22] Band D, Matteson J, Ford L, Schaefer B, Palmer D, Teegarden B, Cline T, Briggs M, Paciesas W, Pendleton G, Fishman G, Kouveliotou C, Meegan C, Wilson R and Lestrade P 1993 *Astrophys. J.* **413** 281–292

[23] Markwardt C B 2009 *Astronomical Data Analysis Software and Systems XVIII ASP Conference Series* **411** 251

[24] Cane H V, Richardson I G and von Rosenvinge T T 2010 *Journal of Geophysical Research* **115** A08101

[25] Kahler S W 2013 *The Astrophysical Journal* **769** 110

[26] Ding L, Jiang Y, Zhao L and Li G 2013 *The Astrophysical Journal* **763** 30

[27] Richardson I G, von Rosenvinge T T, Cane H V, Christian E R, Cohen C M S, Labrador A W, Leske R A, Mewaldt R A, Wiedenbeck M E and Stone E C 2014 *Solar Physics* **289** 3059–3107

[28] Reames D V 2013 *Solar Physics* **astro-ph.SR(3)** 201–993

[29] Mason G M, Li G, Cohen C M S, Desai M I, Haggerty D K, Leske R A, Mewaldt R A and Zank G P 2012 *The Astrophysical Journal* **761** 104

[30] Li G and Lee M A 2015 *The Astrophysical Journal* **810** 82

[31] Zank G P, Rice W K M and Wu C C 2000 *J. Geophys. Res.* **105** 25079–25096

[32] Drury L O 1983 *Rep. Prog. Phys.* **46** 973–1027

[33] Schwadron N A, Lee M A, Gorby M, Lugaz N, Spence H E, Desai M I, Török T, Downs C, Linker J, Lionello R, Mikić Z, Riley P, Giacalone J, Jokipii J R, Kota J and Kozarev K 2015 *The Astrophysical Journal* **810** 97

[34] Schwadron N A, Lee M A, Gorby M, Lugaz N, Spence H E, Desai M I, Török T, Downs C, Linker J, Lionello R, Mikić Z, Riley P, Giacalone J, Jokipii J R, Kota J and Kozarev K 2015 *Journal of Physics: Conference Series* **642** 012025

[35] Viñas A F and Scudder J D 1986 *Journal of Geophysical Research (ISSN 0148-0227)* **91** 39

[36] Li G, Zank G P, Verkhoglyadova O, Mewaldt R A, Cohen C M S, Mason G M and Desai M I 2009 *Astrophys. J.* **702** 998–1004

[37] Dröge W 1994 *Astrophys. J. Suppl. Ser.* **90** 567–576

[38] Battarbee M, Laitinen T and Vainio R 2011 *Astronomy and Astrophysics* **535** A34

[39] Battarbee M, Vainio R, Laitinen T and Hietala H 2013 *Astronomy and Astrophysics* **558** A110

[40] Vainio R, Pönni A, Battarbee M, Koskinen H E J, Afanasiev A and Laitinen T 2014 *Journal of Space Weather and Space Climate* **4** A08

[41] Zank G P, Li G, Florinski V, Hu Q, Lario D and Smith C W 2006 *Journal of Geophysical Research* **111** A06108–16

[42] Tylka A J and Lee M A 2006 *Astrophys. J.* **646** 1319–1334

[43] Giacalone J 2005 *Astrophys. J.* **628** L37–L40

Hard X-ray morphology of the X1.3 April 25, 2014 partially occulted limb solar flare

Frederic Effenberger, Fátima Rubio da Costa and Vahé Petrosian

Department of Physics and KIPAC, Stanford University, Stanford, CA 94305, USA

E-mail: feffen@stanford.edu

Abstract. At hard X-ray energies, the bright footpoint emission from solar flare loops often prevents a detailed analysis of the weaker loop-top source morphology due to the limited dynamic range available for X-ray imaging. Here, we study the X1.3 April 25, 2014 flare with the *Reuven Ramaty High-Energy Solar Spectroscopic Imager* (*RHESSI*). This partially occulted limb flare allows the analysis of the loop-top emission in isolation. We present results on the flare light curve at different energies, the source morphology from X-ray imaging and a detailed spectral analysis of the different source components by imaging spectroscopy. The loop-top source, a likely site of particle acceleration, shows a clear composition of different emission components. The results indicate the opportunities that detailed imaging of hard X-rays can provide to learn about particle acceleration, transport and heating processes in solar flares.

1. Introduction

Hard X-rays provide one of the most direct diagnostics for energetic particle processes in solar flares [1]. Studies in the past have shown that in some flares at least two distinct types of sources, namely from the coronal solar flare loop-top and chromospheric footpoints, can be distinguished [2, 3, 4, 5, 6, 7, 8, 9]. It is suggested by theory and observations that the coronal region at the loop-top is the main acceleration site for electrons (see, e.g. the review [10]). However, due to the limited dynamical range of *RHESSI* [11], it is often hard to clearly distinguish a coronal source, once strong footpoint emission is present. Partially occulted flares, in which the footpoints are a few degrees behind the solar limb, offer the opportunity to observe the coronal sources in isolation. Krucker & Lin [12] performed a statistical study of partially occulted flares during solar cycle 23. They concluded that thin-target radiation of electrons is a plausible scenario for the coronal emission. Some of the analyzed flares also show significant structure, indicating multiple coronal source regions. Kappa-distribution models [13, 14, 15] have been discussed as possible models for coronal X-ray sources, in particular if the sources are separated into different components, including an "above-the-loop-top" source.

Here, we focus on the X1.3 April 25, 2014 partially occulted limb flare, which shows a rich loop-top source morphology. Results on the flare light curve in different energies, the source structure from X-ray imaging, and a detailed spectral analysis of the different source components from imaging spectroscopy are presented. We note that this flare has also recently been studied in the context of CME initiation and a magnetic breakout scenario. See the paper by Chen et al. [16] for further details on these aspects.

Journal of Physics: Conference Series **767** (2016) 012005 doi:10.1088/1742-6596/767/1/012005

2. Temporal evolution

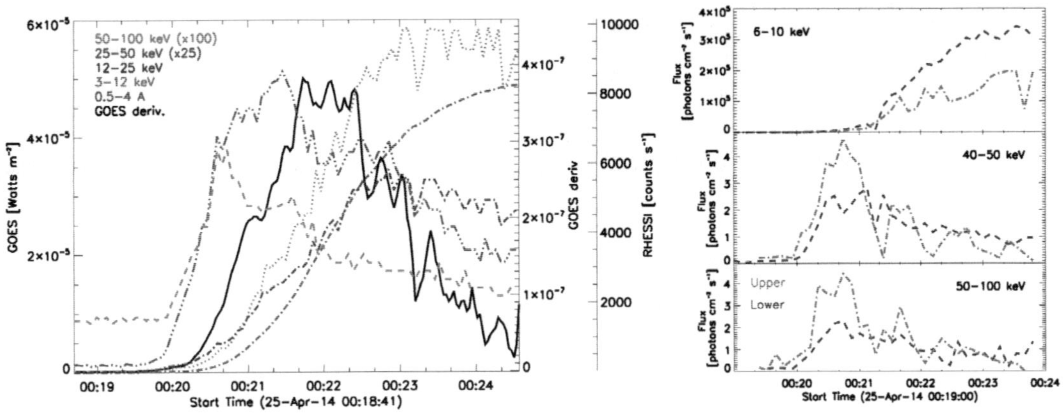

Figure 1. Upper left panel: Light curves of *RHESSI* count rates at four energy ranges (green, blue, purple and magenta), *GOES* high energy flux (0.4 to 4 Å, grey, dash-dotted) and its time derivative (black, solid). The counts in the two high energy channels are multiplied by 25 and 100, respectively, to make them comparable in magnitude. Right panel: Time evolution of photon flux in the two coronal regions selected (see the right panel of Figure 2), for three different energy channels.

We analyzed the temporal evolution of the hard X-ray flux for this flare and compared it with the *GOES* soft X-ray evolution in the high energy channel (0.4 to 4 Å). The left panel of Figure 1 gives an overview of the *RHESSI* light curve in four energy channels and the correlation with the *GOES* flux and its time derivative.

It can be seen that the two lowest energy channels are delayed with respect to the *GOES* derivative and are roughly correlated with the slow rise in *GOES* flux. The two high energy channels 25-50 keV and 50-100 keV, on the other hand, have an earlier and quicker rise to their maxima. In particular the 50-100 keV shows a prominent peak at the early onset and then a fast decay. A cross-correlation analysis showed the best correlation between the 25-50 keV channel and the *GOES* derivative, with a correlation coefficient of 0.81 and a lag of 12 s.

We note that a statistical study of more partially occulted events is currently in progress and a significant number of flares show better correlations and often small or zero lag. The implications of these results, in particular in the context of the so-called Neupert effect [17, 18], will be discussed elsewhere.

The right panel of Figure 1 shows the photon flux evolution in two different regions of the coronal source, namely an upper and lower part. The regions are defined in the right panel of Figure 2. We see that there is a faster rise at high energies in the upper part of the coronal source resulting in a stronger maximum. The lower energy fluxes slowly rise to larger values for the lower source. We now turn to the details of the source structure in X-ray images.

3. Hard X-ray Imaging

RHESSI's unique imaging capability allows a detailed study of the spatial structure of the hard X-ray emission. In Figure 2, we present the evolution over four subsequent time intervals of 12 s length. The X-ray contours, obtained using the CLEAN algorithm [19], at three different energies are overlaid over AIA 131 Å background images closest in time to the *RHESSI* integration time interval (start times given above the images). An expanding upper loop structure can be detected in the AIA images, associated to a high energy (50-100 keV) upper source moving outwards.

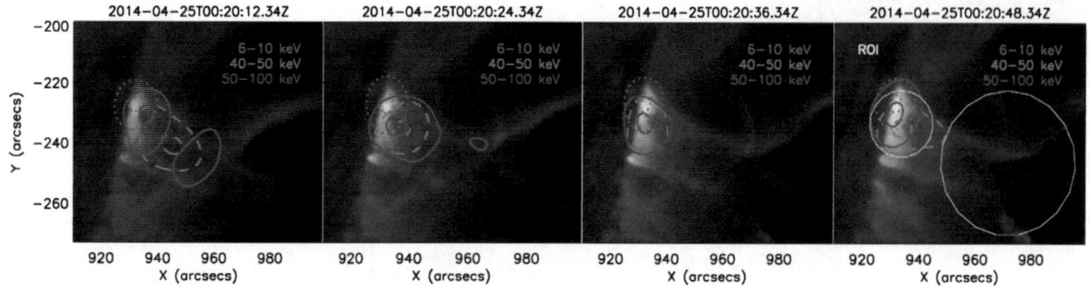

Figure 2. Time evolution of the X1.3 April 25, 2014 flare in AIA 131 Å with *RHESSI* contours at different energy ranges (77% and 97% of the maximum intensity at each time, applying the CLEAN algorithm): 6-10 keV (dotted blue), 40-50 keV (dashed green) and 50-100 keV (solid purple). The rightmost panel also indicates the two selected ROI for the imaging spectroscopy analysis of Section 4.

The high energy (up to 100 keV), non-thermal loop-top emission comes mainly from the upper region of the X-ray source, while the thermal 6-10 keV emission is much stronger in the lower parts of the loops. We note that no quasi-footpoint emission is detected, confirming that the flare is partially occulted and that we are seeing a simultaneously observed loop-top and a second higher above-the-loop-top source (see, e.g., [20, 21, 22]). These uncommon observations allow us to separate the upper and lower coronal emission for further studies. To this end, we performed an imaging spectroscopy study discussed next.

4. Imaging spectroscopy

Thanks to the high energy and spatial resolution of *RHESSI*, it is possible to obtain X-ray spectra of different separated regions. In order to study the spectral evolution of the two sources, we defined two regions of interest (ROI), indicated as white circles in the right panel of Figure 2. We obtained photon spectra of the two different regions resulting from the CLEAN images of these regions.

The red and blue data points of Figure 3 give the results for the upper and lower sources, respectively, at four different times. The spectra are fitted to a thin-target model with a power law electron energy spectrum. The resulting electron spectral index δ is reported in the bottom part of the figures. In general, the upper source has a harder high energy spectrum.

The fit derived from the thin target model is in good agreement with the data in all cases. We note that a thick-target model is often not an appropriate model for coronal sources, due to the low column depth of gas in the source (see, however, [23]). In particular, the relevant time scales of particle escape and energy loss in the loop source are often such that particles tend to lose only a fraction of their energy before they escape. This was found in the study by Chen & Petrosian [24] for an X3.9 and an M2.1 on disk flares, where these time scales could be determined non-parametrically with an inversion procedure.

A thin-target kappa function fit [13] to the photon spectra gives, as should be expected, almost identical values for κ and δ. This is due to the fact that κ is defined to be the parameter in the number density distribution while δ is the index of the electron flux density distribution, see the discussion in [14]. We do not show these fit results since they look very similar to Figure 3.

To gain an additional perspective on the spectral properties of electrons, we obtained electron images and spectra directly using a regularized imaging inversion method [25]. The resulting electron images (not shown) display a similar upward motion of the upper source as in the photon images. The electron spectra for the same ROI are shown in Figure 4, and were fitted

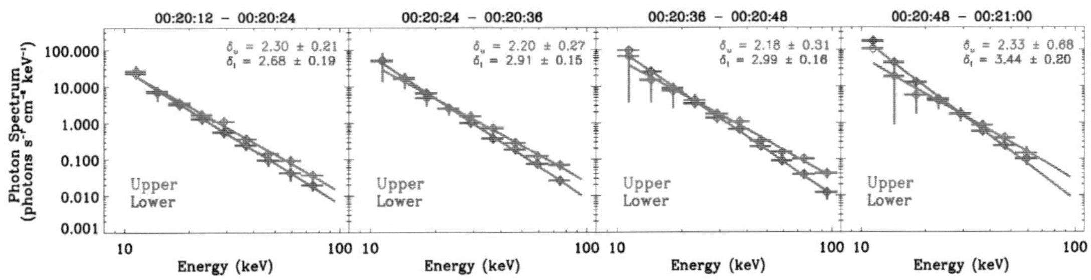

Figure 3. Photon spectra at the same four time ranges as in Figure 2 for the two separate ROI (red: upper source, blue: lower source). The spectra have been fitted to a thin-target model based on a broken power-law. The parameter values of the fit are given in the lower parts of the panels.

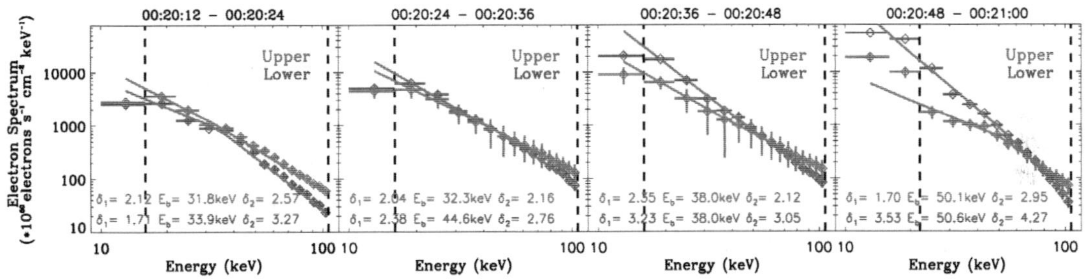

Figure 4. Similar to Figure 3, but now as electron spectra resulting from the regularized imaging inversion with the same ROI. The spectra have been fitted to a broken power-law, with the parameter values of the fit reported in the lower parts of the panels.

to a broken power-law. We see that there is significant electron flux at high energies, with clear power-laws extending to up to ~100 keV, for both source regions. Considering that the errors in the fitting parameters are of the same order as in the previous thin-target fit, the upper source has generally slightly harder spectral indices and higher fluxes at high energies, pointing at a possibly more effective acceleration mechanism in this part of the loop. Most of the spectra are compatible with a single power-law, in particular for the two time intervals in the middle.

Taking into account the confidence intervals for the photon spectra thin target fitting parameters, as stated in Figure 3, we find that the resulting electron spectra are in reasonable agreement for both methods. This confirms the notion of thin-target emission for the entire coronal source region and gives independent support to the different spectral characteristics of the two parts of the source.

5. Conclusions

We have analyzed the X1.3 April 25, 2014 partially occulted limb flare in hard X-rays with *RHESSI* imaging and spectroscopy. The coronal emission shows a rich morphology with at least two different coronal source regions. The upper source is moving upwards and detaching from the lower source during the flare evolution over a short time interval of tens of seconds (see also [26, 27] for similar observations of other flares).

We found a different electron spectral behavior in the upper and lower source derived from a forward thin-target fit and regularized imaging inversion. The differences between the results obtained with the two methods are mostly within the error bounds. The upper source has somewhat smaller (harder) spectral indices than the lower source. Different scenarios could

Journal of Physics: Conference Series **767** (2016) 012005 doi:10.1088/1742-6596/767/1/012005

explain this result. Assuming the reconnection region lies between these two source regions, the effective distance to the reconnection region could imply different turbulence levels. This in turn can result in harder (higher turbulence level) or softer (lower turbulence level) electron spectra for the accelerated population [20, 28].

For further insights into the differences of coronal source modeling and electron properties, a larger sample of partially occulted flares needs to be studied. Detailed comparisons with simulations of particle acceleration and transport will help to uncover the acceleration mechanisms at work in different parts of the flare loop.

Acknowledgments
The authors thank S. Krucker for suggesting the combined *RHESSI* and *GOES* light-curve analysis. F.E. thanks the organizers and participants of the 15th Annual International Astrophysics Conference for the stimulating discussions. This work was supported by NASA grants NNX14AG03G and NNX13AF79G.

References

[1] Fletcher L, Dennis B R, Hudson H S *et al.* 2011 *Space Sci. Rev.* **159** 19–106
[2] Frost K J and Dennis B R 1971 *ApJ* **165** 655
[3] Roy J R and Datlowe D W 1975 *Sol. Phys.* **40** 165–182
[4] McKenzie D L 1975 *Sol. Phys.* **40** 183–191
[5] Hudson H S 1978 *ApJ* **224** 235–240
[6] Masuda S, Kosugi T, Hara H, Tsuneta S and Ogawara Y 1994 *Nature* **371** 495–497
[7] Petrosian V, Donaghy T Q and McTiernan J M 2002 *ApJ* **569** 459–473
[8] Krucker S, White S M and Lin R P 2007 *ApJ* **669** L49–L52
[9] Simões P J A and Kontar E P 2013 *A&A* **551** A135
[10] Petrosian V 2012 *Space Sci. Rev.* **173** 535–556
[11] Lin R P, Dennis B R, Hurford G J *et al.* 2002 *Sol. Phys.* **210** 3–32
[12] Krucker S and Lin R P 2008 *ApJ* **673** 1181–1187
[13] Kašparová J and Karlický M 2009 *A&A* **497** L13–L16
[14] Oka M, Ishikawa S, Saint-Hilaire P, Krucker S and Lin R P 2013 *ApJ* **764** 6
[15] Bian N H, Emslie A G, Stackhouse D J and Kontar E P 2014 *ApJ* **796** 142
[16] Chen Y, Du G, Zhao D, Wu Z, Liu W, Wang B, Ruan G, Feng S and Song H 2016 *ApJ* **820** L37
[17] Neupert W M 1968 *ApJ* **153** L59
[18] Veronig A M, Brown J C, Dennis B R, Schwartz R A, Sui L and Tolbert A K 2005 *ApJ* **621** 482–497
[19] Hurford G J, Schmahl E J, Schwartz R A *et al.* 2002 *Sol. Phys.* **210** 61–86
[20] Liu W, Petrosian V, Dennis B R and Jiang Y W 2008 *ApJ* **676** 704-716
[21] Liu W, Chen Q and Petrosian V 2013 *ApJ* **767** 168
[22] Oka M, Krucker S, Hudson H S and Saint-Hilaire P 2015 *ApJ* **799** 129
[23] Fleishman G D, Xu Y, Nita G N and Gary D E 2016 *ApJ* **816** 62
[24] Chen Q and Petrosian V 2013 *ApJ* **777** 33 (*Preprint* 1307.1837)
[25] Piana M, Massone A M, Hurford G J, Prato M, Emslie A G, Kontar E P and Schwartz R A 2007 *ApJ* **665** 846–855
[26] Sui L and Holman G D 2003 *ApJ* **596** L251–L254
[27] Sui L, Holman G D and Dennis B R 2004 *ApJ* **612** 546–556
[28] Petrosian V and Liu S 2004 *ApJ* **610** 550–571

Journal of Physics: Conference Series **767** (2016) 012006 doi:10.1088/1742-6596/767/1/012006

Faraday tomography: a new, three-dimensional probe of the interstellar magnetic field

Katia Ferrière

IRAP, Université de Toulouse, CNRS, 9 avenue du Colonel Roche, BP 44346, F-31028 Toulouse Cedex 4, France

Abstract.
Much of our present observational knowledge of the interstellar magnetic field of our Galaxy comes from two different sources, both of which involve mechanisms operating at radio wavelengths: the first one is the Faraday rotation of linearly-polarized radio waves propagating through the magneto-ionic interstellar medium, and the second one is the diffuse synchrotron emission from our Galaxy. I will review what these two classical probes have taught us about the strength, the direction/orientation, and the spatial distribution of the interstellar magnetic field. I will then present a recent method, known as Faraday tomography or rotation measure synthesis, which relies on a combination of Faraday rotation and synchrotron emission and which makes it possible to probe the interstellar magnetic field in three dimensions.

1. Introduction

The interstellar magnetic field (ISMF), $\vec{B}$, is an important component of the interstellar medium (ISM) of our Galaxy. It plays a crucial role in a variety of physical processes, including the acceleration, propagation, and confinement of cosmic rays, the spatial distribution, dynamics, and energetics of interstellar matter, the vital process of star formation... However, its properties remain poorly understood.

One of the main difficulties is the lack of direct measurements, apart from Zeeman-splitting measurements. In addition, all existing observational methods provide only partial information: Zeeman-splitting measurements yield $B_\parallel$, the line-of-sight component of $\vec{B}$, in neutral (molecular and atomic) clouds. Faraday rotation measures (RMs), in combination with a model for the thermal-electron density, lead to $B_\parallel$ in ionized regions. Synchrotron emission data, in combination with assumptions on the relativistic-electron density, lead to $\vec{B}_\perp$, the plane-of-sky projection of $\vec{B}$, in the general, cosmic-ray filled ISM. Finally, linear polarization of starlight and of dust thermal emission yields the orientation of $\vec{B}_\perp$ in the general, dusty ISM.

In most cases, magnetic field observations are difficult to interpret, because the observed emission (or Faraday rotation) along any line of sight is generally produced by a number of structures, whose contributions are often hard to disentangle and locate along the line of sight. One way to overcome this intrinsic problem is to perform Faraday tomography, a recent method also known as rotation measure synthesis, which exploits the Faraday rotation of the Galactic diffuse synchrotron emission.

Because understanding Faraday rotation and synchrotron emission is prerequisite to understanding Faraday tomography, I will start with a brief overview of Faraday rotation in

Section 2 and synchrotron emission in Section 3. I will then proceed with a discussion of Faraday tomography in Section 4.

2. Faraday rotation

Faraday rotation is the rotation of the polarization direction of a linearly-polarized radio wave that propagates through a magneto-ionic medium, where it interacts with the free (thermal) electrons gyrating about magnetic field lines. The polarization direction rotates by an angle $\Delta\theta = \mathrm{RM}\ \lambda^2$, where λ is the observing wavelength and RM is the so-called rotation measure, given by

$$\mathrm{RM} = \mathcal{C} \int_0^L n_e\ B_\parallel\ ds\ , \tag{1}$$

with $\mathcal{C}$ a numerical constant, n_e the thermal-electron density, $B_\parallel$ the line-of-sight component of the magnetic field (positive/negative when the field points toward/away from the observer), and L the path length between the source and the observer. In practice, the RM of a given radio source can be determined by measuring the polarization direction of the incoming radiation at at least two different wavelengths.

RMs have now been derived for about 1 100 Galactic pulsars [1] and 42 000 extragalactic radio point sources (see Figure 1). Used in conjunction either with a model for the spatial distribution of thermal electrons or, in the case of Galactic pulsars, with their distances and dispersion measures (DMs), these numerous RMs have made it possible to gather a wealth of information on the strength, direction, and spatial configuration of the ISMF in the magneto-ionic ISM. Here are the key points that have emerged from Faraday rotation studies.

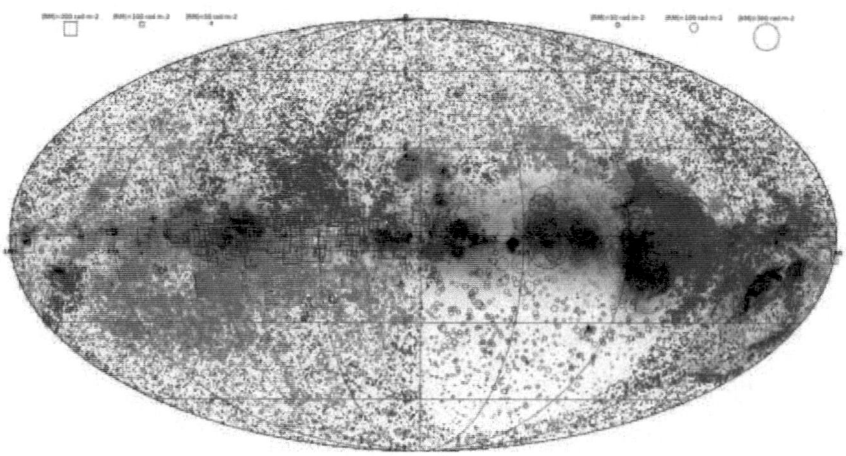

Figure 1. All-sky map in Galactic coordinates (with the Galactic center in the middle, longitude increasing to the left, and latitude increasing upward) of the RMs of $\simeq 42\,000$ extragalactic radio point sources from the NVSS ($\delta > -40°$) and S-PASS ($\delta < 0°$) surveys. Positive/negative RMs, wich correspond to a magnetic field pointing on average toward/away from the observer, are plotted in blue/red. Figure credit: Dominic Schnitzeler.

1) The ISMF has a regular component, $\vec{B}_{\rm reg}$, and a randomly fluctuating component, $\vec{B}_{\rm fluct}$. In the large-scale vicinity of the Sun, $B_{\rm reg} \simeq 1.5\ \mu$G and $B_{\rm fluct} \sim 5\ \mu$G [2]. $B_{\rm reg}$ increases toward the Galactic center, to $\gtrsim 3\ \mu$G at Galactocentric radius $R = 3$ kpc [3], i.e., with an exponential scale length $\lesssim 7.2$ kpc. Moreover, $B_{\rm reg}$ decreases away from the Galactic plane, albeit at a very uncertain rate; for reference, the exponential scale height inferred from the RMs of extragalactic radio point sources is ~ 1.4 kpc [4].

2) In the Galactic disk, $\vec{B}_{\rm reg}$ is nearly horizontal and generally dominated by its azimuthal component. In the large-scale vicinity of the Sun, $\vec{B}_{\rm reg}$ runs clockwise at an angle $\simeq -8°$ to the azimuthal direction [5], which is very close to the pitch angle $\simeq -7°$ inferred from starlight polarization [6]. $\vec{B}_{\rm reg}$ reverses direction at least a couple of times with decreasing Galactocentric radius, but the exact number and radial locations of the field reversals are still highly controversial [3; 5; 7–10]. These field reversals have often been interpreted as evidence that $\vec{B}_{\rm reg}$ is bisymmetric (azimuthal wavenumber $m = 1$), while an axisymmetric ($m = 0$) field would be expected from dynamo theory. In reality, [11] showed that neither the axisymmetric nor the bisymmetric picture is consistent with the existing pulsar RMs, and they concluded that $\vec{B}_{\rm reg}$ must have a more complex pattern.

3) In the Galactic halo, $\vec{B}_{\rm reg}$ could have a significant vertical component. Near the horizontal position of the Sun, [12] obtained $(B_{\rm reg})_z \simeq -0.14\ \mu$G above the Galactic midplane ($z > 0$) and $(B_{\rm reg})_z \simeq +0.30\ \mu$G below the midplane ($z < 0$), whereas [13] obtained $(B_{\rm reg})_z \simeq 0.00\ \mu$G toward the north Galactic pole and $(B_{\rm reg})_z \simeq +0.31\ \mu$G toward the south Galactic pole. In contrast to the situation prevailing in the Galactic disk, the horizontal component of $\vec{B}_{\rm reg}$ shows no sign of reversal with decreasing Galactocentric radius.

4) $\vec{B}_{\rm reg}$ exhibits some symmetry properties with respect to the Galactic midplane. At low latitudes (basically, in the disk), $\vec{B}_{\rm reg}$ appears to be roughly symmetric in z [7; 14], while at high latitudes (in the halo), the RM sky features a rather striking antisymmetry/symmetry in z in the inner/outer Galactic quadrants [5; 15; 16], which suggests that $\vec{B}_{\rm reg}$ is roughly antisymmetric in z inside the solar circle [5; 16, but see also [14]]. Finding $\vec{B}_{\rm reg}$ to be symmetric in the disk and antisymmetric in the inner halo is consistent with the predictions of dynamo theory and with the results of galactic dynamo calculations [e.g., 17–19].

3. Synchrotron emission

Synchrotron emission is produced by relativistic electrons gyrating about magnetic field lines. The synchrotron emissivity at frequency ν due to a power-law energy spectrum of relativistic electrons, $f(E) = K_e\, E^{-\gamma}$, is given by

$$\mathcal{E}_\nu \;=\; \mathcal{F}(\gamma)\; K_e\; B_\perp^{\frac{\gamma+1}{2}}\; \nu^{-\frac{\gamma-1}{2}}\;, \tag{2}$$

where $\mathcal{F}(\gamma)$ is a known function of the electron spectral index and $B_\perp$ is the strength of the sky-projected magnetic field.

The spatial distribution of the Galactic synchrotron emissivity was modeled by [20], based on the all-sky 408 MHz radio continuum map of [21] (see Figure 2). Several authors then used this synchrotron distribution model to derive the ISMF distribution in our Galaxy. To do so, the vast majority of them resorted to the standard double assumption that (1) relativistic electrons represent a fixed fraction of the cosmic-ray population and (2) cosmic rays and magnetic fields are in (energy or pressure) equipartition. Relying on the cosmic-ray ion and electron spectra directly measured by the *Voyager* spacecraft, [22] verified that, near the Sun, cosmic rays and magnetic fields are indeed close to (pressure) equipartition, with a total magnetic field strength $B \simeq 5\ \mu$G. She also found that the total field strength has a radial scale length $\simeq 12$ kpc and a vertical scale height near the horizontal position of the Sun $\simeq 4.5$ kpc.

Because synchrotron emission is linearly polarized perpendicular to $\vec{B}_\perp$, information can also be gained on the orientation of $\vec{B}_\perp$. Evidently, if the observing frequency is low enough to be affected by Faraday rotation, the received polarized signal must somehow be "de-rotated" in order to retrieve the true field orientation. In addition, if the ISMF has a randomly fluctuating component, the contributions from isotropic magnetic fluctuations to the polarized emission cancel out, leaving only the contribution from the ordered (i.e., regular + anisotropic random)

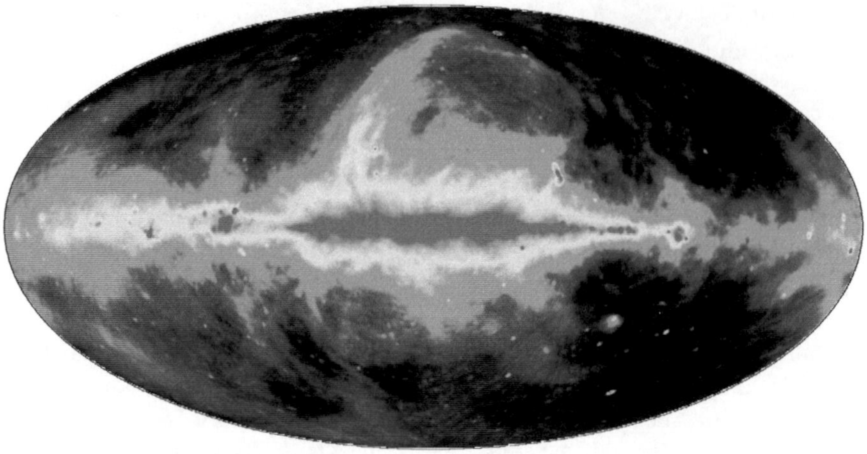

Figure 2. All-sky map in Galactic coordinates (with the Galactic center in the middle) of the 408 MHz total intensity from the Bonn 408-MHz All-Sky Survey, using data obtained with the Jodrell-Bank 76 m, Effelsberg 100 m and Parkes 64 m radio telescopes. Figure credit: MPIfR/Patricia Reich.

ISMF, $\vec{B}_{\rm ord}$. Thus, while the *total* synchrotron intensity yields the strength of the *total* sky-projected ISMF, the *polarized* synchrotron intensity yields the strength and the orientation of the *ordered* sky-projected ISMF.

In the large-scale vicinity of the Sun, the ratio of ordered to total ISMF strengths turns out to be $\simeq 0.6$ [23]. Together with $B \simeq 5\ \mu$G, this ratio implies an ordered ISMF strength $B_{\rm ord} \simeq 3\ \mu$G. Besides, the radio polarization vectors confirm that the large-scale ordered ISMF in the Galactic disk is horizontal.

4. Faraday tomography

An important limitation of the methods described in Sections 2 and 3 is that they provide only line-of-sight integrated quantities, with no details on how the integrants vary along the line of sight. For instance, the synchrotron intensity measured in a given direction tells us only about the total amount of synchrotron emission produced along the entire line of sight through the Galaxy, with no information on the local value of $\mathcal{E}_\nu$ (Eq. 2). Similarly, the RM of a given radio source (Eq. 1) tells us only about the total amount of Faraday rotation incurred along the line of sight between the source and the observer, with no information on the local value of $n_e\, B_\|$. For the latter, some large-scale trends can in principle be inferred if the line of sight encounters a sizeable number of well-separated Galactic pulsars with known distances and RMs. In practice, however, there are only $\simeq 1\,100$ Galactic pulsars with measured RMs [1], and their distances remain quite uncertain.

An alternative, more powerful and more promising, approach to probe the three-dimensional structure of the ISMF is now being increasingly utilized. This approach is also based on Faraday rotation, but instead of considering the Faraday rotation of the linearly-polarized radiation from a background radio source (as explained in Section 2), the idea is to exploit the Faraday rotation of the synchrotron radiation from the Galaxy itself (discussed in Section 3).

As a reminder, in the case of a background radio source, i.e., when the regions of radio emission and Faraday rotation are spatially separated, the Faraday rotation angle, $\Delta\theta$, increases linearly with wavelength squared, λ^2, and one may define RM as being the slope of the linear relation between $\Delta\theta$ and λ^2, such that $\Delta\theta = {\rm RM}\ \lambda^2$ (see Section 2). Hence, RM is a purely observational quantity, which can be meaningfully measured only for a background radio source and which can then be related to the physical properties of the foreground Faraday-rotating medium through Eq. (1).

In contrast, when the radio source is the Galaxy itself, the regions of radio emission and Faraday rotation are spatially mixed. In that case, $\Delta\theta$ no longer increases linearly with λ^2 and the very concept of RM becomes meaningless. However, one may resort to the more general concept of Faraday depth (FD), defined as

$$\Phi = \mathcal{C} \int_0^z n_e \, B_\parallel \, ds \,, \tag{3}$$

where $\mathcal{C}$, n_e, and $B_\parallel$ have the same meaning as in Eq. (1) and z is the line-of-sight distance from the observer [24; 25]. Φ has basically the same formal expression as RM (Eq. 1), but it differs from RM in the sense that it is a truly physical quantity, which can be defined at any point of the ISM, independent of any background radio source. Φ simply corresponds to the line-of-sight depth measured in terms of Faraday rotation – in much the same way as optical depth corresponds to line-of-sight depth measured in terms of opacity.

When radio emission and Faraday rotation are mixed along the line of sight, the polarized intensity measured at a given wavelength λ is the superposition of the polarized emission produced at all line-of-sight distances z, i.e., at all Faraday depths Φ, and Faraday-rotated by an angle $\Delta\theta = \Phi \, \lambda^2$:

$$P(\lambda^2) = \int_{-\infty}^{+\infty} F(\Phi) \, e^{2i\Phi\lambda^2} \, d\Phi \,, \tag{4}$$

where $P(\lambda^2)$ is the complex polarized intensity ($P = Q + iU$) at λ, and $F(\Phi)$ is the complex polarized intensity per unit Φ, also called complex Faraday dispersion function, at Φ [24]. Since the Faraday rotation angle varies with wavelength, the polarized intensities measured at different wavelengths correspond to different combinations of all the line-of-sight contributions and, therefore, provide different pieces of information. Thus, the idea is to measure the polarized intensity at a large number of different wavelengths and to convert its variation with λ^2 into a variation with Φ. Mathematically, this can be done inverting Eq. (4), i.e., by taking the Fourier-like transform of $P(\lambda^2)$ to obtain $F(\Phi)$.

The method is illustrated in Figure 3, which depicts a situation where the line of sight intersects two Faraday-rotating clouds (shaded in light grey), across which Φ increases or decreases according to Eq. (3), and two synchrotron-emitting clouds (shaded in dark blue). The top panel in both Figures 3a and 3b indicates the positions of the four clouds along the line of sight with respect to the observer placed on the far left, as well as the directions of the ISMF (red arrows) in the two Faraday-rotating clouds: in the closer/farther cloud, the ISMF points toward/away from the observer, so that $B_\parallel$ is positive/negative and Φ increases/decreases with increasing z. The corresponding run of Φ with z is plotted in the middle panel, where Φ_1 denotes the Faraday thickness of the closer cloud and Φ_2 the cumulated Faraday thickness of both Faraday-rotating clouds. The bottom panel displays the Faraday dispersion spectrum, $|F(\Phi)|$, with the two peaks representing the polarized emissions from the two synchrotron-emitting clouds. In Figure 3a, where the Faraday-rotating and synchrotron-emitting clouds are spatially separated, the closer and farther synchrotron-emitting clouds lie at FDs Φ_1 and Φ_2, respectively. In Figure 3b, the closer synchrotron-emitting cloud is embedded inside the closer Faraday-rotating cloud, so it has a finite Faraday thickness, i.e., it extends over a range of FDs (up to nearly Φ_1).

In practice, Faraday tomography can be used to separate synchrotron-emitting regions located at different FDs along the line of sight and to estimate their respective polarized synchrotron intensities, which in turn can lead to the strength and the orientation of their $\vec{B}_\perp$ (see Section 3).

Journal of Physics: Conference Series **767** (2016) 012006 doi:10.1088/1742-6596/767/1/012006

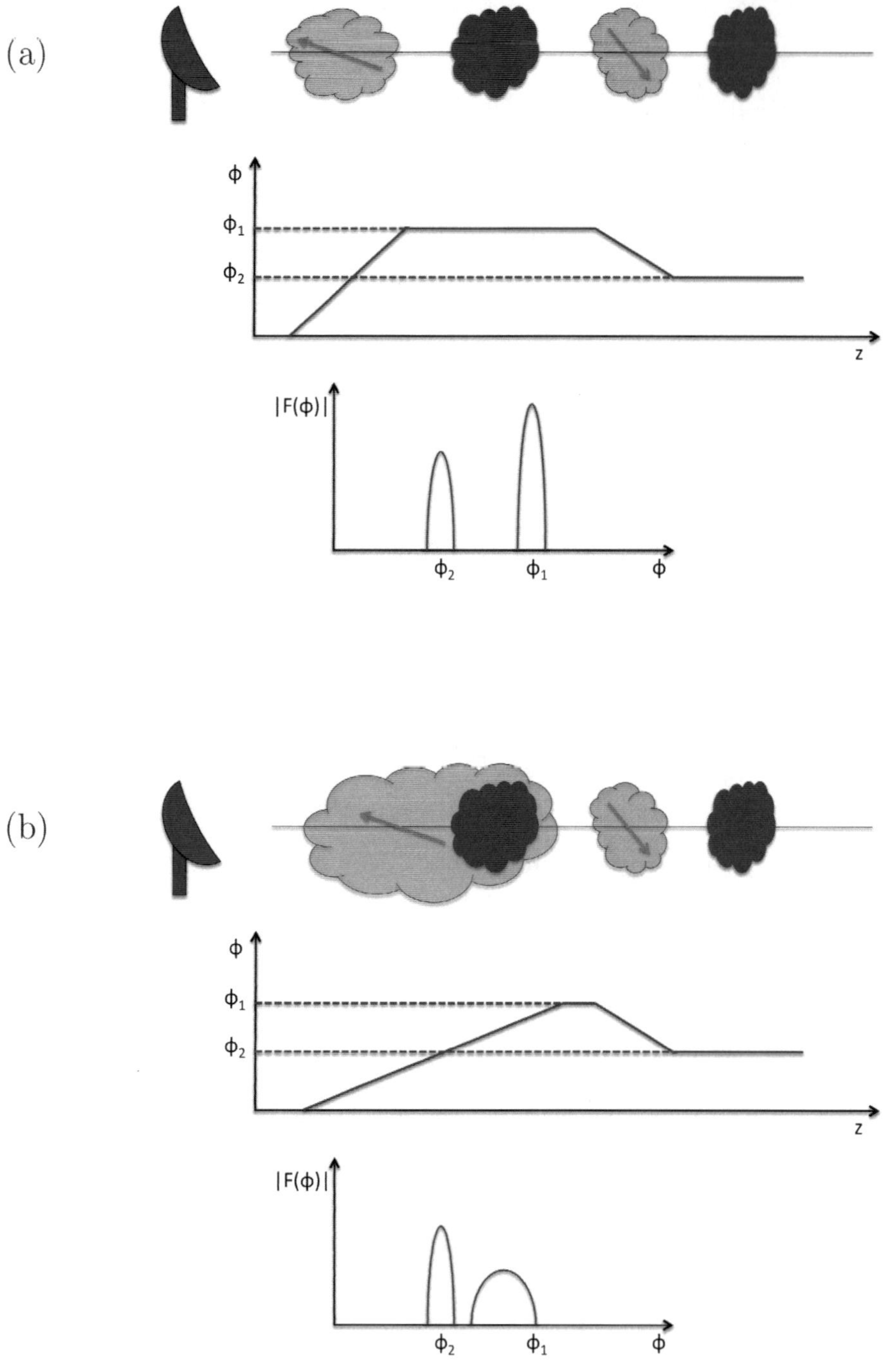

Figure 3. Schematics illustrating the concept of Faraday tomography. The top panel in both (a) and (b) pictures the spatial configuration of the system: two Faraday-rotating clouds (light grey shading, with a red arrow representing the ISMF) and two synchrotron-emitting clouds (dark blue shading), with the observer on the far left. The middle panel shows how the Faraday depth, Φ (given by Eq. 3), varies with line-of-sight distance from the observer, z. The bottom panel provides the Faraday dispersion spectrum, $|F(\Phi)|$. See the main text for more details. Figure credit: Marta Alves.

Faraday tomography can also be used to uncover intervening Faraday screens and to estimate their Faraday thickness, which in turn can lead to their $B_{\|}$. The method is particularly interesting when the uncovered Faraday screens can be identified with known gaseous structures, because it then offers a new way of probing their magnetic field.

As an example of the applicability of the method, [26] performed Faraday tomography of a $140° \times 100°$ field in the northern Galactic hemisphere, based on polarization data from the Global Magneto-Ionic Medium Survey (GMIMS). The observed field turns out to contain a known HI bubble-shell structure, estimated to lie at a distance ~ 150 pc. [26] derived $F(\Phi)$ along many lines of sight covering the observed field, and they assembled all their derived $F(\Phi)$ into a so-called RM-synthesis cube, i.e., a three-dimensional cube of $F(\Phi)$ data, with two axes in the plane of the sky and the third axis along Φ. By scanning this RM-synthesis cube, they found that the background polarized radio emission toward the east and west parts of the HI shell is shifted along the Φ axis, and hence Faraday-rotated, by $\simeq +60$ rad m^{-2} and -50 rad m^{-2}, respectively. Their interpretation was that the east and west parts of the HI shell act as Faraday screens, with respective Faraday thicknesses $\simeq +60$ rad m^{-2} and -50 rad m^{-2}, implying a magnetic field pointing toward/away from the observer, respectively. This interpretation, in turn, suggested that the magnetic field is wrapped around the bubble. In addition, the Faraday thicknesses of both parts of the shell enabled the authors to obtain a rough estimate for the magnetic field strength inside the shell.

Within the next few years, Faraday tomography is likely to lead to a big leap in our knowledge and understanding of the ISMF. Polarization data should be acquired at low radio frequencies in order to enhance the effects of Faraday rotation (which increase as λ^2) and, therefore, to improve sensitivity in FD space. At the same time, broad frequency coverage is needed to achieve fine resolution in FD space [25]. This is why low-frequency, broad-band radio-telescopes, such as the LOw-Frequency ARray (LOFAR) and, in the near future, the Square Kilometer Array (SKA), are ideally suited for the task at hand.

Acknowledgments
I would like to express my gratitude to my colleague, Marta Alves, who kindly accepted to prepare the cartoons illustrating the concept of Faraday tomography in Figure 3.

References

[1] Han J L, Manchester R N, van Straten W and Demorest P B 2016 (*in preparation*)
[2] Rand R J and Kulkarni S R 1989 ApJ **343** 760–772
[3] Han J L, Manchester R N, Lyne A G, Qiao G J and van Straten W 2006 ApJ **642** 868–881 (*Preprint* `arXiv:astro-ph/0601357`)
[4] Inoue M and Tabara H 1981 PASJ **33** 603–+
[5] Han J L, Manchester R N and Qiao G J 1999 MNRAS **306** 371–380 (*Preprint* `arXiv:astro-ph/9903101`)
[6] Heiles C 1996 ApJ **462** 316–+
[7] Rand R J and Lyne A G 1994 MNRAS **268** 497–+
[8] Vallée J P 2005 ApJ **619** 297–305
[9] Brown J C, Haverkorn M, Gaensler B M, Taylor A R, Bizunok N S, McClure-Griffiths N M, Dickey J M and Green A J 2007 ApJ **663** 258–266 (*Preprint* `arXiv:0704.0458`)
[10] Nota T and Katgert P 2010 A&A **513** A65+
[11] Men H, Ferrière K and Han J L 2008 A&A **486** 819–828 (*Preprint* `0805.3454`)
[12] Taylor A R, Stil J M and Sunstrum C 2009 ApJ **702** 1230–1236

[13] Mao S A, Gaensler B M, Haverkorn M, Zweibel E G, Madsen G J, McClure-Griffiths N M, Shukurov A and Kronberg P P 2010 ApJ **714** 1170–1186 (*Preprint* `1003.4519`)

[14] Frick P, Stepanov R, Shukurov A and Sokoloff D 2001 MNRAS **325** 649–664 (*Preprint* `arXiv:astro-ph/0012459`)

[15] Oren A L and Wolfe A M 1995 ApJ **445** 624–641

[16] Han J L, Manchester R N, Berkhuijsen E M and Beck R 1997 A&A **322** 98–102

[17] Ruzmaikin A A, Sokolov D D and Shukurov A M (eds) 1988 *Magnetic fields of galaxies* (*Astrophysics and Space Science Library* vol 133)

[18] Moss D and Sokoloff D 2008 A&A **487** 197–203

[19] Moss D, Sokoloff D, Beck R and Krause M 2010 A&A **512** A61

[20] Beuermann K, Kanbach G and Berkhuijsen E M 1985 A&A **153** 17–34

[21] Haslam C G T, Salter C J, Stoffel H and Wilson W E 1982 A&AS **47** 1

[22] Ferrière K 1998 ApJ **497** 759–+

[23] Beck R 2001 *Space Science Reviews* **99** 243–260 (*Preprint* `arXiv:astro-ph/0012402`)

[24] Burn, B. J. 1966, MNRAS, 133, 67

[25] Brentjens, M. A. & de Bruyn, A. G. 2005, A&A, 441, 1217

[26] Wolleben, M., Fletcher, A., Landecker, T. L., et al. 2010, ApJ, 724, L48

Testing the H-wave Hypothesis for the Origin of the IBEX Ribbon

Horst Fichtner, Adama Sylla, Klaus Scherer
Institut für Theoretische Physik IV, Ruhr-Universität Bochum, Germany
E-mail: hf@tp4.rub.de

Hans-Jörg Fahr, Jochen Zönnchen
Argelander Institut für Astronomie, Universität Bonn, Germany

Abstract. The Interstellar Boundary Explorer (IBEX) spacecraft maps the neutral atom fluxes across the whole sky. Thereby it is indirectly mapping the structure of the outer heliosphere and the (very) local interstellar medium. A particularly interesting feature in the IBEX-Hi all-sky maps of the differential flux of Energetic Neutral Atoms (ENAs) in the 0.7 to 4.3 keV range is the so-called ribbon, i.e. a factor of two to three enhancement in a twenty to thirty degree wide band across the sky. Amongst other hypotheses it has been argued that this ribbon may be related to a neutral density enhancement, a so-called H-wave, in the local interstellar medium. By employing an analytical model of the large-scale structure of the heliosphere it is demonstrated that the H-wave scenario for the ribbon formation leads to results that are fully consistent with the observed location of the ribbon in the full-sky maps as well as in space at all energies detected with IBEX-Hi. As a further extension to previous work and as a necessary prerequisite for the computation of differential ENA fluxes, the evolution of the proton velocity distribution function in the inner heliosheath is semi-analytically computed in terms of a κ-functions with locally determined κ-values.

1. Introduction

The most unexpected feature in the all-sky maps of the differential fluxes of energetic neutral (hydrogen) atoms (ENAs) is the so-called 'ribbon' [1]. In this band the ENA fluxes are two- to three-fold higher than the globally-distributed 'background' fluxes [2]. Even after now seven years since its discovery [3], there is no generally accepted model for the formation of the ribbon.

After at least one ribbon model had to be discarded [4], the remaining ones can be ordered into two groups. According to the first and so far more popular one, the ribbon ENAs are produced from charge exchange processes in the outer heliosheath (OHS), i.e. in the disturbed interstellar flow beyond the heliopause (see, e.g., [5, 6, 7, 8, 9, 10, 11, 12, 13, 14, 15]). The second group consists of those models that assume the ribbon ENAs to be generated in the inner heliosheath (IHS), i.e. between the solar wind termination shock and the heliopause (see, e.g., [16, 17, 18, 19, 20, 21]).

The present paper deals with the second group. It is, first, re-emphasized that - independent of the details of any model belonging to this group - if the ribbon ENAs are produced in the IHS, they must originate in a predestined, favourable region there. Second, it is shown that such models are consistent with the available observations. And since any such model needs to consider the seed population of the ENAs in order to provide differential fluxes, we discuss, third, a model for the evolution of the joint thermal (solar wind) and suprathermal (pick-up) proton velocity distribution in the IHS.

2. A simplified model of the heliosphere

For a number of purposes it is sufficient to consider an analytical representation rather than a numerical simulation of the large-scale flow structure of the heliosphere, for recent examples see [22, 23, 24]. Following the original idea by Parker [25], most of such analytical approaches formulate the overall plasma flow in terms of a velocity potential that consists of two sources, namely a spherically symmetric one at the origin representing the solar wind and a homogeneous one at infinity representing the interstellar flow with velocity $\vec{u}_{ISM}$. Such model can be expected to be a reasonable approximation in the nearly incompressible region between the termination shock and the bow shock [26, 27]. From the specific representation [23, 21] in cylindrical coordinates ρ and z

$$\mathbf{u}(\mathbf{r}) = -\nabla\Phi(\mathbf{r}) \quad ; \quad \Phi(\mathbf{r} = \rho\mathbf{e}_\rho + z\mathbf{e}_z) = |\vec{u}_{ISM}| \left(z + \frac{2}{r}\right) \quad ; \quad r = |\mathbf{r}| = \sqrt{\rho^2 + z^2} \tag{1}$$

follow the flow lines (parameterized with η) as

$$z(\rho) = \left(\eta - \frac{\rho^2}{4}\right) \rho \left(1 - \left[\eta - \frac{\rho^2}{4}\right]^2\right)^{-\frac{1}{2}} \tag{2}$$

The heliopause surface is obtained with $\eta = 1$. Distances are measured in units of the (spherical) termination shock distance R_{TS}. Together with a spherical termination shock one obtains the principal flow geometry shown in Fig. 1. As demonstrated in detail in [22, 23, 21]

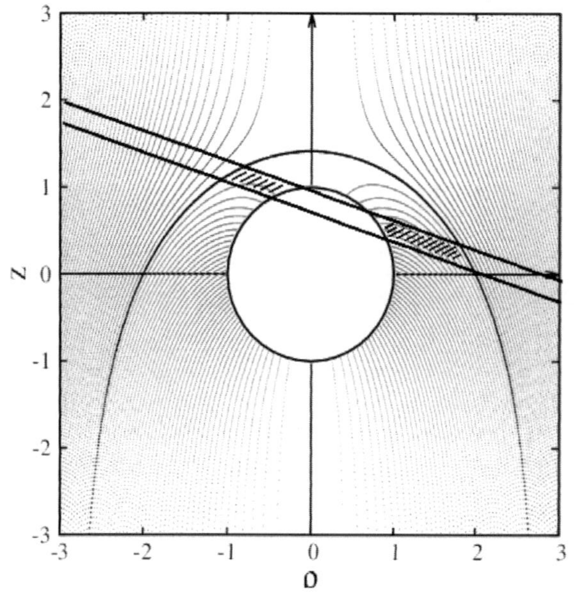

Figure 1. The analytical representation of the large-scale flow structure of the heliosphere and the local interstellar medium (LISM) accoding to Eqs.(1). The upper parabolically shaped thick black line marks the heliopause surface (given by Eq.(2) for $\eta = 1$), the lower one is the termination shock. The colors indicate the flow in the IHS (red) and OHS (blue). The region between the two parallel planes (indicated by the two straight lines, representing the front and rear of the H-wave, for details see text) that is shaded in black must be the source region of ENAs if they originate in the IHS.

via comparisons with numerical simulations and observations, this flow model is remarkably sufficient to principally test models against measurements. Hence, it can be used to test the so-called H-wave hypothesis, which was proposed and first discussed in [20].

3. The H-wave hypothesis

The so-called H-wave model falls into the second group of models explaining the ENA ribbon in the IBEX-Hi all-sky maps with ENAs originating in the IHS. It is, however, the only one that makes a variation in the number density of neutral hydrogen atoms responsible for the increased ENA flux. The basic idea is to acknowledge the inhomogeneity of the local interstellar medium (LISM) also regarding its neutral hydrogen component. From various observations of the general ISM it is known that it is not only inhomogeneous on the large kpc- and pc-scales [28, 29, 30], but also on the sub-pc- down to the 10-100 AU-scales [31] and that this is not only valid for the plasma but also for the neutral component [32, 33]. As argued in [20], the coupling of the plasma and the neutral component (see, e.g., [34, 35, 36]) may lead to the formation of so-called H-waves, i.e. an enhancement in the number density of neutral hydrogen. The existence of such waves has also been predicted by [37]. Such H-wave signature must be expected to decouple from the plasma flow structures in the OHS [20] and to penetrate the heliopause as sketched in Fig. 1 with the two parallel lines indicating the leading and trailing edge of the density enhancement. The (in 3-D) ring-like intersection of such an H-wave with the IHS defines the source region of the ribbon ENAs, because the neutral density enhancement there directly implies a correspondingly enhanced ENA flux. This way the circularity of the ribbon is naturally explained [38].

3.1. Test 1: Geometry of the ribbon

The quantitative study of the H-wave scenario outlined above [21] demonstrated that the indicated ring-like intersection of neutral density enhancement and IHS must indeed be the source region of the ribbon ENAs if the latter are originating inside the heliosphere. The best fit to the observed ribbon in the all-sky flux maps, obtained from the simplified, analytically prescribed termination shock and heliopause that is possibly tilted with respect to the LISM inflow direction, is shown in Fig. 2 at the example of the differential ENA flux at 1.11 keV. The fit parameters comprise the orientation of the H-wave normal $\vec{n}$ and its width in terms of the heliocentric distances $d_{1,2}$ of its leading and trailing edge, respectively. If the upwind direction is given by $\vec{e}_z = (0, 0, 1)$, the best fit normal (unit) vector is given by $\vec{n} = (-0.672, 0.485, 0.560)$ and the best fit H-wave width is $d_2 - d_1 = 0.4\ R_{TS} - 0.1\ R_{TS} = 0.3\ R_{TS}$. Evidently, the described method fits the ribbon excellently. That this also applies to the other IBEX-Hi energy bands up to 4.29 keV is discussed in [21] along with an analysis of the sensitivity of the fit to its parameters.

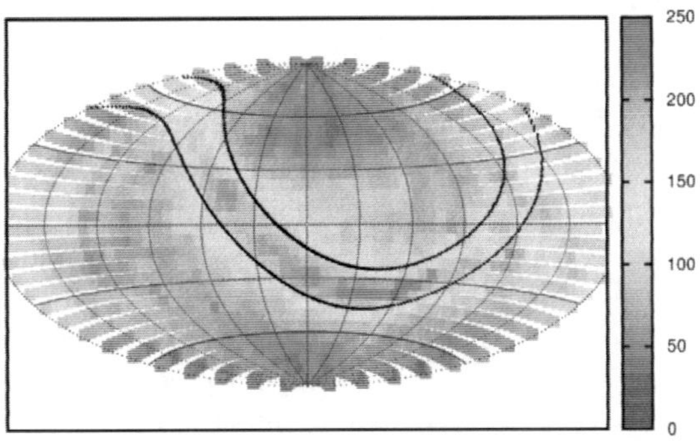

Figure 2. All-sky map of the differential ENA fluxes (ENAs/(cm^2 s sr keV) as observed by IBEX-Hi at 1.11 keV the simulated best-fit ribbon geometry, for details see text. Taken from [21].

It is interesting to note that, independent from the H-wave hypothesis, this finding demonstrates that for geometrical reasons the source region of the ribbon ENAs must be the shaded area indicated in Fig. 1, if they are assumed to originate in the IHS.

3.2. Test 2: Orientiation of the undisturbed LISM magnetic field
Within the H-wave scenario it is clear that the normal vector $\vec{n}$ of the wavefront is not necessarily aligned to the undisturbed local interstellar magnetic field. Since, however, the orientation of the ribbon is undoubtedly related to the latter (see, e.g., [39, 1], another test of the H-wave hypothesis becomes possible. It is straightforward to compute the angle between the best fit normal vector and the upwind direction given by the z-axis as well as the corresponding ecliptic longitude λ_{ecl} and latitude δ_{ecl}. One finds [21] for the angle $\sphericalangle(\vec{n}, -\vec{u}_{ISM}) \approx (55 \pm 5)^{\rm o}$, i.e. a value similar to the 49° estimated by, e.g., [13]. While the values $\lambda_{ecl} \approx 205^{\rm o}$ and $\delta_{ecl} \approx 35^{\rm o}$ are slightly different from those discussed in the literature, these findings translate into the fact that the H-wave front is at least almost perpendicular to the undisturbed LISM magnetic field. This, in turn, is consistent with studies claiming indeed that a tilt of the phase front normal w.r.t. the field direction should be expected to be small [40, 41].

3.3. Test 3: Location in space
One reason for the persistence of two model groups to explain the IBEX ribbon is the fact that the spatial location of the ENAs constituting the associated enhanced fluxes has, as yet, not been identified. Only recently, a first attempt has been made to estimate the heliocentric distance d_{ribbon} of a ribbon element by determining its parallax [42]. While as such this represents a very worthwile enterprise that, in principle, has the potential to discriminate between the two model groups, the accuracy is, so far, still too low to do so in practice. These authors discussed in detail all major error or uncertainty sources and eventually found a value of $d_{ribbon} = 140^{+84}_{-38}$ AU, which obviously does neither favour the IHS- nor the OHS-origin of the ribbon ENAs.

The authors state further that they can neither confirm nor reject that the ribbon ENAs are created at different heliocentric distances depending on energy. It is, however, interesting to note that, if anything, their results would favour a decrease in heliocentric distance with increasing ENA energy (see their Table 1), which cannot be expected from the models of the first group claiming an OHS-origin of the ribbon ENAs, see, e.g. [15]. Evidently, one must hope for a significant increase in the accuracy of the parallax determination by either refining the analysis presented by [42] or by devising a different method to identify the source region of the ribbon ENAs.

Despite these positive and partially inconclusive tests of the H-wave hypothesis, the question whether such waves do exist in the interstellar medium remains to be answered. From a conceptual point of view it appears plausible that H-waves exist because (i) they could be generated at the nearest border of the local cloud the heliosphere is immersed in, i.e. in a distance of about 10000 AU [43] or (ii) they could be a result of turbulence cascading from larger to smaller scales (and corresponding coupling to the neutral component) as expected in view of the 'big power law in the sky' [44]. If the disturbances are first generated in the plasma component they can be transferred via change exchange coupling to the neutral component. While it has been demonstrated by [37] with an analytical study that such coupling indeed results in the generation of H-waves, more detailed and refined analyses still have to be made with numerical simulations in order to clarify the nature of disturbances in the neutral component of the LISM.

4. Proton velocity distribution in the inner heliosheath

Future tests must go beyond these mainly geometric considerations and comprise a computation of the differential fluxes of ENAs at different energies. A central ingredient for this is the knowledge of the proton source distribution in the IHS from which the ENAs are produced as a consequence of charge exchange. Assuming the principal form of the combined velocity distribution of thermal (solar wind) and suprathermal (pick-up) protons to be of κ-type (as done in [45] and [46]) and following the procedure used in [47], one can derive an ordinary differential equation for the variation of the κ-parameter along a given flow line:

$$\frac{d\kappa}{d\rho} = \frac{\kappa(\rho)(2\kappa(\rho)-3)u(\rho)}{3u_\rho(\rho)}\left[\frac{2}{3u}\frac{du}{d\rho} - \frac{10D_0(\rho)}{u(\rho)} + \frac{3n_H\sigma}{10\pi^{3/2}\kappa(\rho)}\right] \tag{3}$$

where D_0 denotes the velocity diffusion coefficient, n_H the neutral hydrogen density, and σ the charge exchange cross section. Typical values are $D_0 = 10^{-9}\text{s}^{-1}$ [48], $n_H = 0.1\text{cm}^{-3}$ [49], and $\sigma_{ex} = 10^{-15}\text{cm}^2$ [50]. The solution of this differential equation is shown along three different flowlines in the IHS ($\eta = 0.4, 0.65$ and 0.95 in Eq.(2). Fig. 3 illustrates that velocity diffusion

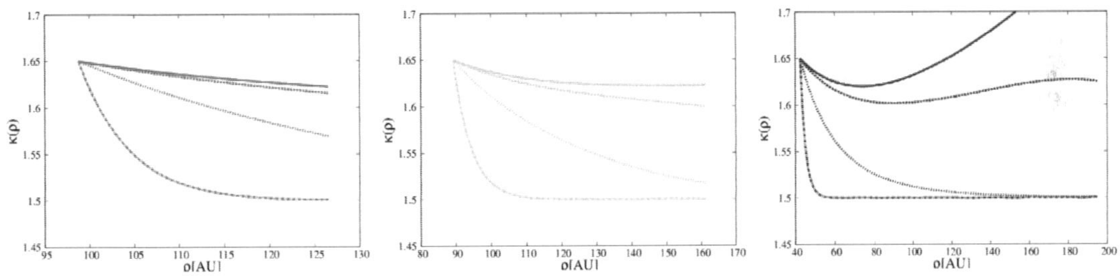

Figure 3. The variation of κ between the termination shock and the plane $z = 0$ along three flow lines given by Eq.(2) with $\eta = 0.4$ (left), 0.65 (middle) and 0.95 (right panel). In each panel the curves are computed from Eq.(3) for a velocity diffusion coefficient $D_0 = 0\ \text{s}^{-1}$ (solid line), 10^{-10}s^{-1} (dashed line), 10^{-9}s^{-1} (dotted line), and 10^{-8}s^{-1} (dash-dotted line).

overcompensates any increase of κ due to a loss of suprathermal protons via charge exchange. Given that the velocity diffusion coefficient is varied over two orders of magnitude this finding appears to be robust. Evidently, $1.62 < \kappa < 1.65$ for most regions in the upwind heliosphere ($z \geq 0$) for the case of vanishing diffusion. For the cases with $D_0 \neq 0$ one has $1.5 < \kappa < 1.65$ everywhere with a preference for the lower values. One the one hand this corroborates a value of $\kappa < 1.65$ in the upwind IHS as used earlier (e.g., [45]), on the other hand it clearly shows that the assumption of a constant κ should not be made.

5. Conclusions

In this brief report we have reviewed the H-wave hypothesis and discussed three tests, namely the resulting geometry of the ENA ribbon, the associated orientation of the local interstellar magnetic field, and the location of the ribbon ENA source region in space. With all tests we found the hypothesis to be consistent with the available data. A future test should consist in an analysis as to the existence of H-waves in the interstellar medium.

We continued with presenting an approximate solution of the combined thermal and suprathermal proton transport equation in the inner heliosheath. This solution, expressed as a κ-function, gives the range of κ-values to be expected, therewith confirming both that previsouly used values were of the right order and that the previously made asumption of a location-independent κ should not be made.

With such distribution functions of the seed population of the ENAs in the inner heliosheath, the next step of the quantitative analysis can be carried out, namely the computation of all-sky ENA flux maps that include the ribbon feature.

Acknowledgments

The work is directly related to the project FI 706/21-1 with the Deutsche Forschungsgemeinschaft (DFG) whose support we acknowledge.

References

[1] McComas D J, Allegrini F, Bzowski M, Dayeh M A, DeMajistre R, Funsten H O, Fuselier S A, Gruntman M, Janzen P H, Kubiak M A, Kucharek H, Möbius E, Reisenfeld D B, Schwadron N A, Sokół J M and Tokumaru M 2014 *Astrophys. J. Supp.* **213** 20
[2] Schwadron N A, Allegrini F, Bzowski M, Christian E R, Crew G B, Dayeh M, DeMajistre R, Frisch P, Funsten H O, Fuselier S A, Goodrich K, Gruntman M, Janzen P, Kucharek H, Livadiotis G, McComas D J, Moebius E, Prested C, Reisenfeld D, Reno M, Roelof E, Siegel J and Vanderspek R 2011 *Astrophys. J.* **731** 56
[3] McComas D J, Allegrini F, Bochsler P, Bzowski M, Collier M, Fahr H, Fichtner H, Frisch P, Funsten H O, Fuselier S A, Gloeckler G, Gruntman M, Izmodenov V, Knappenberger P, Lee M, Livi S, Mitchell D, Möbius E, Moore T, Pope S, Reisenfeld D, Roelof E, Scherrer J, Schwadron N, Tyler R, Wieser M, Witte M, Wurz P and Zank G 2009 *Space Sci. Rev.* **146** 11–33
[4] Grzedzielski S, Bzowski M, Czechowski A, Funsten H O, McComas D J and Schwadron N A 2010 *Astrophys. J.l* **715** L84–L87 (*Preprint* 1001.3917)
[5] Chalov S V, Alexashov D B, McComas D, Izmodenov V V, Malama Y G and Schwadron N 2010 *Astrophys. J.l* **716** L99–L102 (*Preprint* 1003.4826)
[6] Heerikhuisen J, Pogorelov N V, Zank G P, Crew G B, Frisch P C, Funsten H O, Janzen P H, McComas D J, Reisenfeld D B and Schwadron N A 2010 *Astrophys. J.l* **708** L126–L130
[7] Heerikhuisen J and Pogorelov N V 2011 *Astrophys. J.* **738** 29
[8] Strumik M, Ben-Jaffel L, Ratkiewicz R and Grygorczuk J 2011 *Astrophys. J.l* **741** L6
[9] Möbius E, Liu K, Funsten H, Gary S P and Winske D 2013 *Astrophys. J.* **766** 129
[10] Schwadron N A and McComas D J 2013 *Astrophys. J.* **764** 92
[11] Zirnstein E J, Heerikhuisen J, McComas D J and Schwadron N A 2013 *Astrophys. J.* **778** 112
[12] Burlaga L F, Ness N F, Florinski V and Heerikhuisen J 2014 *Astrophys. J.* **792** 134
[13] Heerikhuisen J, Zirnstein E J, Funsten H O, Pogorelov N V and Zank G P 2014 *Astrophys. J.* **784** 73
[14] Isenberg P A 2014 *Astrophys. J.* **787** 76 (*Preprint* 1404.2170)
[15] Zirnstein E J, Heerikhuisen J, Funsten H O, Livadiotis G, McComas D J and Pogorelov N V 2016 *Astrophys. J.l* **818** L18
[16] Krimigis S M, Mitchell D G, Roelof E C and Decker R B 2010 *American Institute of Physics Conference Series* (*American Institute of Physics Conference Series* vol 1302) ed Le Roux J, Zank G P, Coates A J and Florinski V pp 79–85
[17] Fahr H J, Siewert M, McComas D J and Schwadron N A 2011 *Astron. Astrophys.* **531** A77
[18] Kucharek H, Fuselier S A, Wurz P, Pogorelov N, Borovikov S, Lee M A, Moebius E, Reisenfeld D, Funsten H, Schwadron N and McComas D 2013 *Astrophys. J.* **776** 109
[19] Siewert M, Fahr H J, McComas D J and Schwadron N A 2013 *Astron. Astrophys.* **551** A58
[20] Fichtner H, Scherer K, Effenberger F, Zönnchen J, Schwadron N and McComas D J 2014 *Astron. Astrophys.* **561** A74
[21] Sylla A and Fichtner H 2015 *Astrophys. J.* **811** 150 (*Preprint* 1508.07815)
[22] Isenberg P A, Forbes T G and Möbius E 2015 *Astrophys. J.* **805** 153 (*Preprint* 1504.00585)
[23] Röken C, Kleimann J and Fichtner H 2015 *Astrophys. J.* **805** 173 (*Preprint* 1412.7199)
[24] Kleimann J, Röken C, Fichtner H and Heerikhuisen J 2016 *Astrophys. J.* **816** 29 (*Preprint* 1509.02699)
[25] Parker E N 1961 *Astrophys. J.* **134** 20
[26] Czechowski A, Fahr H J, Lay G and Hilchenbach M 2001 *Astron. Astrophys.* **379** 601–610
[27] Fahr H J and Siewert M 2015 *Astron. Astrophys.* **576** A100
[28] de Avillez M A and Breitschwerdt D (eds) 2004 *From Observations to Self-Consistent Modelling of the ISM in Galaxies. JENAM 2002 Workshop, Porto, Portugal, 3-5 September 2002. Reprinted from Astrophys. Space Sci. 289, Nos. 3-4, 2004. Kluwer Academic Publishers, Dordrecht*

[29] Hennebelle P and Audit E 2007 *SINS - Small Ionized and Neutral Structures in the Diffuse Interstellar Medium* (*Astronomical Society of the Pacific Conference Series* vol 365) ed M Haverkorn & W M Goss p 133 (*Preprint* `arXiv:astro-ph/0701685`)
[30] Breitschwerdt D, de Avillez M A, Fuchs B and Dettbarn C 2009 *Space Sci. Rev.* **143** 263–276
[31] Haverkorn M and Goss W M (eds) 2007 *SINS - Small Ionized and Neutral Structures in the Diffuse Interstellar Medium* (*Astronomical Society of the Pacific Conference Series* vol 365)
[32] Heiles C 1997 *Astrophys. J.* **481** 193
[33] Cho J and Lazarian A 2003 *Mon. Not. Roy. Astron. Soc.* **345** 325–339 (*Preprint* `arXiv:astro-ph/0301062`)
[34] Diver D A, Potts H E and Teodoro L F A 2006 *New Journal of Physics* **8** 265
[35] Shaikh D and Zank G P 2010 *Physics Letters A* **374** 4538–4542 (*Preprint* `1009.1859`)
[36] Kellum M E and Shaikh D 2011 *Journal of Plasma Physics* **77** 577–588 (*Preprint* `1012.0940`)
[37] Krasnobaev K V 1989 *Soviet Astronomy Letters* **15** 429
[38] Funsten H O, DeMajistre R, Frisch P C, Heerikhuisen J, Higdon D M, Janzen P, Larsen B A, Livadiotis G, McComas D J, Möbius E, Reese C S, Reisenfeld D B, Schwadron N A and Zirnstein E J 2013 *Astrophys. J.* **776** 30
[39] McComas D J, Lewis W S and Schwadron N A 2014 *Reviews of Geophysics* **52** 118–155
[40] Lerche I 1978 *Astrophys. Space Sci.* **57** 233–239
[41] Zaqarashvili T V, Khodachenko M L and Rucker H O 2011 *Astron. Astrophys.* **529** A82 (*Preprint* `1101.3913`)
[42] Swaczyna P, Bzowski M, Christian E R, Funsten H O, McComas D J and Schwadron N A 2016 *Astrophys. J.* **823** 119 (*Preprint* `1603.09134`)
[43] Redfield S and Linsky J L 2000 *Astrophys. J.* **534** 825–837
[44] Spangler S 2007 *American Astronomical Society Meeting Abstracts #210* (*Bulletin of the American Astronomical Society* vol 39) p 162
[45] Heerikhuisen J, Pogorelov N V, Florinski V, Zank G P and le Roux J A 2008 *Astrophys. J.* **682** 679–689 (*Preprint* `0803.2538`)
[46] Zirnstein E J and McComas D J 2015 *Astrophys. J.* **815** 31
[47] Fahr H J, Fichtner H and Scherer K 2014 *Journal of Geophysical Research (Space Physics)* **119** 7998–8005
[48] Kallenbach R, Hilchenbach M, Chalov S V, Le Roux J A and Bamert K 2005 *Astron. Astrophys.* **439** 1–22
[49] Zank G P, Heerikhuisen J, Wood B E, Pogorelov N V, Zirnstein E and McComas D J 2013 *Astrophys. J.* **763** 20
[50] Lindsay B G and Stebbings R F 2005 *Journal of Geophysical Research (Space Physics)* **110** 12213–+

The Fisk & Gloeckler model for the nose region of the heliosheath: Another model for Ed Stone to test

L A Fisk and G Gloeckler

Department of Climate & Space, University of Michigan
2455 Hayward Street, Ann Arbor, MI 48109-2143 USA

Email: lafisk@umich.edu

Abstract. Ed Stone has always been receptive to new ideas, including controversial ones, provided that the idea can be tested by observations. This paper will discuss the latest controversial idea for Ed to test: the Fisk & Gloeckler model for the nose region of the heliosheath, which concludes that *Voyager* 1 remains in the heliosphere and is not in the local interstellar medium, and that the nose region of the heliosheath is a very different place from what others have considered, and it can be argued, a more interesting place than others have imagined. The test of the model is that within this year, *Voyager* 1 should encounter another current sheet, due to the reversal of the magnetic field of the Sun, thereby establishing conclusively that *Voyager* 1 remains in the heliosphere. If/when the current sheet is crossed, it will be appropriate to extend the Fisk & Gloeckler model. A preview will be provided of these extensions and the interesting properties and physical properties that result for: (1) the region inside the *Heliocliff*; (2) the *Heliocliff*; (3) the region between the *Heliocliff* and the actual *Heliopause*; and (4) the *Heliopause*.

1. Introduction

Ed Stone has always been receptive to a new thought. That is a talent that in others can diminish with age. With Ed, however, the approach has always been the same. Give him something that he can test. How can he use his data to test your idea, however outlandish? That is the right partnership between theory and observation. Ed's talent for what is the proper scientific method should be a model for all of us, and for all generations of scientists.

With this attribute of Ed Stone in mind, this paper will present our latest controversial idea: that *Voyager 1* remains in the heliosheath and is not in the local interstellar medium. Thus, the nose region of the heliosheath is a very different place from what others have considered, and frankly, in our judgment, a more interesting place than others have imagined.

We do assume that Ed approves of our approach. We are putting forth a new and interesting model, and most important we have a definitive test of whether our model is correct. We can all let the data decide.

The model of Fisk & Gloeckler for the nose region of the heliosheath has been presented and published numerous times, including at previous meetings of this conference [1-5]. Thus, in this paper we will simply summarize the basic concepts of the model. Also, as is well known, the test of our model is that we predict that *Voyager 1* will encounter another current sheet, due to the magnetic field reversal of the Sun, within the next few months, and since changes in magnetic polarity are a feature of the solar magnetic field, not the local interstellar magnetic field, if this current sheet is observed, as predicted, we will establish that *Voyager 1* is still inside the *Heliopause*.

We have not been idle waiting for the current sheet crossing. Rather we have developed many other aspects and predictions of our model, which we will preview in this paper. You should conclude that if the current sheet is crossed, and our model for the nose region of the heliosheath is correct, there are numerous additional consequences, which will have to be fully developed, and we will have a treasure of interesting physics to talk about.

2. The basics of the Fisk & Gloeckler model

We begin by reviewing briefly the essence of our model for the nose region of the heliosheath, which is illustrated in Figure 1, a simplified version of the detailed calculations presented in Fisk and Gloeckler [1]. The model is based upon the observed solar wind flow, as determined from the particle anisotropy observations [6-8], which need to be used to determine the solar wind flow in the absence of a working plasma detector on *Voyager 1*. The solar wind flow is observed to be primarily in the radial and azimuthal directions, particularly with increasing distance into the heliosheath, and to become small in all directions as it approaches what we call the *Heliocliff,* where the ACRs are observed to escape [9,10]. The *Heliocliff* is the location the *Voyager* investigators consider to be the *Heliopause*.

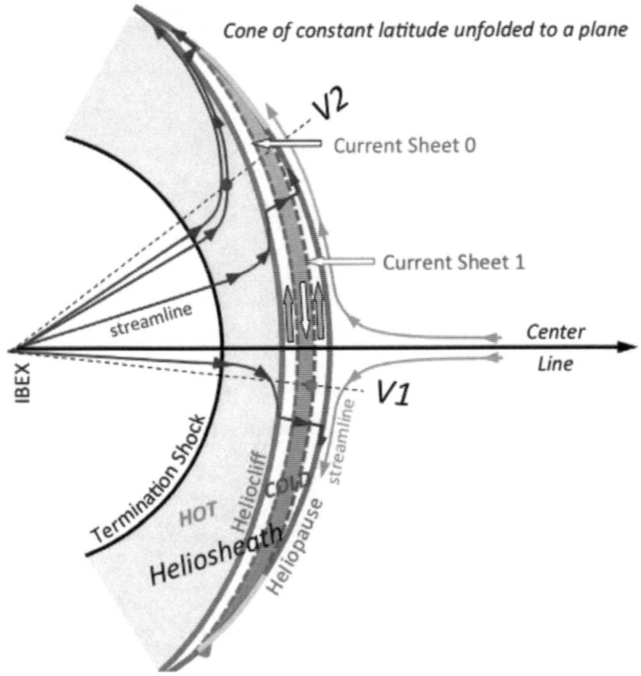

Figure 1. A schematics of the basic features of the Fisk & Gloeckler model for the nose region of the heliosheath (after Fisk and Gloeckler [1]).

As can be seen in Figure 1, the streamlines of the solar wind cross the termination shock, and then turn initially azimuthal, so the solar wind can flow out of the nose region of the heliosheath. The azimuthal flow creates a centerline region, which as is demonstrated in Fisk and Gloeckler [1] causes the flow speed to decrease. The azimuthal speed goes to zero first, and then the streamlines flow radially outward at a very low speed, across the *Heliocliff*, out to the actual *Heliopause*, which is beyond where *Voyager 1* is now located. The solar wind flows across the *Heliopause*, which is a rotational discontinuity.

The model naturally yields the so-called stagnation region, identified by the *Voyager 1* investigators (e.g. [11]), where the solar wind flow becomes very small. In this region, the solar wind is compressed and heated in our model. We refer to the region where the solar wind is compressed and heated as the *Hot Heliosheath*. Beyond the *Heliocliff*, where the solar wind and more energetic particles can escape across the actual *Heliopause*, the solar wind is cold and dense, and we call this the *Cold Heliosheath*. Gloeckler and Fisk [12] demonstrate that *Hot* and *Cold Heliosheaths* are necessary to explain the full spectrum of the *IBEX* ENA observations.

3. The *Hot Heliosheath*

As is discussed in detail in Fisk & Gloeckler [1], an essential feature of our model is that the pickup ions and ACRs, which are the mobile particles, are decoupled from the solar wind. The pressure in the pickup ions and ACRs is essentially constant, balancing the pressure of the local interstellar medium. If the pickup ions and ACRs are coupled to the solar wind, then the solar wind density must also be constant and cannot be compressed. However, if we decouple the pickup ions and ACRs from the solar wind, that constraint no longer applies, and the solar wind can be compressed, yielding the observed stagnation region and the *Hot Heliosheath*.

There is a known problem with the behavior of the magnetic field in the stagnation region, where the solar wind slows down in all directions, particularly in directions normal to the magnetic field. Even assuming a reasonable amount of reconnection, we should expect by Faraday's law that the magnetic field strength will increase dramatically. Yet it is observed by Burlaga and Ness [11] not to.

We can resolve this discrepancy by assuming that the pickup ions and ACRs are decoupled from the solar wind, thus providing corroborating evidence for the decoupling. If the pickup ions and ACRs are decoupled they can be considered mobile energetic particles, with a large pressure, little mass, and a ram pressure that can be ignored, and thus by using standard plasma physics, the pickup ions and ACRs create a current:

$$\mathbf{J}_{ep} = c\frac{\mathbf{B}\times\nabla_{\perp}P_{ep,\perp}}{B^2}. \tag{1}$$

Here, $\mathbf{B}$ is the magnetic field convected into the heliosheath across the termination shock; P_{ep} is the pressure in the pickup ions and ACRs; c is the speed of light.

The current in equation (1) is a magnetization current — the response of the pickup ions and ACRs to the application of $\mathbf{B}$. The magnetization current creates a magnetic field that offsets the applied $\mathbf{B}$-field and creates the observed magnetic field, which we refer to as $\mathbf{H}$, which is standard notation in E&M textbooks. Thus, the current in equation (1), and $\mathbf{B}$ and $\mathbf{H}$ are related by

$$\frac{c}{4\pi}\nabla\times\mathbf{B} = \frac{c}{4\pi}\nabla\times\mathbf{H} + c\frac{\mathbf{B}\times\nabla_{\perp}P_{ep,\perp}}{B^2}. \tag{2}$$

Then, ignoring the curvature of the magnetic field, and noting from pressure equilibrium that

$$\nabla P_{ep,\perp} = -\nabla\left(H^2/8\pi\right), \tag{3}$$

we find that

$$\nabla H = \frac{\nabla B}{\left(1+H/B\right)}. \tag{4}$$

Note that the current in equation (1) is identical to the current of modulated galactic cosmic rays. In the cosmic ray case, the pressure in the cosmic rays is small, and thus this current is usually ignored. Also, the physics here is exactly the same as the physics of the ring current of the Earth. The ring current is a separate particle population with a current that alters the intrinsic magnetic field of the

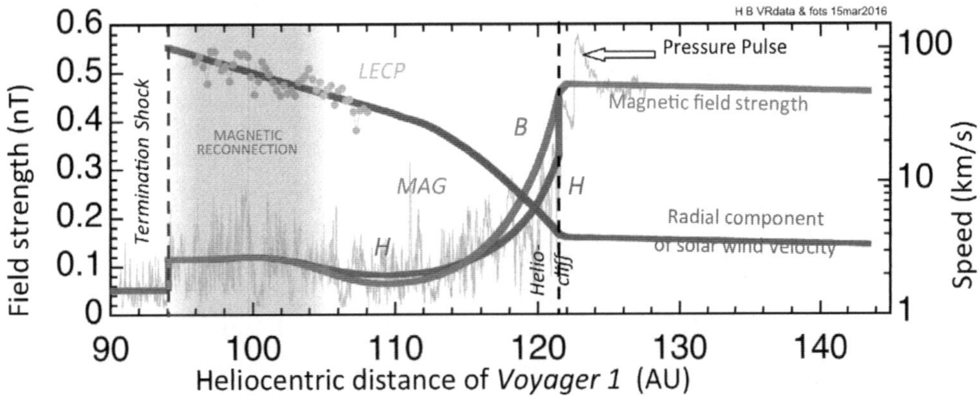

Figure 2. The underlying magnetic field that is convected into the *Hot Heliosheath* by the solar wind (B, shown in red), as determined from the observed magnetic field (H, shown in green). Also shown is the observed and predicted radial component of the solar wind velocity (shown in blue). See text for details.

Earth, just as the current in the pickup ions and ACRs alters the intrinsic, frozen-in magnetic field convected into the heliosheath with the solar wind.

We can then use equation (4) and the observed magnetic field H in the *Hot Heliosheath* to determine the underlying value of B. Shown in Figure 2, in green, is the observed magnetic field, H, smoothed to remove short-period fluctuations. Then from equation (4), we determine B, shown in red. Note that the calculated B is much larger than the observed H in the stagnation region inside the *Heliocliff*, consistent with the expectation, from Faraday's Law, that B should be larger than the observed field. Note also that $B=H$ beyond the *Heliocliff*. The pickup ions and ACRs escape beyond the *Heliocliff* and thus there is no magnetization current or difference between B and H.

Finally, if we know the underlying solar wind magnetic field, which obeys Faraday's law, and use the data to make a reasonable assumption about the amount of reconnection, we can constrain the solar wind speed across the *Heliocliff*, and into the *Cold Heliosheath*, to around 5 kms^{-1}. This is the solar wind speed that we have used to estimate when the next current sheet should be observed by *Voyager 1* sometime this year.

It is worth emphasizing how many features of our model are interdependent and verifiable by observations. In order to create the *Hot Heliosheath*, the existence of which is verified by *IBEX* ENA observations [12], we need to decouple the pickup ions and ACRs from the solar wind, which permits us to explain the observations of the magnetic field magnitude in the *Hot Heliosheath* and to predict when the current sheet crossing should occur in the *Cold Heliosheath*.

4. The *Heliocliff*

Consider next the controlling physics of the *Heliocliff*, which can be seen in the simplified illustration of the nose region of the heliosheath in Figure 3. Immediately outside the *Heliocliff*, the magnetic field intersects the actual *Heliopause*, on the flanks, which provides for escape of the ACRs, and entry of GCRs, as is observed at the *Heliocliff*. In our model the entry of the GCRs is facilitated because the heliosheath magnetic field is oriented in approximately the general direction of the draped interstellar magnetic field. The fields merge and provide easy access to the GCRs. This should not be the case when we cross the next current sheet into a region of opposite magnetic polarity.

Consider a magnetic field line in the *Hot Heliosheath* just inside the *Heliocliff*. This field line remains in the heliosheath, executing the spiral pattern through the tail region of the heliosheath. This field line is then convected across the *Heliocliff*, and intersects the *Heliopause* on the flanks. Note, however, that in our model the *Heliocliff* is blunt. The bluntness results from the centerline region

Journal of Physics: Conference Series **767** (2016) 012008 doi:10.1088/1742-6596/767/1/012008

created by the diverging azimuthal flows illustrated in Figure 1. The bluntness is confirmed by the *IBEX* ENA observations [4].

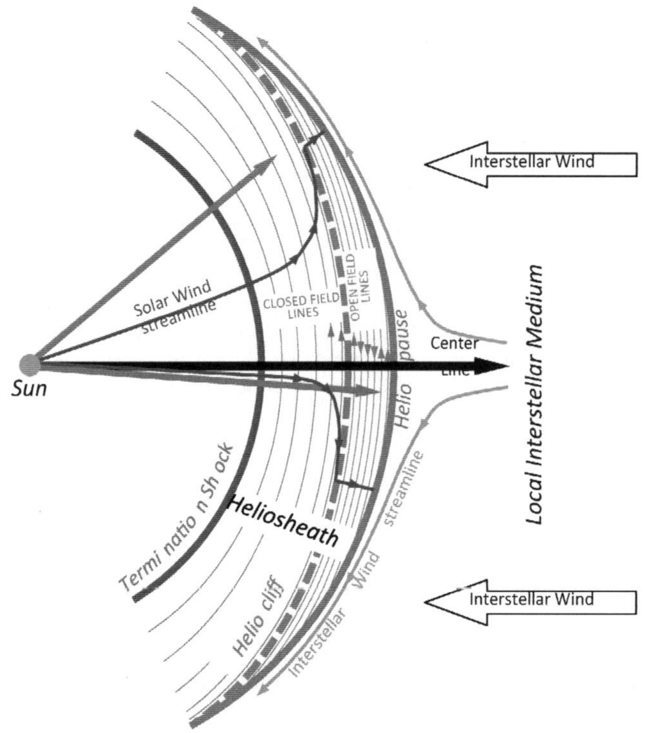

Figure 3. A simplified schematic of the nose region of the heliosheath.

Consider the length of a magnetic field line in the *Cold Heliosheath*, between its two *Heliopause* crossings. Because the *Heliocliff* is blunt, the length of the field line is rapidly reduced as the field line is convected outward with the radial flow speed of the solar wind. Since the density of the interstellar gas is larger than the initial density of the solar wind in the *Cold Heliosheath*, the shortening of the field line is the equivalent of a piston that rapidly compresses the density of the solar wind to equal the density of the interstellar medium.

The rapid increase in solar wind density will result in a rapid increase in the solar wind pressure. The mobile particles will perform their function of maintaining constant pressure by escaping across the *Heliopause*, to which the magnetic field is now attached. The pressure increase is rapid and the *Heliopause* is a long distance away, in which case only the higher energy mobile particles can reduce the pressure by escaping. Any particle previously considered a mobile particle, e.g., the pickup ions and ACRs, that cannot escape can no longer be considered a mobile particle that maintains pressure equilibrium. Since all the actual mobile particles escape, without the mobile particles, the observed magnetic field is simply the **B**-field of the solar wind, as seen in Figure 2.

5. The *Cold Heliosheath*

Now consider the *Cold Heliosheath*, which is bounded by the *Heliocliff* and the *Heliopause*. The higher energy particles, including the higher energy solar wind thermal particles, have escaped. The solar wind is compressed, but now cold, hence the term *Cold Heliosheath*. Also, the magnetic field is observed to be roughly constant, ignoring for the moment pressure waves, in which case the solar wind flow speed must decrease like $1/r$ from its estimated value at the *Heliocliff* out to the *Heliopause*.

The *Cold Heliosheath* is where *Voyager 1* should encounter the next current sheet resulting from the magnetic field reversal of the Sun, which, using the predicted solar wind velocity at the *Heliocliff*, should occur sometime in 2016. We can speculate on what conditions we will find in the region beyond the next current sheet.

In our model for the nose region of the heliosheath, mobile particles readily escape across the *Heliopause*, and GCRs readily enter when the heliosheath magnetic field is aligned with and merges with the interstellar magnetic field. This is the present situation observed by *Voyager 1*, which has been in a single sector since before crossing the *Heliocliff*; the mobile particles escape and the GCRs have entered. At the next current sheet crossing, the magnetic field will reverse direction, and thus it has been argued that the escape of the mobile particles and entry of GCRs will be restricted.

Consider what happened to the last current sheet when it crossed the *Heliocliff*, ~11 years ago. The same physical process should occur at the *Heliocliff* as with the sector *Voyager 1* is now in. There will be a compression of the solar wind along the magnetic field due to the blunt *Heliocliff* versus the spherical *Heliopause*, as described in the previous section, and the mobile particles need to escape in order that the pressure equals the interstellar pressure along the magnetic field lines, which now intersect the *Heliopause*. The escape may not be as easy as in the current sector, but will certainly have occurred by the time *Voyager 1* encounters the next current sheet. Thus, we should not expect to see ACRs at the next current sheet. What could be different, modified by the different escape, is the spectrum of the remaining particles that have not escaped. Also, the GCRs should not be as enhanced as they are in the current sector. Unlike the mobile particles, which simply have to escape across the *Heliopause*, for the GCRs to enter they need to penetrate through the region beyond the *Heliopause*, where there is reconnection and magnetic islands formed between the draped interstellar magnetic field and the oppositely directed heliosheath magnetic field.

Another interesting feature of the *Cold Heliosheath* is the striking lack of turbulence. Burlaga et al. [13] observe a very low level of compressive turbulence, much reduced compared to the compressive turbulence upstream of the *Heliocliff*. This reduction in compressive turbulence is to be expected because the *Cold Heliosheath* no longer contains mobile energetic particles, the pickup ions and ACRs that yield a magnetization current that determines the observed magnetic field (equations 1-4). In the *Hot Heliosheath*, the observed magnetic field, H, is determined by the perpendicular pressure of the mobile particles. The mobile particles are continuously attempting to maintain pressure isotropy, and pressure equilibrium along and across the magnetic field, i.e. the mobile particles are attempting to maintain pressure equilibrium both locally and remotely with one value for the pressure. Since the magnetic field does not have a uniform direction, but rather is entwined and random walking, pressure equilibrium on the scale of the entwining is not possible, and extensive compressive turbulence will result. In the *Cold Heliosheath*, the mobile particles are absent, and pressure equilibrium can be established locally, greatly reducing the compressive turbulence.

Another interesting feature of the *Cold Heliosheath* is pressure waves. Burlaga et al. [14] observe pressure waves beyond the *Heliocliff*; also in Burlaga et al. [13], the paper on turbulence beyond the *Heliocliff*, there is evidence of pressure waves. In some cases, these pressure waves are associated with the plasma waves observed by Gurnett et al. [15], which are observed preceding the pressure waves. Gurnett et al. and Burlaga et al. interpret the pressure wave as propagating into the interstellar medium (since they consider the *Heliocliff* to be the *Heliopause*), preceded by a shock, and thus the plasma waves are generated by electrons propagating upstream from a shock, as is observed in the supersonic solar wind.

Pressure waves can certainly propagate into the *Cold Heliosheath* of our model for the nose region of the heliosheath. Indeed, it is easier to propagate pressure waves into the *Cold Heliosheath* than into the interstellar medium since the solar wind flows outward in the *Cold Heliosheath* as opposed to the interstellar medium where the plasma flows inward.

A pressure wave in the *Cold Heliosheath* will be subject to pressure loss along the magnetic field, due to escaping mobile particles, just as in the underlying solar wind. Since most of the mobile particles have escaped, balancing the pressure in the underlying solar wind, the dissipation of the pressure wave will not be immediate, but should occur, reducing the outward speed of the pressure wave so that it basically moves with the outward radial speed of the solar wind. The loss of pressure along the magnetic field does not change the magnetic field, and a region of enhanced magnetic field, moving with the solar wind, is created in the *Cold Heliosheath*. *Voyager 1*, which is moving faster than the solar wind, will overtake the region of enhanced magnetic field, as opposed to the Burlaga et al. [14] interpretation that the wave overtakes *Voyager 1*.

The loss of pressure from the pressure wave is faster closer to the *Heliopause* than at the centerline in Figure 3, where *Voyager 1* is located. The disturbance thus moves faster along the centerline than it does closer to the flanks and develops a curved shape relative to the magnetic field direction, which is parallel to the *Heliocliff*. The magnetic field lines that *Voyager 1* observes in advance of the pressure wave thus intersect the pressure wave closer to the flanks. Mobile particles, in particular electrons, can propagate along the magnetic field lines that intersect the pressure wave, exciting the plasma waves observed by Gurnett et al. [15].

6. The *Heliopause*

The *Heliopause* is a rotational discontinuity at which the heliosheath magnetic field rotates and then merges or reconnects with the draped interstellar magnetic field. Thus, beyond the *Heliopause* there should be a turbulent boundary layer. Thus, we should know when we have crossed the actual *Heliopause*.

We can estimate the distance to the *Heliopause* by using our knowledge of the *Cold Heliosheath* and a property of a rotational discontinuity: it propagates relative to the plasma with a normal speed equal to the Alfven speed. Thus, the normal component of the escaping solar wind, u_n, must equal

$$u_n = \frac{B_n}{\sqrt{4\pi\rho_{\text{int}}}} . \tag{5}$$

Here, B_n is the normal component of the magnetic field across the *Heliopause*, and ρ_{int} is the mass density at the *Heliopause*, which is the interstellar mass density.

Referring to Figure 3, we can see that all of the magnetic flux in the *Cold Heliosheath* that passes through the centerline between the *Heliocliff* and the *Heliopause* must escape across the *Heliopause*. Thus, we can calculate the distance to the *Heliopause* by requiring that

$$B_{CH}\left(r_{hp} - r_{hc}\right) = \left(\sqrt{4\pi\rho_{\text{int}}}\right) u_n r_{hp} \Delta\phi . \tag{6}$$

Here, B_{CH} is the magnetic field in the *Cold Heliosheath*; '*hp*' denotes *Heliopause*; '*hc*' denotes *Heliocliff*; and $\Delta\phi$ is the angle along the *Heliopause* on one side of the centerline, where the magnetic field escapes.

We take B_{CH} to be the magnetic field at the *Heliocliff*, or 0.46 nT, and constant with radial distance. With a constant magnetic field, the product of the radial solar wind speed, and the radius is also constant, and thus we can evaluate $u_n r_{hp}$ at the *Heliocliff*, or $u_n r_{hp}$ =5.3 kms^{-1}*122 AU. We take $\Delta\phi$ to be 1 radian and the interstellar number density to be 0.08 cm^{-3}, in which case equation (6) becomes

$$r_{hp} = 122 + \frac{1}{.46}(0.013)(5.3)(122) = 140 \text{ AU} . \tag{7}$$

It should be noted that *Voyager 1* will cross a *Heliopause* at 140 AU within the next two years.

7. Concluding remarks

In summary, there are several conclusions to draw from this paper: The Fisk & Gloeckler model is a comprehensive model for the nose region of the heliosheath. It provides explanations for the governing physics of the *Hot Heliosheath*, the *Heliocliff*, the *Cold Heliosheath*, and the *Heliopause*. The various features of the model are interdependent and verifiable by observations. In some cases, the explanations contained in the model need further development to be sure they are correct. In all cases the model describes a heliosheath that is a very different place from what others have considered, and it can be argued, a more interesting place than others have imagined.

Thus, if we can establish through observations of the upcoming current sheet that *Voyager 1* has not crossed the *Heliopause*, and is still in the *Cold Heliosheath*, and that the *Heliopause* still lies ahead, we will have a treasure of interesting physics to talk about. We will have made one of the most interesting and profound discoveries of the space program. All brought to you by *Voyager*, the mission that Ed Stone has so ably led these many years.

Acknowledgements

This work was supported in part by NASA Grant NNX10AF23G and by NSF Grant AGS-1043012.

References

[1] Fisk L A and Gloeckler G 2014 *ApJ* **789** 41
[2] Fisk L A and Gloeckler G 2015 *J. Phys.: Conference Series* **577** 012009
[3] Gloeckler G and Fisk L A 2014 *GRL* **41** 5325
[4] Gloeckler G and Fisk L A 2015a, *ApJ Letts.* **806** L27
[5] Gloeckler G and Fisk L A 2015b *J. Phys.: Conference Series* **642** 012011
[6] Krimigis S M, Roelof E C, Decker R B and Hill M E 2011 *Nature* **247** 359
[7] Decker R B, Krimigis S M, Roelof E C and Hill M E 2012 *Nature* **489**, 124
[8] Stone E C and Cummings A C 2011 *Proc. 32nd Int. Cosmic Ray Conf.* vol 1 (Beijing: IUPAP) p 29
[9] Krimigis S M et al. 2013 *Sci.* **341** 144
[10] Stone E C et al. 2013 *Sci.* **341** 150
[11] Burlaga L F and Ness N F 2012 *ApJ* **749** 13
[12] Gloeckler G and Fisk L A 2016 these proceedings
[13] Burlaga L F, Florinski V and Ness N F 2015 *ApJ Letts.* **804** L31
[14] Burlaga L F, Ness N F, Gurnett D A, and Kurth W S 2013 *ApJ* **778** L3
[15] Gurnett D A, Kurth W S, Burlaga L F, and Ness N F 2013 *Sci.* **341** 1489

Journal of Physics: Conference Series **767** (2016) 012009 doi:10.1088/1742-6596/767/1/012009

The ribbon of the rings: the stability of the rings

V Florinski

Department of Space Science and Center for Space Plasma and Aeronomic Research, University of Alabama in Huntsville, Huntsville, AL 35899, USA

E-mail: vaf0001@uah.edu

Abstract. The astounding result of the Voyager 1 observations in the outer heliosheath region is the Kolmogorov-like spectrum of magnetic fluctuations, consistent with the ambient spectrum of interstellar turbulence deduced from remote sensing. Such quiet conditions appear to be incompatible with the ionization of the neutral solar wind hydrogen atoms by charge exchange, which would generate abundant turbulent fluctuations due to an instability of the resulting proton ring. Here we revisit the problem of stability of ring and shell distributions embedded in a warm plasma characteristic of the outer heliosheath. We show that ion rings with parallel velocity dispersion between a few km/s and the core thermal speed are stable, while those outside these bounds are subject to magnetic-fluctuation-producing left- and right-handed instabilities. Wave generation by the instability is too efficient to be transported to shorter scales and converted into heat by the turbulent cascade.

1. Introduction

One of the most surprising results of the IBEX mission was the discovery of a circular band in space emitting large quantities of energetic neutral atoms or ENAs [12]. A number of possible explanations for the band, dubbed the IBEX ribbon, were proposed in the following years. Theories and models placing the ribbon source outside the heliopause (mainly in the outer heliosheath, OHS) and using charge exchange of pickup ions (PUIs) of solar-wind origin on interstellar hydrogen atoms [7] have been very successful in explaining the observed properties of the ribbon. In these models the ribbon (a vast ring in space) is produced by ring-like distributions of protons gyrating in place in directions where the magnetic field is normal to the line of sight from IBEX and waiting for a chance to capture an electron from a passing interstellar hydrogen atom (hence the title).

The outstanding problem with these models is the requirement to maintain anisotropic ring- or shell-like distributions of pickup protons for several years, long enough to undergo a charge transfer collision. Such distributions are though to be unstable, having the tendency to scatter onto an isotropic spherical shell and excite magnetic fluctuations in large quantities [5, 11]. No magnetic fluctuations of pickup-ion origin were observed in the OHS by Voyager 1 [1]. We note, however, that Voyager 1 is not located in the ribbon, and the fluctuations could still exist in other directions.

Summerlin and co-workers [16] proposed that a ring distribution could be stabilized by a warm proton core. We will show here that this stabilization only occurs in a relatively narrow range of ring parallel temperatures, while outside this range the rings and shells are still unstable. Models show that the proton distributions generated from ENAs of solar-wind origin could be hotter than the upper limit of the stability range. The situation is exacerbated by the presence of the second population of PUIs produced from inner heliospheric ENAs that have very high temperatures.

In this report we summarize the plasma physics of the instability generated by a warm gyrating proton ring. We also employ kinetic (hybrid) plasma simulations to study the scattering of proton rings injected into a warm plasma characteristic of the OHS. The simulations do confirm the existence of a region of stability in between two unstable regions. Finally, we argue that the existence of unstable ring population in the OHS is not compatible with Voyager 1's observations of the interstellar turbulence spectrum.

2. Stability analysis

Charge transfer collisions between interstellar H atoms and solar-wind protons generate a stream of energetic hydrogen atoms called neutral solar wind (NSW). These atoms can undergo a second charge-exchange interaction outside the heliopause, producing a gyrating population of pickup protons. Suppose the NSW has a Maxwellian distribution with a bulk radial speed $v_{\rm sw}$ (the solar-wind speed) and thermal velocity spread $\delta v = (2k_B T_{\rm SW}/m_p)$, where $T_{\rm SW}$ is the temperature of the solar-wind protons, m_p is the proton mass and k_B is the Boltzmann constant. Because NSW atoms are produced over a range of heliocentric distances, from a few AU all the way to the termination shock, their actual distribution in the OHS would be a superposition of Maxwellians with different temperatures. Observations show that the core solar wind proton temperature decreases between 1 and 10 AU, and afterwards remains roughly constant at $\sim (1-2) \times 10^4$ K until the termination shock because of turbulent heating by PUI-excited waves [15]. The effective temperature of the entire NSW distribution therefore will be in the tens of thousands of K, and generally higher in the fast solar wind.

The distribution of secondary PUIs produced from NSW hydrogen by electron capture is obtained by transforming the atoms' velocities into the OHS plasma rest frame and averaging over the gyrophase. For perpendicular pickup, the resulting ring distribution is easily shown to be

$$f_r(v_\parallel, v_\perp) = \frac{n_r}{\pi^{3/2}\delta v^3} I_0\left(\frac{2v_\perp v_{\rm sw}}{\delta v^2}\right) \exp\left(-\frac{v^2 + v_{\rm sw}^2}{\delta v^2}\right), \tag{1}$$

where n_r is the number density of the ring protons, and I_0 is the modified Bessel function. Generally, in the solar wind $\delta v \ll v_{\rm sw}$, in which case the expression (1) is very similar to the Gaussian toroid of [16], which is mathematically tractable.

Consider now a Gaussian toroidal distribution of pickup protons with the major radius $v_{\rm sw}$. We take $v_{\rm sw} = 450$ km s^{-1}, the core plasma temperature in the OHS of 20,000 K, the OHS magnetic field of $B_0 = 3\,\mu$G, and the ratio of the PUI density to the core density of 10^{-4}. From linear Vlasov theory, the normal modes $\omega(\mathbf{k})$ are the eigenvalues of the Maxwell tensor

$$\Lambda_{ij} = (\omega^2 - k^2c^2)\delta_{ij} + c^2 k_i k_j + 4\pi i\omega\sigma_{ij}, \tag{2}$$

where $\mathbf{k}$ is the wavevector, and σ_{ij} is the conductivity tensor which depends on the properties of the particles' distribution function. For a Gaussian toroid the expressions for σ_{ij} for arbitrary wave propagation angles may be found in [6].

Figure 1 shows the imaginary part of the complex frequency as a function of wavenumber for the Alfvén-ion cyclotron (AIC) wave propagating in the direction of $\mathbf{B}_0$ (it may be shown that maximum growth is attained for $\mathbf{k} \times \mathbf{B}_0 = 0$) for $\delta v = 10$ km s^{-1}, 22.5 km s^{-1}, 50 km s^{-1}, and 100 km s^{-1}, corresponding to solar wind temperatures of 6×10^3 K, 2.4×10^4 K, 1.5×10^5 K, and 6×10^5 K, respectively. Growing modes are obtained when the PUI ring effective temperature (proportional to the minor radius of the toroid) exceeds the temperature of the core OHS plasma. In that case the ring excites waves in the low-k range where there are few resonant core protons and the cyclotron damping is weak. The ring is also unstable at very *low* temperatures, below a

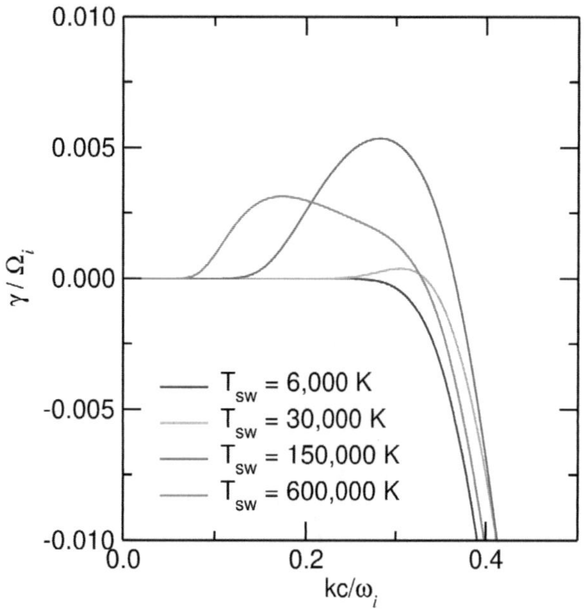

Figure 1. Imaginary part of the complex frequency vs. wavenumber times the proton inertial length for different NSW temperatures. A positive value of γ indicates an unstable mode.

few hundred K [16, 6]. However, this is well below the range of temperatures in the solar wind, and we will not discuss the cold ring instability here.

For oblique pickup angles the resulting ring-beam distribution can resonate with the magnetosonic-whistler wave, in addition to the Alfvén wave. As shown in [6], both of these resonances are associated with instabilities. Generally, the former instability has the higher growth rate of the two when the temperature of the ring is lower than the temperature of the core; otherwise the Alfvén mode is the dominant instability. In the direction of Voyager 1 the pickup angle is close to 70° [2] and the PUI ring-beam may be shown to be always unstable thus generating magnetic fluctuations [6]. The apparent absence of such fluctuations in the Voyager 1 magnetometer measurements [1] remains an unsolved puzzle. We will return to this topic in Section 4.

3. Particle-mesh simulations

Recently [13] performed two sets of numerical kinetic simulations of a tenuous but energetic proton ring distribution embedded in a warm plasma characteristic of the OHS. The first set used a hybrid model, which uses a fluid-like approximation for the electron component, in one dimension along the mean magnetic field. The second set used a particle description for both species and two spatial dimensions. The two sets of simulation had nearly identical wave growth and PUI scattering rates, which demonstrates that including only ion dynamics and fluctuationa traveling along $\mathbf{B}_0$ is sufficient to properly model the relevant instabilities. For this reason, we only use the hybrid simulations here, since they are much less computationally expensive.

To test the linear stability conclusions of the preceding section, we performed two numerical simulations of proton rings and toroids using the 1D hybrid code. The first had a small, but finite temperature $T_\parallel = T_\perp = 200$ K, and the second had a large parallel temperature $T_\parallel = 125,000$ K with $T_\perp = 17,500$ K. The two simulations used 10^6 and 10^5 particles per cell, respectively. Both were run for 1000 proton Larmor orbital periods (about 2.5 days).

Figure 2 compares the time evolution of the magnetic fluctuation intensity $\langle(\mathbf{B}-\mathbf{B}_0)^2\rangle$ and the pitch-angle width $\langle\mu^2\rangle$ (μ is the pitch angle cosine) for the two cases. The 200 K ring appears to have a very short wave growth period, but the fluctuation intensity barely rises above the statistical noise level. The thin ring undergoes some minimal scattering initially (this is unavoidable in kinetic simulations with finite numbers of particles), but remains essentially cold at the end of the run. By contrast, the 125,000 K toroid generates intense fluctuations for the first few hundred Larmor periods after which the instability saturates (the exponential wave growth interval is a telltale sign of an instability in case of the hot distribution). The warm toroid spreads in pitch angle even more; by the end of the run it has reached $\langle\mu^2\rangle \sim 0.1$ (for comparison, an isotropic distribution has $\langle\mu^2\rangle = 1/3$).

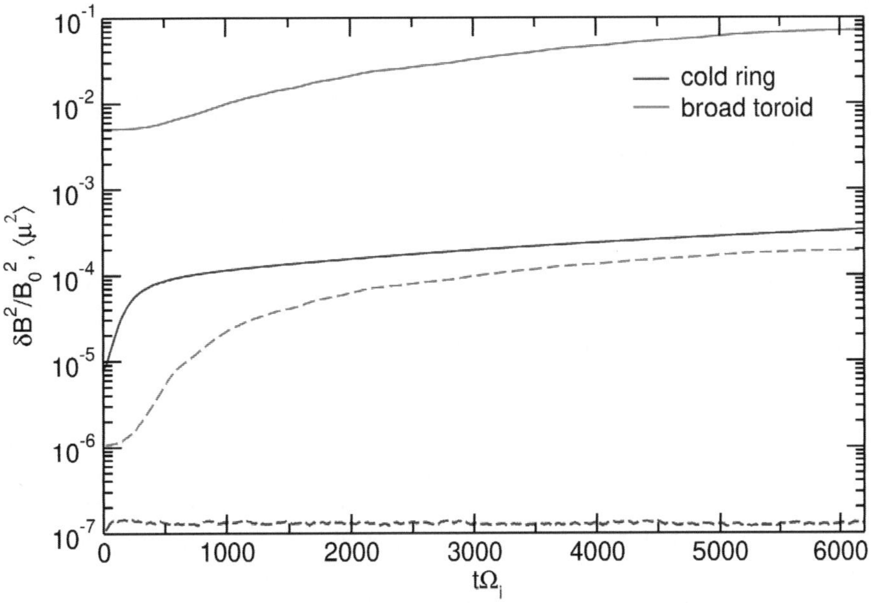

Figure 2. Pitch angle variance $\langle\mu^2\rangle$ (solid lines) and magnetic fluctuation intensity $\langle(\mathbf{B}-\mathbf{B}_0)^2\rangle$ (dashed lines) as functions of time for initial $T_\parallel = T_\perp = 200$ K (blue) and $T_\parallel = 125,000$ K, $T_\perp = 17,500$ K (red). The cold ring is stable and the hot ring is unstable.

The numerical model results shown here are in agreement with the linear theory predictions. The 200 K is nominally unstable [16], but this type of instability is very quick to saturate and the resulting turbulence does not scatter the ring ions appreciably. The 125,000 K ring, however, is strongly unstable (cf. the 150,000 K curve in Figure 1), and the turbulence it generates is significant. While this result might appear counter-intuitive, it has been obtained using two completely different methods, and its veracity is not in doubt. As we discuss below, it has profound implications for the IBEX ribbon models.

In addition to toroids, [6] analyzed partial shell proton distributions away from the ribbon direction, i.e., at oblique pickup angles. These shells consist of ions with pickup angles larger than the local pickup angle that travel along the magnetic field lines from the direction of the ribbon. A simple model with straight magnetic field lines and zero temperature NSW yields the PUI distribution function at a heliocentric distance r as

$$f_r(v,\mu) = \frac{n_{\rm NSW} n_p \sigma r}{4\pi v_{\rm SW}^2 \mu (1-\mu^2)} \exp\left[n_H \sigma r \left(1 - \frac{\mu_0}{\mu}\sqrt{\frac{1-\mu^2}{1-\mu_0^2}}\right)\right] \delta(v - v_{\rm sw}) H(\mu_0 - \mu), \quad (3)$$

where σ is the charge exchange cross-section, μ_0 is the local pickup angle, n_H and n_p are the number densities of interstellar neutral and ionized hydrogen, respectively, $n_{\rm NSW}(r)$ is the number density of the neutral solar wind, and H is the Heaviside step function. For the Voyager 1 direction ($\mu_0 = 69°$) the distribution (3) is unstable and tends to scatter toward a half-shell [6]. The ENAs generated from the half-shell PUI distribution are not directed toward Earth and therefore do not contribute to the distributed flux. However, the level of magnetic fluctuations in the hybrid simulation was some three orders of magnitude above the background inferred from Voyager 1 observations. If indeed the instability is operating in the OHS, an efficient sink of magnetic fluctuation is required to dissipate the wave energy released by the PUI scattering process.

4. Discussion

According to [16], a pickup proton ring with a temperature in excess of a few hundred K is stable in the OHS plasma. This, however, ignores the second instability that has a temperature threshold close to the core plasma temperature in the OHS, $\sim 15,000 - 25,000$ K, based on MHD models [17]. The stable region therefore extends roughly from 200 K to 20,000 K. [13] have found, using Monte Carlo simulations of the neutral solar wind, that the latter is generally hotter than the upper limit on stability. Exacerbating the problem is the second PUI population produced from heliosheath ENAs. That population is significantly denser than the NSW, and has a toroid-like velocity distribution with a temperature of several million degrees K. According to [13], adding the heliosheath PUI population doubles the wave growth rate in the particle-mesh simulations. Therefore, contrary to the conclusions of [5] and [16], ribbon PUI stability requires low, and not high, ion ring temperatures.

Models have appeared recently where scattering is a beneficial factor concentrating ENA emissions in the ribbon direction [14, 9]. These models are compatible with our results for high-temperature toroids. The problem is that observations show extremely low magnetic fluctuation levels in the OHS that all but rule out a strong local source of turbulence in the direction of Voyager 1 [1]. PUI signature waves are also rarely seen in the solar wind as a consequence of rapid spectral transfer to smaller scales by the turbulence cascade [3]. However, we found [6] that for the OHS conditions the rate of energy spectral transfer is much smaller than the rate of energy production by the instability. The ratio can be estimated as

$$\frac{\dot{E}_{\rm turb}}{\dot{E}_{\rm prod}} \simeq \frac{2k^{-5/2}}{\eta n_{\rm NSW}\sigma v_{\rm SW}^3}\left(\frac{du^2}{dk}\right)^{3/2} \tag{4}$$

for a Kolmogorov cascade, where $\eta \sim 0.01 - 0.02$ is the fraction of the PUI energy transferred to the waves [6], $n_{\rm NSW}$ is the number density of NSW atoms [8], and u^2 is the turbulent energy per unit mass [1]. Using the published numbers, one arrives at a ratio of only 0.01. It must accepted, therefore, that the ambient turbulent cascade in the OHS is not the mechanism responsible for the absence of the waves.

If the OHS ENA model of the IBEX ribbon is correct, one must choose between two possibilities. The first is that the ion rings are stable and the waves are not generated. Then the PUIs either scatter very slowly or not at all (the ribbon's properties in that case could be explained by focusing effects from magnetic field draping around the heliopause [4] or the large scale magnetic irregularities that are part of the ambient wave spectrum in the LISM [10]). The second possibility is that there is efficient scattering on self-excited fluctuations [14, 9], but the waves are removed by turbulent processes and eventually dissipated as heat. Given the above estimate of the dissipation rate, the first possibility appears to be more realistic. In a follow up publication we will show that the NSW could be cold enough to prevent the Alfvén wave instability from operating in the OHS.

Acknowledgments
This research was supported by NASA grants NNX12AH44G and NNX14AF43G, and by NSF grant AGS-0955700. Resources supporting this work were provided by the NASA High-End Computing (HEC) Program through the NASA Advanced Supercomputing (NAS) Division at Ames Research Center.

References
[1] Burlaga L F, Florinski V and Ness N F 2015 *Astrophys. J. Lett.* **804** L31
[2] Burlaga L F and Ness N F 2014 *Astrophys. J.* **784** 146
[3] Cannon B E, Smith C W, Isenberg P A et al. 2014 *Astrophys. J.* 787 133
[4] Chalov S V, Alexashov D B, McComas D et al 2010 *Astrophys. J. Lett.* **716** L99
[5] Florinski V, Zank G P, Heerikhuisen J, Hu Q and Khazanov I 2010 *Astrophys. J.* **719** 1097
[6] Florinski V, Heerikhuisen J, Niemiec J and Ernst A 2016 *Astrophys. J.* in press
[7] Heerikhuisen J, Pogorelov N V, Zank G P, et al. 2010 *Astrophys. J. Lett.* **708** L126
[8] Heerikhuisen J, Zirnstein E, Funsten H, Pogorelov N V and Zank G P 2014 *Astrophys. J.* **784** 73
[9] Isenberg P A 2014 *Astrophys. J.* **787** 76
[10] Giacalone J and Jokipii J R 2015 *Astrophys. J. Lett.* **812** L9
[11] Liu K, Möbius E, Gary S P and Winske D 2012 *J. Geophys. Res.* **117** A10102
[12] McComas D J, Allegrini F, Bochsler P, et al. 2009 *Science* **326** 959
[13] Niemiec J, Florinski V, Heerikhuisen J and Nishikawa K-I 2016 *Astrophys. J.* in press
[14] Schwadron N A and McComas D J 2013 *Astrophys. J.* **764** 92
[15] Smith C W, Isenberg P A, Matthaeus W H and Richardson J D 2006 *Astrophys. J.* **638** 508
[16] Summerlin E J, Viñas A F, Moore T E, Christian E R and Cooper J F 2014, *Astrophys. J.* **793** 93
[17] Zank G P, Heerikhuisen J, Wood B E et al 2013 *Astrophys. J.* **763** 20

Journal of Physics: Conference Series **767** (2016) 012010 doi:10.1088/1742-6596/767/1/012010

Following the interstellar magnetic field from the heliosphere into space with polarized starlight

P C Frisch[1], A B Berdyugin[2], V Piirola[2], A M Magalhaes[3], D B Seriacopi[3], T Ferrari[3], F P Santos[4], N A Schwadron[5], H O Funsten[6], D J McComas[7], C E Heiles[8]

[1] Department of Astronomy and Astrophysics, University of Chicago, Chicago IL 60637 USA
[2] Tuorla Observatory and Finnish Centre for Astronomy with ESO, University of Turku, Finland
[3] Inst. de Astronomia, Geofisica e Ciencias Atmosfericas, Universidade de Sao Paulo, Sao Paulo, Brazil
[4] Center for Interdisciplinary Exploration and Research in Astrophysics and Department of Physics and Astronomy, Northwestern University, Evanston, IL 60208 USA
[5] Space Science Center, University of New Hampshire, NH 03824 USA
[6] Los Alamos National Laboratory, Los Alamos, NM 87544 USA
[7] Princeton Plasma Physics Lab, Princeton University, Princeton NJ 08543 USA
[8] Astronomy Department, University of California, CA 94720 USA

Abstract. Starlight linearly polarized by aligned interstellar dust grains provides the necessary data for tracing the structure of the very local interstellar magnetic field (ISMF). Two methods have been developed to recover the ISMF direction from polarized starlight, using data from an ongoing polarization survey. Both methods rely on the probability distribution function for polarized light. Method 1 calculates the ISMF direction from polarization position angles regardless of the data accuracy, while Method 2 relies on high-probability data points. The ISMF direction $B_{\rm IBEX}$ recovered by Method 1 corresponds to the closest ISMF to the heliosphere, traced by the center of the IBEX Ribbon arc. Method 2 reveals a new direction for the more distant ISMF, $B_{\rm new}$, toward ℓ=41.1° ± 4.1° and b= 25.8° ± 3.0°, which differs by 30.4° ± 5.6° from the IBEX ISMF direction. Polarizations of filament stars that are located within 25° of a pole of $B_{\rm new}$, where background polarizations would be minimal, show the highest statistical probabilities of tracing the filament ISMF. The IBEX ISMF direction orders the kinematics of interstellar clouds within 15 pc, and $B_{\rm new}$ must therefore dominate beyond 15 pc. These new data are consistent with the location of the Sun in the rim of an expanding superbubble shell associated with the evolved Loop I superbubble.

1. Introduction

Large-scale coherent magnetic features are found throughout the interstellar medium where they order synchrotron emission, rotate polarized radio emission, and organize the alignment of interstellar dust grains. At heliosphere scales of $\sim$ 0.001 pc the interstellar magnetic field (ISMF) direction is traced by the center of the Ribbon arc of energetic neutral atoms (ENAs) discovered by the Interstellar Boundary Explorer (IBEX) [1, 2, 3, 4, 5, 6]. Linearly polarized starlight provides the only currently available means for following the IBEX Ribbon ISMF from the heliosphere out into the low density interstellar gas in the immediate solar neighborhood. This linear polarization arises as starlight traverses a dichroic interstellar medium created by charged dust grains that are aligned in the presence of a magnetic field [7, 8]. A survey is underway to map the structure of the local ISMF using polarized starlight, and connect the interstellar field to the magnetic field traced by the IBEX Ribbon. Previous results from this survey have been presented [9, 10, 11] and the new results are part of a broader study of the structure of the ISMF within 40 pc (Frisch et al. 2016 *in preparation*).

Two methods for obtaining the ISMF direction from the polarized starlight have been developed for use in this survey, one of which is sensitive to weak polarizations that may not alone be statistically significant [9, 10], and the other that builds the likely magnetic field directions from significant polarizations only and is presented for the first time in this paper (Section 2.2).

Three interstellar magnetic features stand out in the survey polarization data. The ISMF direction traced by the IBEX ENA Ribbon, $B_{\rm IBEX}$ [3], dominates the most weakly polarized stars within 40 pc and 90° of the heliosphere nose [9, 10]. These ENAs were created by charge exchange between interstellar neutral hydrogen atoms and energetic ions in directions that are perpendicular to the interstellar magnetic field draping over the heliosphere. The heliosphere nose direction corresponds to the upwind direction of interstellar neutral helium flowing into the heliosphere [12]. A nearby filamentary-shaped polarization feature was identified while searching for the ISMF direction associated with the IBEX Ribbon, leading to a more precise identification of the IBEX ISMF direction in the polarization data ([10, 11], Section 3). The filament shows geometrical properties related to the heliosphere, suggesting that the filament is the heliospheric version of the dusty bowshocks or bowwaves common around external stars [13]. A third magnetic field direction, $B_{\rm new}$, is traced by stronger polarizations.

These polarization survey data trace aligned interstellar dust grains in the cluster of local interstellar clouds. Additional features of the polarization data underlying the results in this paper are mentioned in Appendix A. The dominant local ISMF direction, which matches the IBEX Ribbon ISMF direction, orders the kinematics of nearby interstellar gas [10] and controls the flux of TeV galactic cosmic rays into the heliosphere [14]. Early measurements of the nearby interstellar magnetic field were puzzling, with disagreement between studies [15, 16]. A better understanding of the partially ionized magnetoionic medium through which the Sun is now traveling should emerge from the data provided by this survey.

2. Recovering Magnetic Field Directions from Optical Polarization Data

Optically polarized starlight forms as light traverses a dichroic medium created when optically asymmetic interstellar dust grains become aligned with respect to the interstellar magnetic field. Dissipation of grain angular momentum aligns the principle axis of inertia with the angular momentum vector, so that the spinning charged grains become aligned with respect to the ISMF direction [7, 8]. The consensus view is that grains are aligned so that the axis of lowest optical opacity, and therefore the polarization position angle PA of the linearly polarized starlight E-vector, is parallel to the ISMF direction. This assumption is the basis of our efforts to infer the ISMF direction from optically polarized starlight.

Approximately 1% of the mass of an interstellar cloud is carried by the dust grains. Typical column densities in the immediate solar neighborhood are $N(H^o) \leq 10^{18.5}$ cm^{-2} [17], leading to interstellar polarization strengths of 0.01%–0.02% using standard relations between color excess and $N(H^o)$, and polarization strength and color excess [10]. Sensitive measurements of the linear polarizations of nearby stars therefore provide a viable tool for tracing the spatial configuration of the local interstellar magnetic field.

Low local column densities, a patchy ISM distribution [17] that also includes dense clouds [18], and an inhomogeneous population of dust grains [17], lead to detections of significant polarizations that depend on both polarimeter sensitivity and the interstellar material. These survey polarization data are a combination of early (20th century) polarization data with relatively large mean errors and recent (21st century) polarization measurements, some with mean errors $3\sigma < 0.01\%$.

2.1. Method 1: Minimize weighted polarization position angles in an ensemble of data
The first method that we established for finding the ISMF direction from a set of interstellar polarization PA data is based on creating a merit function that tests for goodness-of-fit of the PA data to an ISMF direction, and then evaluating that merit function for each possible ISMF direction on the sky to determine the ISMF direction that provides the best match to the weighted polarization data [9, 10]. The sine of a polarization position angle will be zero when evaluated in a coordinate system where the pole is rotated to coincide with the true interstellar magnetic pole. The advantage of this merit function is that data with all levels of signal-to-noise are utilized to determine the local ISMF direction, which is useful because of the low local column densities. Data are weighted with a probability distribution function (PDF) suitable for polarization position angles. Additional information about the statistical techniques of Method 1 can be found in Appendix B and Frisch et al. [10].

The best-fitting ISMF direction to polarization data for stars within 90° of the heliosphere nose and 40 pc (termed here the "studied region") was obtained using Method 1 for two separate fittings [10]. The first fit included all stars in this spatial region while the second fit excluded the filament stars (Section 3). The second fit provided the best match to the direction of the ISMF traced by the IBEX Ribbon and yielded an ISMF direction toward $\ell,b = 36.2°, 49.0°$ ($\pm 16°$). This fitting is sensitive to the large number of low-probability data points, that together place significant constraints on the merit function. The PDF for survey polarizations that trace the direction of the IBEX Ribbon ISMF is shown in Figure 1, right. The weighted center of the IBEX Ribbon is toward ecliptic coordinates $\lambda, \beta = 219.2° \pm 1.3°, 39.9° \pm 2.3°$ [3], or galactic coordinates $\ell, b = 34.8° \pm 4.3°, 56.6° \pm 1.2°$ [10]. These two ISMF directions are in agreement. There are a large number of data points with large mean errors, $P/dP < 2.0$ (for polarization P with mean error dP), and low probabilities of matching the IBEX Ribbon ISMF direction, but these low-probability data points form smooth shoulders on the distribution with higher probabilities than the predicted probabilities from a statistically random probability distribution.

2.2. Method 2: Map most probable magnetic poles across the sky
An alternate method for obtaining the ISMF direction from linear polarization is to build the result directly from the PDF (Figure 1) by comparing polarization position angles rotated into the new coordinate system, $PA_{\rm coord}$, where the pole is aligned with a possible ISMF direction, and then counting the number of data points with statistical probabilities $p_{\rm s}$ larger than an arbitrary minimum value, $p_{\rm s} > p_{\rm min} = 0.5$ for example. The coordinate system rotation with the most number of rotated polarization position angles with $PA_{\rm coord} > p_{\rm min}$ then defines the best-fitting ISMF direction. Effectively Method 2 draws a horizontal line through the PDF in Figure 1 and then selects the data points above that line for the analysis. Method 2 therefore ignores the more numerous low-probability data points, $p_{\rm min} < 0.5$, while identifying the ISMF directions traced by the strongest polarizations in the sample. Method 2 is particularly useful for yielding new insights into the filament of dust around the heliosphere ([10, 11], Section 3).

3. Evidence for dust filament around heliosphere

An unexpected result of the polarization survey was the identification of a polarization filament with properties related to the heliosphere geometry [11]. Filament stars were originally selected from the polarization data based on a marginal correlation between polarization position angles and star distance [9] that did not survive testing with a larger set of polarization data [10]. The larger data set instead showed that the filament polarizations are seen toward sixteen stars that

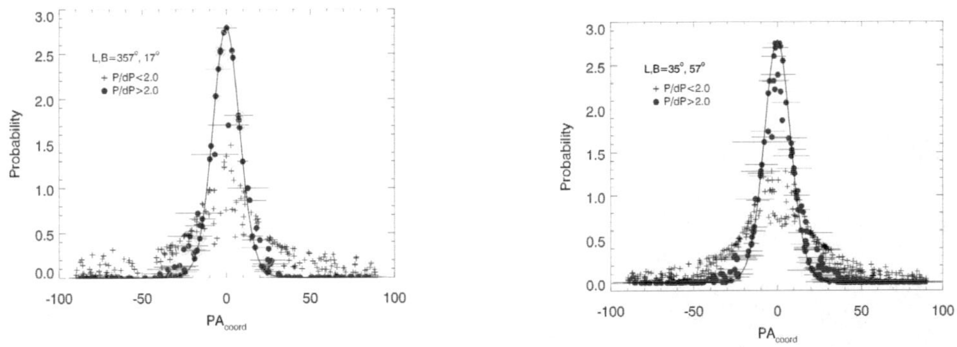

Figure 1. Probabilities that polarization position angles are aligned with the "true" ISMF direction. $PA_{\rm coord}$ is the polarization position angle expressed in a coordinate system where the magnetic pole is aligned with the true ISMF pole so that $PA_{\rm coord}$=0 for perfectly aligned grains. Left: Polarization position angles are expressed in a coordinate system aligned with the dust filament magnetic pole (ℓ=357°, b=17°). Solid lines show the underlying statistical probability distribution function for $P_{\rm o}$ = 3.5 (Appendix B). Blue dots (black crosses) indicate data where $P/dP>$ 2.0 ($<$ 2.0). Cyan-colored circles surround the filament stars. Right: The coordinate system pole and ISMF direction correspond to the ISMF direction traced by the center of the IBEX Ribbon ISMF arc (ℓ=35°, b=57°).

cover an angular extent of $\sim$ 100° and include three stars within 10 pc. Calculating the ISMF direction traced by the filament polarizations using Method 1 yielded a field direction toward galactic coordinates $\ell, b = 357.3°, 17.0°$. This direction is within $6.7° \pm 11.2°$ of the heliosphere nose direction given by the velocity vector of interstellar neutral helium entering the heliosphere and suggests that the filament is associated with the heliosheath region. Figure 2 shows the filament polarizations plotted in galactic coordinates in a stereographic projection (left), and a linear projection (right) that display other landmarks of the heliosheath region including the IBEX Ribbon [1, 19, 3].

Method 1 was originally applied to all polarization survey position angle data for the region under study, giving an ISMF direction offset by $32.4° \pm 16.6°$ from IBEX Ribbon ISMF direction. After the filament star subsample was identified and omitted from the calculations, the ISMF directions indicated by the polarization data and IBEX Ribbon agreed, differing by $7.6° \pm 16.6°$. The improved agreement between the polarization ISMF direction and the IBEX Ribbon ISMF found with omission of the filament stars from the sample supports a conclusion that the filament is a real magnetic structure in space.

Another filament property suggesting a heliosphere connection is that the axis of the filament is perpendicular to the heliosphere symmetry plane, the B-V plane, which is created by the velocity V of interstellar gas with respect to the Sun, the interstellar magnetic field direction B, and the Sun. Heliosphere symmetries about the B-V plane arise from Lorentz forces on charged particles that deflect charged plasma and particles in directions perpendicular to the magnetic field and particle velocities (e.g. [21, 22]). A linear fit to the filament axis yields an angle of $80° \pm 14°$ between the filament axis and the direction defined by the interstellar He velocity vector and IBEX ISMF pseudo-vector [11].

The scenario suggested here is that the filament polarizations are seen where the ion flow becomes fully deflected in the outer heliosheath regions and the flows of primary and secondary interstellar particles are beginning to diverge. In this region the ISMF direction and the gas velocity become parallel and are close to the heliopause.

Journal of Physics: Conference Series **767** (2016) 012010 doi:10.1088/1742-6596/767/1/012010

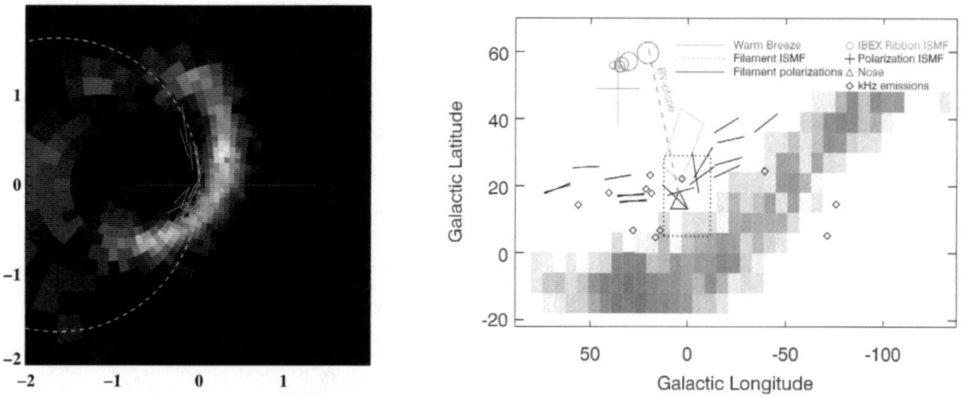

Figure 2. Left: A stereographic projection of the polarizations of the filament stars together with the brightest parts of the 1.1 keV IBEX Ribbon plotted in galactic coordinates. The scaling is nonlinear with a distance of 90° between labels 0 and 1, and 127° between labels 0 and 2. Longitude increases counterclockwise around the Ribbon arc. Right: The polarizations of the filament stars (black bars) are compared to markers of the outer heliosheath region or beyond. Uncertainties on the ISMF traced by the filament, which is toward the inflowing direction of interstellar helium wind, are outlined with a dotted box. The ISMF directions from the centers of the IBEX ribbon arc for the five energy bands of IBEX-HI [3] are plotted as red open circles, with the size of the circle increasing with energy. The black diamonds shows the locations of the outer heliosheath 2–3 kHz emissions [20]. The ISMF direction found from Method I applied to the polarization data is the large black cross. Dashed and solid gray lines show the locations of the B-V plane and the axis of the filament, respectively, which differ by an angle of $80° \pm 14°$. The IBEX Ribbon is shown for 1.1 keV data. See [11] for additional information.

The motion of the Sun through the surrounding low density weakly ionized interstellar cloud drives a flow of neutral or weakly charged interstellar gas and dust particles through the heliosphere. Enhanced charging of small grains in the hot heliosheath regions trap small grains with large charge-to-mass ratios in the plasma deflected around the heliosphere, where the interstellar grains that form the filament would be present [23, 24, 25]. Decades of *in situ* measurements of the mass distribution of interstellar dust grains by spacecraft confirm the exclusion of low-mass grains from the heliosphere [23, 26, 27] and indicate they have a silicate composition [28, 29], as is required by grain alignment scenarios [8] and predicted by abundance patterns of interstellar gas in the Local Interstellar Cloud (LIC) surrounding the heliosphere [30].

Low rates of collision between the interstellar grains and gas prevent collisional disruption of the coupling between grains with high charge-to-mass ratios and the ISMF. Voyager 1 entered the outer heliosheath at a distance of 124 AU [31], and found a non-turbulent ordered ISMF with a strength of $\sim 4.7\ \mu$G for the interstellar field compressed against the heliopause [32]. The temporal variation of the ISMF direction detected by the magnetometer on Voyager 1 (presently located 30° north, at higher ecliptic latitude, of the heliosphere nose) is converging to the ISMF direction traced by the center of the IBEX ENA Ribbon [33]. The laminar characteristic found

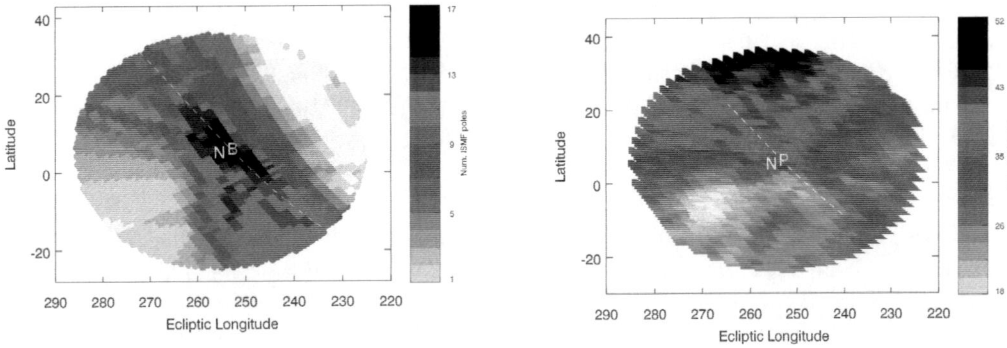

Figure 3. Left: The range of possible ISMF direction allowed by the polarizations of the filament stars is plotted in ecliptic coordinates. The number of polarizations with magnetic poles directed toward each location was calculated from Method 2 using probabilities larger than 0.5 for the PDF (Figure 1). The dashed line shows the best fit to the filament axis from [11], and the "B" marks the direction of the filament polarization obtained from Method 1. The "N" shows the heliosphere nose direction. The filament polarizations provide a well-defined parameter space for the filament ISMF direction. Right: Same as the left figure except that all stars within 40 pc and 85° of the heliosphere nose in the polarization catalog are included in the search for the number of data points that predict a magnetic pole for each location. Figure 1, left, explains the absence of the filament feature in the larger data set above-right since few stars have position angles directed toward the heliosphere nose.

by Voyager 1 for the ISMF in the outer heliosheath signifies the orderly draping of the ISMF over the heliosphere [34, 35] that would be required for a filamentary structure to form. Bending of the interstellar magnetic field-lines around the heliopause create the condition where the ISMF direction and therefore polarization position angles are parallel to the deflected currents that flow away from the heliosphere nose. Such a scenario might explain the existence of a polarizing dust filament around the heliosphere.

Theoretical simulations of interstellar dust grains interacting with a 3D MHD heliosphere predict that small charged dust grains, excluded from the heliosphere by the Lorentz force, create filamentary-like structures depending on the phase of the solar magnetic activity cycle [25]. These simulations show that dust grains interacting with the heliosphere experience forces that depend on solar cycle phase. The influence of the solar magnetic cycle has been shown to extend into the outer heliosheath and beyond, so that the filament might be a time-varying feature [36, 37].

A reasonable argument that the filament is a spurious effect could be made based on the fact that the nearest of the filament stars is 2500 times further away than the bowshock or bowwave of the heliosphere, at roughly 500 AU [38]. This is a valid concern because of the possibility that light from the filament stars is already polarized when it arrives in the heliosphere vicinity. The current data indicate that the filament stars with the largest statistical probabilities for tracing the filament ISMF direction, $p_s > 1.5$ (left figure of Figure 1), tend to be located within 25° of the direction of B_{new} and have low statistical probabilities for tracing B_{new}. The background polarization caused by B_{new} should therefore be weak because interstellar polarizations are weakest when sampling sightlines toward the ISMF direction. Polarizations arising from the background field B_{new} should be weaker by factors of ~ 2 or more for those filament stars with the largest probabilties of tracing the filament ISMF direction.

Journal of Physics: Conference Series **767** (2016) 012010 doi:10.1088/1742-6596/767/1/012010

4. Probability distribution function for the IBEX ISMF direction

The direction of the ISMF field in the very local ISM was obtained by analyzing the best-fitting ISMF direction to the polarization data set, excluding the filament stars, using Method 1. Method 1 allows the large number of data points where the polarization is not detected at a statistically significant level to contribute to the result. Figure 1, right, shows the PDF for the polarization position angle values calculated with respect to the pole of the IBEX ISMF direction. It is apparent that even measurements with low statistical probability look like they are also sampling the IBEX ISMF direction. This is in contrast to the same distribution calculated for the filament stars (left, Figure 1) where the position angles of low probability data points do not look regularly distributed or as if they were sampling the filament ISMF direction.

The ISMF direction obtained with Method 1 is in agreement with the ISMF direction defined by the weighted mean center of the IBEX Ribbon ([3], see Section 3), and the large number of low probability measurements (Figure 1, right) is consistent with the very low column densities nearby.

The Ribbon models are still under debate [39]. Several variants of these models yield observational signatures measured near the Sun (e.g., at Earth) with energetic neutral atom fluxes predominantly from directions perpendicular to the local interstellar magnetic field draping over the heliosphere. The Ribbon properties have been simulated using MHD models of the ISMF draping over the heliosphere, with the production of ENAs seeded by secondary pickup ions that form upstream of the heliopause and produce preferred directions of ENA emission after charge-exchange with a neutral hydrogen atom [2, 4, 6]. These simulations predict an offset as large as $\sim 9^\circ$ between the center of the Ribbon arc and the ISMF direction far upstream of the heliopause.

4.1. Testing whether Method 1 and Method 2 give the same result for ISMF direction in the galactic center hemisphere

As a test we apply Method 2 to the whole polarization data set to see whether the result is the same as found from Method 1. The two methods inherently sample different parts of the PDF, with Method 1 also responding to low probability measurements and Method 2 counting only high-probability data points by definition (see Figure 1). Both methods utilize stars at any location in the region of study for the evaluation of the best-fitting ISMF. Figure 4, left, shows the number of stars within 40 pc and 85° (stet) of the heliosphere nose, with $P/dP > 2.0$ and a probability > 1.0 for polarization position angle to trace a magnetic pole at each location in the plotted region.

The longitude and latitude of the most probable magnetic pole is considered the location of the maximum number of probable position angles. For $p_s > 1.0$, the maximum number of position angles at any location is 34, and this maximum is found at five separate locations. Averaging the coordinates of these five locations together gives ℓ=41.1° ± 4.4° and b= 25.8° ± 3.0° for this new magnetic field direction, $B_{\rm new}$ (see Frisch et al. 2016, *in preparation*, for additional information). The separation between this location and the IBEX Ribbon ISMF is 30.4° ± 5.6°. This maximum likelihood ISMF direction is a prominent feature in Figure 4 that suggests that the ISMF far beyond the heliosphere has a different orientation than the ISMF shaping the heliosphere. If the difference between $B_{\rm IBEX}$ and $B_{\rm new}$ is indicative of magnetic turbulence in the local ISM then the turbulence would on average represent changes in direction of $\sim 0.8^\circ$ per parsec. The IBEX ISMF is the ISMF around the heliosphere, and $B_{\rm new}$ which is traced by statistically stronger data, is likely to trace the ISMF much further away but within 40 pc. $B_{\rm new}$ is unsettlingly close to the direction of the solar apex motion [10]. and the topic of a separate study (Frisch et al. 2016, *in preparation*).

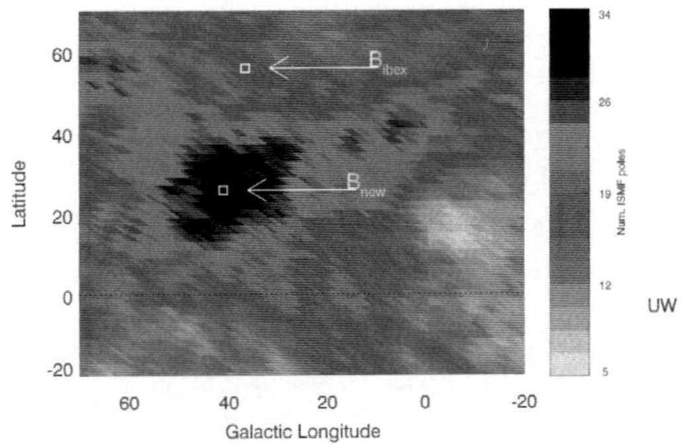

Figure 4. A section of the range of possible ISMF directions for the local ISMF direction as found by Method 2. The region that is displayed shows the most probable ISMF direction. The data in the figure represent measurements where $P/dP > 2.0$ and the probability of a polarization position angle tracing a magnetic pole at each location is $p_s > 1.0$ (see Figure 1). A new ISMF direction B_{new} appears as an excess of data points tracing a magnetic pole at ℓ=41.1° ± 4.1° and b= 25.8° ± 3.0° that is separated by 30.4° ± 5.6° from B_{IBEX}.

A large semi-circular feature of unknown origin also appeared in the fourth galactic quadrant when Method 2 was applied to the position angle data in the studied region. This feature consists of preferred directions for the directions of polarization position angles, and therefore for magnetic field poles. Additional information about the polarization survey data is presented in Appendix A.

5. Influence of ISMF on local ISM

By three measures the dominant ISMF direction traced by the IBEX Ribbon orders the kinematics of nearby interstellar gas. The only interstellar cloud for which we have *in situ* measurements of both the cloud velocity and the ISMF direction is the LIC that surrounds the heliosphere. IBEX data on the interstellar helium velocity and the Ribbon ISMF direction show that the LIC velocity through the LSR is perpendicular to ISMF direction, with a relative angle of 87.3° ± 3.0° [40, 10].

A comparison between the individual kinematical groups used to model the interstellar gas within ~ 15 pc [41] and the IBEX ISMF direction B_{IBEX} shows that interstellar cloud velocities through the LSR have a speed that increases with the angle between the IBEX ISMF direction and cloud velocity vector (Figure 5, [10]). This increase is not seen when cloud LSR velocity is compared with B_{new}. B_{new} traces the strongest polarizations that are more likely to form further from the Sun, while the Redfield and Linsky model places the fifteen clouds within 15 pc. It is not surprising that B_{IBEX} rather than B_{new} orders the cloud speeds.

Longstanding measurements of nearby interstellar absorption lines towards the Sco-Cen Association region have indicated that the very local ISM flows away from the center of the Loop I superbubble [17]. Wolleben (2007, [42]) developed a quantitative model for the synchrotron emission as a composite of two superbubbles, with the "S1" shell being the most likely to affect the very local interstellar material. Figure 6 shows the fifteen local interstellar clouds [41] that are flowing away from the center of the Loop I superbubble (also see [43]). Berdyugin and Piirola [44] have shown that high-latitude polarized starlight traces the S1 shell, while Santos et al. [45] have shown that the entire region of Loop I is well-defined by polarization data. The Wolleben model places the nearside of the S1 shell at the solar location. The survey polarization data

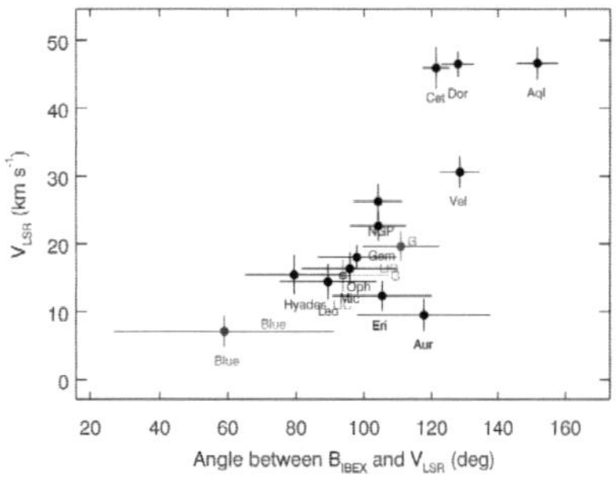

Figure 5. The angle between the LSR velocity of nearby interstellar clouds and the direction of the IBEX ISMF is shown plotted against the cloud speed. The ISMF direction is more disrupted for clouds accelerated to higher LSR speeds. Cloud names and velocities are based on the fifteen cloud local ISM model presented in [41]. The figure is from [10].

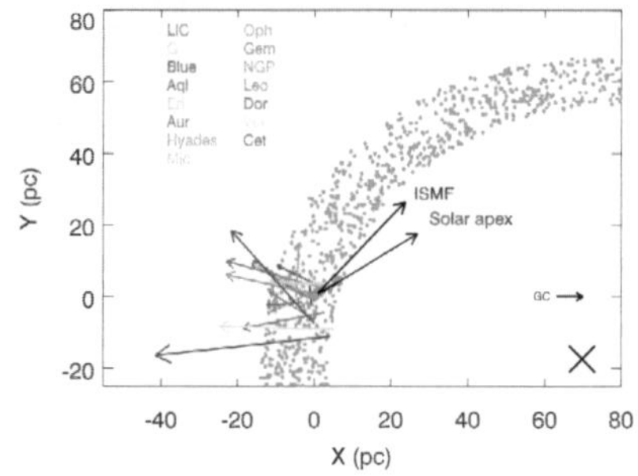

Figure 6. Cloud LSR velocities (arrows) are plotted over the S1 shell (gray dots) that describes synchrotron features associated with the Loop I superbubble centered at the "X", using the model of [42]. LSR cloud velocities are color coded. Black arrows show the directions of the IBEX ISMF direction and the solar apex motion through the LSR.

further support this interpretation, as does the fact that $B_{\rm new}$ and the IBEX Ribbon ISMF are directed toward different regions of the North Polar Spur that is the most prominent feature of Loop I. An alternate model that couples the local interstellar gas and ISMF models the detachment of a magnetic flux tube from the walls of the Local Bubble [46].

6. Conclusion

A survey is underway to probe the structure of the interstellar magnetic field within 40 pc using polarized starlight. Two methods are used to recover the ISMF direction from these data. Both methods rely on the statistical probabilities that a polarization position angle traces the "true" ISMF direction. Method 1 utilizes all suitable polarization data, regardless of uncertainties, while Method 2 selects only data with high probabilities for tracing a given field direction. Three ISMF directions are currently found from these data. The IBEX Ribbon ISMF direction that represents the ISMF closest to the heliosphere was found with Method 1, a new ISMF direction that is $30.4^\circ \pm 5.6^\circ$ away from $B_{\rm IBEX}$ was found with Method 2, and a third magnetic field direction toward the heliosphere nose is found to be traced by a filament of polarized starlight. Understanding the paleoheliosphere and astrospheres requires a full understanding of the ISMF direction in the surrounding interstellar medium. These polarization data are filling a gap in knowledge of the ISMF in the solar neighborhood.

Acknowledgments

This research has been supported by the NASA Explorer program through support for the IBEX mission and by the ERC Advanced Grant Hot-Mol ERC-2011-AdG-291659. AMM is grateful for support from FAPESP (grant no. 2010/19694-4) and CNP (Research Grant). DBS is grateful for support from CAPES (Phd scholarship). We are grateful to the Institute for Astronomy at the University of Hawaii (IfA) for the allocation of IfA time for observations at the T60 and UH88 telescopes, and to LNA, Brazil, for the allocation of observing time at the Observatorio do Pico dos Dias (OPD).

Appendix A. Polarization survey data

A survey is underway to map the structure of the local ISMF within 40 pc of the Sun using linearly polarized starlight. The results reported in this paper utilize data in the literature, data previously acquired specifically for this survey [9, 10, 11], and data recently acquired with the University of Hawaii UH88 telescope at Mauna Kea and the T60 telescope at Haleakala, and the Observatorio do Pico dos Dias (OPD), LNA, Brazil. The data utilized in the results reported in this paper were collected during the 20th and 21st centuries [9, 10, 11] and include new data (Frisch et al., 2016, *in preparation*).

Polarization data collected after the year 2000 are more accurate than data collected earlier due the higher accuracies of modern polarimeters [47, 48, 49, 50]. Polarization measurements of 755 stars provide the basis for the analysis in this paper. The mean value of P/dP for the 334 measurements collected after 2000 is 3.90. The mean value for 20th century measurements of P/dP is 1.60. The mean uncertainties for the 21st and 20th century data are 0.015% and 0.026% respectively.

Appendix B. Equations for evaluating polarization probabilities

A merit function has been utilized to extract the "true" ISMF direction from an ensemble of polarization position angles [10]. The goal of the merit function is to minimize the mean sine of the polarization position angles while weighting the data using a statistical distribution suitable for polarization data. The merit function is:

$$F_{\rm II}(B_{\rm i}) = {\rm N}^{-1} \sum_{\rm n=1}^{\rm N} f_{\rm n}(B_{\rm i}) = |\frac{\sin(\theta_{\rm n}(B_{\rm i}))}{G_{\rm n}}| \qquad \text{(B.1)}$$

The contribution of each individual star n to the merit function for ISMF direction $B_{\rm i}$ is $f_{\rm n}(B_{\rm i})$. The quantity $\theta_{\rm n}(B_{\rm i})$ is the polarization position angle $PA_{\rm n}$ for star n, which is calculated with respect to the i$^{\rm th}$ possible interstellar magnetic field direction $B_{\rm i}$. The sum is over N stars with the statistical weighting function for each star given by:

$$G_{\rm n}(\theta_{\rm obs};\ \theta_{\rm o}, P_{\rm o}) \ = \ \frac{1}{\sqrt{\pi}}\ \{\frac{1}{\sqrt{\pi}} + \eta_{\rm o}\ \exp(\eta_{\rm o}^2)\ [1 + {\rm erf}(\eta_{\rm o})]\}\exp(-\frac{{\rm P}_{\rm o}^2}{2}) \qquad \text{(B.2)}$$

using the probability distribution for polarization position angles in Naghizadeh and Clarke (1993,[51]). The observed position angle is $\theta_{\rm obs}$, the "true" position angle is $\theta_{\rm o}$, and $P_{\rm o} = \frac{P_{\rm true}}{\sigma}$, mean error σ=dP, $\eta_{\rm o} \ = \ \frac{P_{\rm o}}{\sqrt{2}}\ \cos\ [2(\theta_{\rm obs} - \theta_{\rm o})]$, and the Gaussian error function $erf(Z) \ = \ \frac{2}{\sqrt{\pi}} \int_0^Z exp(-t^2)\ dt$.

This probability distribution function is non-gaussian because of the bivariate nature of the position angle and the fact that polarizations are always positive while the underlying Stokes parameters can be positive or negative; the PDF reverts to a Gaussian at large probabilities (> 6). Constraints are placed on some parameters to avoid introducing spurious results related to poor data or variations in observing techniques. See Frisch et al. (2015, [10]) for additional information.

References

[1] McComas D J, Allegrini F, Bochsler P et al. 2009 *Science* **326** 959
[2] Schwadron N A, Bzowski M, Crew G B et al. 2009 *Science* **326** 966
[3] Funsten H O, DeMajistre R, Frisch P C et al. 2013 *ApJ* **776** 30
[4] Heerikhuisen J, Pogorelov N V, Zank G P et al. 2010 *ApJ* **708** L126–L130
[5] Heerikhuisen J and Pogorelov N V 2011 *ApJ* **738** 29
[6] Zirnstein E J, Heerikhuisen J, Funsten H O, Livadiotis G, McComas D J and Pogorelov N V 2016 *ApJ* **818** L18

[7] Lazarian A 2007 *J. Quant. Spec. Radiat. Transf.* **106** 225–256
[8] Andersson B G, Lazarian A and Vaillancourt J E 2015 *ARA&A* **53** 501–539
[9] Frisch P C, Andersson B G, Berdyugin A et al. 2012 *ApJ* **760** 106
[10] Frisch P C, Berdyugin A, Piirola V et al. 2015 *ApJ* **814** 112
[11] Frisch P C, Andersson B G, Berdyugin A et al. 2015 *ApJ* **805** 60
[12] Schwadron N A, Moebius E, Leonard T et al. 2015 *ApJS* **220** 25
[13] Peri C S, Benaglia P, Brookes D P, Stevens I R and Isequilla N L 2012 *A&A* **538** A108
[14] Adams, F. C., Christian, E. R. et al 2014 *Science* **343** 988
[15] Tinbergen J 1982 *A&A* **105** 53–64
[16] Leroy J L 1993 *A&AS* **101** 551
[17] Frisch P C, Redfield S and Slavin J 2011 *ARA&A* **49**
[18] Peek J E G, Heiles C, Peek K M G, Meyer D M and Lauroesch J T 2011 *ApJ* **735** 129
[19] Schwadron N A, Allegrini F, Bzowski M et al. 2011 *ApJ* **731** 56–77
[20] Gurnett D A, Kurth W S, Allendorf S C and Poynter R L 1993 *Science* **262** 199
[21] Pogorelov N V, Heerikhuisen J, Mitchell J J, Cairns I H and Zank G P 2009 *ApJ* **695** L31–L34
[22] Opher M, Bibi F A, Toth G, Richardson J D, Izmodenov V V and Gombosi T I 2009 *Nature* **462** 1036–1038
[23] Frisch P C, Dorschner J M, Geiss J et al. 1999 *ApJ* **525** 492–516
[24] Kimura H and Mann I 1999 *Earth, Planets, and Space* **51** 1223–1232
[25] Slavin J D, Frisch P C, Müller H R, Heerikhuisen J, Pogorelov N V, Reach W T and Zank G 2012 *ApJ* **760** 46
[26] Landgraf M, Baggaley W J, Grün E, Krüger H and Linkert G 2000 *J. Geophys. Res.* **105** 10343–10352
[27] Sterken V J, Strub P, Krüger H, von Steiger R and Frisch P 2015 *ApJ* **812** 141
[28] Westphal A J and the Stardust Team 2014 *Science* **345** 786,791
[29] Altobelli N, Postberg F and et al 2016 *Science* **352** 6283
[30] Slavin J D and Frisch P C 2008 *A&A* **491** 53–68
[31] Gurnett D A, Kurth W S, Burlaga L F and Ness N F 2013 *Science* **341** 1489–1492
[32] Burlaga L F and Ness N F 2014 *ApJ* **784** 146
[33] Schwadron N A, Richardson J D, Burlaga L F, McComas D J and Moebius E 2015 *ApJ* **813** L20
[34] Pogorelov N V, Heerikhuisen J, Zank G P, Borovikov S N, Frisch P C and McComas D J 2011 *ApJ* **742** 104
[35] Opher M and Drake J F 2013 *ApJ* **778** L26
[36] Pogorelov N V, Suess S T, Borovikov S N, Ebert R W, McComas D J and Zank G P 2013 *ApJ* **772** 2
[37] Zirnstein E J, Funsten H O, Heerikhuisen J and McComas D J 2016 *A&A* **586** A31
[38] Zank G P, Heerikhuisen J, Wood B E, Pogorelov N V, Zirnstein E and McComas D J 2013 *ApJ* **763** 20
[39] McComas D J, Lewis W S and Schwadron N A 2014 *Reviews of Geophysics* **52** 118–155
[40] Schwadron N A, Adams F C, Christian E R, Desiati P, Frisch P, Funsten H O, Jokipii J R, McComas D J, Moebius E and Zank G P 2014 *Science* **343** 988–990
[41] Redfield S and Linsky J L 2008 *ApJ* **673** 283–314
[42] Wolleben M 2007 *ApJ* **664** 349–356
[43] Frisch P C 2010 *ApJ* **714** 1679–1688
[44] Berdyugin A, Piirola V and Teerikorpi P 2014 *A&A* **561** A24
[45] Santos F P, Corradi W and Reis W 2011 *ApJ* **728** 104
[46] Cox D P and Helenius L 2003 *ApJ* **583** 205–228
[47] Piirola V, Berdyugin A and Berdyugina S 2014 *Society of Photo-Optical Instrumentation Engineers (SPIE) Conference Series* (*Society of Photo-Optical Instrumentation Engineers (SPIE) Conference Series* vol 9147) p 8
[48] Wiktorowicz S J and Matthews K 2008 *PASP* **120** 1282–1297
[49] Magalhães A M, Pereyra A, Melgarejo R, de Matos L, Carciofi A C, Benedito F F C, Valentim R, Vidotto A A, da Silva F N, de Souza P P F, Faria H and Gabriel V S 2005 *Astronomical Polarimetry: Current Status and Future Directions* (*Astronomical Society of the Pacific Conference Series* vol 343) ed Adamson A, Aspin C, Davis C and Fujiyoshi T pp 305–+
[50] Bailey J, Kedziora-Chudczer L, Cotton D V, Bott K, Hough J H and Lucas P W 2015 *MNRAS* **449** 3064–3073
[51] Naghizadeh-Khouei J and Clarke D 1993 *A&A* **274** 968

Limits on the interstellar magnetic field imposed by observational constraints indicate that *Voyager 1* remains in the inner heliosheath

G. Gloeckler and L. A. Fisk

Department of Climate and Space Sciences and Engineering, University of Michigan, 2455 Hayward St, Ann Arbor, MI 48109-2143 USA

gglo@umich.edu

Abstract. The unmeasured interstellar magnetic field strength is an essential parameter for models of the heliosphere moving relative to the Local Interstellar Cloud (LIC) that surrounds it. Here we use recent measurements of the differential energy spectra of energetic neutral hydrogen (ENH) combined with *Voyager 1* (*V1*) measurements of energetic particles to calculate the total pressure and its 1-σ limit in the inner heliosheath, the region between the termination shock at ~90 AU and the heliopause that *V1* presumably crossed at 122 AU. Balancing pressure across the heliopause, we find the magnetic pressure in the outer heliosheath at ~ 135 AU where *V1* is currently located. From this magnetic pressure we find an extraordinarily large magnetic field strength of 0.84±0.08 nT compared to the 0.464±0.009 nT field strength currently measured by *V1*. We conclude that *V1* is currently not measuring the interstellar field, but rather the solar magnetic field. *Voyager 1* is thus not in interstellar space but still inside a new, unanticipated, and most unusual region of heliosphere that it entered in late August 2012. We recalculate total pressures in the ordinary heliosheath (~90 - 122 AU), in the new, co-called cold inner heliosheath (122 – 145 AU), in the outer heliosheath and in the unperturbed LIC. Pressure balance again requires that all these pressures must be equal. Constrained by the requirement of pressure equality between all regions and minimum reduced χ^2 fits to the ENH spectrum and the differential intensities of particles measured by *V1* in the ordinary, co-called hot heliosheath, we find the common pressure to be $(3.45\pm0.39)\times10^{-12}$ dyne-cm^{-2}. The magnetic field strength in the outer heliosheath is calculated to be 0.90±0.089 nT, and in the unperturbed LIC it is 0.86±0.11 nT. To achieve common pressure for all regions imposes firm limits on the LIC neutral hydrogen density, n(HI) = 0.10±0.02 cm^{-3}. Pressure balance between the LIC and the hot cloud surrounding it places firm limit on the proton density in the unperturbed LIC, n(HII) = 0.036±0.012 cm^{-3}.

1. Introduction

Currently, the *Voyager 1* spacecraft, hereafter referred to as *V1*, is generally believed to be in interstellar space [1-3], in the outer heliosheath beyond the heliopause, which separates the heliosphere from the local interstellar medium. In this region, the putative outer heliosheath, particles that had been accelerated in the inner heliosphere have disappeared [2,3], galactic cosmic rays (GCRs) reached maximum intensity [2,3], suggesting easy escape of the former across the heliopause and entry of the latter. But has *V1* actually crossed the heliopause and is it at this moment in the local interstellar medium? Here we provide conclusive evidence to the contrary. First, we compute the total average pressure in the inner heliosheath (between 94 AU when *V1* crossed the termination shock, and 122 AU, the assumed location of the heliopause [1-3]) using recently published differential intensity spectra of Energetic Neutral Hydrogen (ENH) at low [4] and high [5,6] energies, and direct observations of the magnetic field [8] and > 40 keV particles [3] by *V1*, and find it to be $3.45 \bullet 10^{-12}$ dyne-cm^{-2}. Most of this pressure ($3.22 \bullet 10^{-12}$ dyne-cm^{-2}) comes from the thermal pressure of pickup ions, about half from locally created pickup ions and half from transmitted pickup ions that were produced in the supersonic solar wind, convected across the termination shock that heated and compressed them. The total pressure in the outer heliosheath just outside the heliopause is the sum of the particle (interstellar plasma and locally created pickup ions) thermal pressures and magnetic pressure due to the draped magnetic field. Pressure balance requires that the total pressure in the outer heliosheath near the heliopause must be the same as the total average pressure in the inner heliosheath. ENHs created in the outer heliosheath will also be detected by *IBEX* and thus contribute to the measured *IBEX* spectrum. Constrained by the *IBEX* observations, we compute the outer heliosheath particle pressure, finding it to be only $5.35 \bullet 10^{-13}$ dyne-cm^{-2}. A magnetic field strength of 0.84±0.08 nT is required to achieve pressure balance across the heliopause. However, *V1* measures only 0.464±0.009 nT [7], a field strength that is lower than the 4-σ lower limit of the field required from pressure balance. Not only is the direction of the magnetic field observed by *V1* [8] unchanged from the direction of the heliospheric magnetic field, which is in a substantially different direction from the expected direction of the interstellar magnetic field inferred from both *IBEX* observations of energetic neutral atoms (ENAs) [9,10, and references therein], as well as from observations of the polarization of interstellar grains [11], not only is the magnetic field strength unusually smooth and nonturbulent [7], but now we also find it to be far too weak. It must thus be concluded that *V1* remains in the inner heliosheath.

Fisk and Gloeckler [12,13] have developed a detailed model for the nose region of the heliosphere, the region of the heliosphere in the direction of motion of the Sun through the local interstellar medium, the region that the *Voyagers* are exploring. Our model [12,13] is based on the assumption that the solar wind can be compressed. The high plasma densities measured by Gurnett *et al.* [1] are thus not interstellar densities but rather densities of the compressed solar wind. The model yields flow patterns for the solar wind that are consistent with the convective solar wind velocities inferred from >40 keV anisotropy observations, the technique used to determine the solar wind velocity in the absence of a working plasma detector on *V1* [14,15]. The model also yields the likely distance to the heliopause, as well as the likely properties of the heliopause. For our purposes here, the details of the model of Fisk and Gloeckler [12,13] are not needed, other than to note that it is possible to construct a complete model for the nose region of the heliosheath in which the solar wind is compressed. *V1* is currently in a most unusual region of the inner heliosheath, the so-called cold inner heliosheath, from which particles below ~40 keV accelerated in the hot inner heliosheath (between the termination shock and heliocliff at 122 AU) have disappeared and galactic cosmic rays (GCRs) have reached maximum intensity [2,3], suggesting easy escape of the former across the heliopause and entry of the latter.

IBEX, looking in the *V1* direction, will detect ENHs created in (*a*) the hot, the ordinary inner heliosheath (~90 to 122 AU), (*b*) the cold inner heliosheath (122 AU to the heliopause at ~145 AU), and (*c*) the outer heliosheath (145 to roughly 175 AU). Three quite distinct charged particle populations, interacting with the neutral gas moving at ~26 km/s toward the Sun, produce Energetic Neutral Atoms and, in particular, ENHs: (*i*) the thermal population and their common tail [16], (*ii*) locally created pickup ions, and (*iii*) transmitted pickup ions, which are pickup ions created in the supersonic solar wind [17] upstream of the termination shock. Transmitted pickup ions are convected away from the Sun by the thermal plasma (solar wind), are compressed and heated as they cross the termination shock and enter the inner heliosheath. As detailed below, transmitted (population (*iii*)) together with local (population (*ii*)) pickup ions provide by far the mayor contribution (43% and 41%, respectively) of the total pressure ($3.45 \bullet 10^{-12}$ dyne-cm^{-2}) in the inner hot heliosheath. The solar wind, ACRs, and magnetic field contribute only ~7, ~7, and ~2%, respectively.

In the hot heliosheath the distribution functions of each of the three populations will have strong common tails (-5 power laws with exponential rollovers at high speeds), e.g., ACR tails on the transmitted pickup ion distribution, accelerated by the pump mechanism [16] in this highly turbulent [8] region.

In the cold heliosheath, where the measured magnetic field is incredibly smooth and the pump mechanism extremely inefficient, tails are not produced, and only the core distributions remain. Furthermore, particles with energy above ~ 30 keV (e.g. ACRs) have escaped [2,3], and, as we will show below, the transmitted pickup ion distribution rolls over sharply at $2.55 \cdot 10^{8}$ cm s^{-1} (28.9 keV). In the cold heliosheath, the solar wind contributes 55%, the magnetic field 25%, transmitted pickup ions 17%, and locally created pickup ions only 3%.

In the outer heliosheath near the heliopause the total pressure must be the same as the average pressure in the hot inner heliosheath, which must be the same as in the cold inner heliosheath. The total pressure in the outer heliosheath is the sum of the unknown and as yet unmeasured magnetic pressure and the thermal pressures of the interstellar plasma (interstellar wind) and locally created pickup ions. The ram pressures are negligibly small because the radial component of the bulk speed of the interstellar plasma vanishes at the heliopause. As detailed below, we find that the particles contribute a small amount (6%) to the total pressure of $3.45 \cdot 10^{-12}$ dyne-cm^{-2}, 5.7% and 0.3% for locally produced pickup ions and the interstellar plasma, respectively. Magnetic pressure is the primary balancing force, providing 94% of the total pressure. The strength of the draped magnetic field in the outer heliosheath is 0.90 nT.

Next, we derive the magnetic field strength in the unperturbed LIC, again using pressure balance with respect to the heliosphere, and independently with respect to the hot cloud that surrounds the LIC. To achieve pressure balance with the heliosphere requires a magnetic pressure of $2.94 \cdot 10^{-12}$ dyne-cm^{-2}, or 85% of the total balancing pressure of $3.45 \cdot 10^{-12}$ dyne-cm^{-2}. The only particle population in the unperturbed LIC is the interstellar plasma with its common tail. The sum of the ram and thermal particle pressures (13% and 2%, respectively) accounts for 15% of the total pressure. The magnetic field strength in the unperturbed LIC is 0.86 nT.

The LIC thermal pressure is only a small fraction (2 to 4%) of the hot cloud pressure surrounding the LIC. Thus, to balance the measured [18] thermal pressure of $3 \cdot 10^{-12}$ dyne-cm^{-2} in the hot cloud requires a LIC magnetic field of ~ 0.87 nT [19], which is remarkably close the value of we derive for the LIC magnetic field. To make the two values of the LIC field identical requires a LIC proton density of 0.036 cm^{-3}.

2. Computations of Pressures

At each location of the heliopause surface the total pressure just inside must be equal to that just outside the heliopause that separates the solar wind and magnetic field from the local interstellar field and plasma. In the subsonic inner heliosheath (between the termination shock, where the solar wind becomes subsonic, and the heliopause) the distribution between magnetic, thermal, and ram pressures will vary with heliocentric distance. For example, just downstream of the termination shock, where the radial speed component of the solar wind is large, the ram pressure of the solar wind will be relatively high, while the thermal pressure due to locally created pickup ions will be relatively small because so few are created just downstream of the shock. However, the sum of these separate pressures at any distance must be the same throughout the entire inner heliosheath, and the total average pressure of the entire inner heliosheath must thus be the same as the pressure in the inner heliosheath at the heliopause.

The average total pressure in the inner heliosheath is the sum of (*a*) the average thermal pressures, P_{th}, of (*1*) thermal ions and electrons (e.g. bulk solar wind) and (*2*) pickup ions, (*b*) the average ram pressure of ions, P_{ram} (the ram pressures due to electrons are negligible), and (*c*) the average magnetic pressure, P_{mag}. These pressures, as well as the number densities, n, are computed using equation (1),

$$n = 4\pi \int_0^\infty f(v) v^2 \, dv; \quad P_{th} = \frac{4\pi m_p}{3} \int_0^\infty f(v) v^4 \, dv; \quad P_{ram} = \rho U_r^2; \quad P_{mag} = \frac{B^2}{8\pi}, \tag{1}$$

where $f(v)$ is the phase space density (psd) as a function of v, the particle speed in the solar wind or interstellar plasma reference frame, m_p is the proton mass, ρ the mass density, and U_r the radial component of the solar wind or interstellar plasma bulk speed.

In order to compute particle pressures we need to know the distribution functions, $f_{i,k}(v)$, for each major particle population, where j designates the charged particle's mass, (e.g. j =1 for protons, j =4 for He, etc.), and k refers to one of the three major particle populations in the inner heliosheath (solar wind, transmitted pickup ions [17], and locally created pickup ions), and the two most dominant particle populations (interstellar thermal plasma and locally created pickup ions) in the outer heliosheath where *V1* now presumably resides. Above ~ 40 keV (2.8×10^8 cm s^{-1}) *V1* measures the distribution functions [3], and above ~8.8×10^8 cm s^{-1} the composition [3] directly, and partial pressures above these high speeds are readily computed. Unfortunately, no direct particle measurements from *V1* are available below 40 keV. Therefore we use the measured time-averaged velocity distribution function (psd *versus* particle speed, v_r) of energetic neutral hydrogen (ENH) measured by *IBEX* [4], *Cassini* [5], and *SOHO* [6], shown as filled circles in figure 1, to determine the proton velocity distributions $f_{1,k}(v)$ in the inner heliosheath for speeds below 2.8×10^8 cm s^{-1}, where most of the total particle pressure resides, mostly in pickup ions.

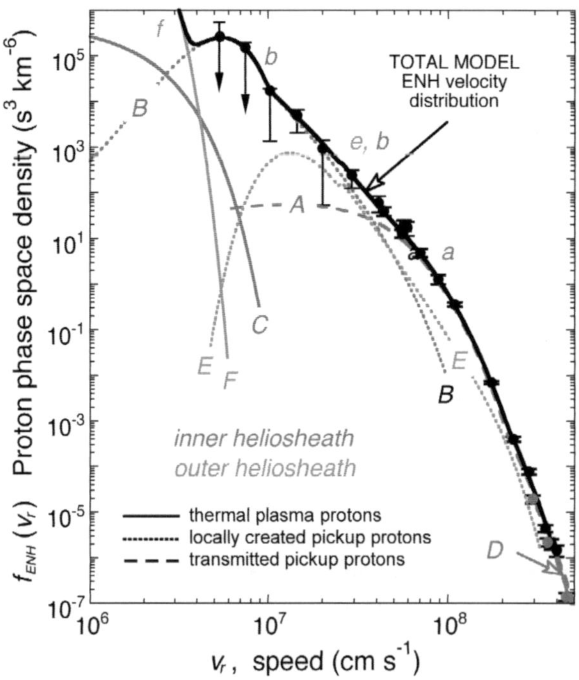

Figure 1. Time-averaged phase space density of energetic neutral hydrogen (ENH), $f_{ENH}(v_r)$, measured by *IBEX* [4] from ~5×10^6 to ~10^8 cm s^{-1}, *Cassini* [5] from ~1.1×10^8 to ~3.1×10^8 cm s^{-1}, and *SOHO* [6] from ~3.1×10^8 to ~4.4×10^8 cm s^{-1} (filled circles), versus the radial component of the speed, v_r, of the hydrogen atoms. The sum of the five labeled curves (*A*, *B*, *C*, *E*, and *F*) ENH phase space density spectra of particle populations in the inner heliosheath and the outer heliosheath near the heliopause (~122 to ~152 AU) yield the total spectrum (bold curve) of ENH below ~4.4×10^8 cm s^{-1} from the emission region (inner and outer heliosheaths) where these ENH are created. Curve *D* is the low end of the ACR ENH spectrum derived from the directly measured [3] proton spectrum by *V1* in the inner heliosheath. In the text we describe how the six spectra were computed. Curve *A* is the velocity distribution of ENH created by transmitted pickup protons. The transmitted pickup ion population is created in the supersonic solar wind upstream of the termination shock (TS), and convected with the solar wind across the TS. In crossing the TS transmitted pickup ions are compressed and heated. Convected through the highly turbulent [8] inner heliosheath, common spectrum [16] ACR suprathermal tails develop on these pickup ion distributions, and at the same time the cutoff speed of the pickup ion distribution is reduced.

2.1. Standard Model of the Heliosphere

Here we use the standard model of the heliosphere with the standard inner heliosheath that extends from the termination shock to the heliopause, and the outer heliosheath beyond the heliopause. This is the model that is generally accepted [1-3,7] with the heliopause at a heliocentric distance of 122 AU that *V1* presumably crossed in late August 2012. If one accepts this model then *V1* is now in the outer heliosheath near the heliopause, exploring interstellar space for almost 4 years.

We begin by computing the distribution functions of protons $f_{1,k}(v)$ using equation (2).

$$f_{1,k}(v) = f_0 \exp\left(-\left(\frac{v_{low}}{v}\right)^{\alpha_{low}}\right)\left(1+\frac{1}{\kappa}\left(\frac{v}{v_{th}}\right)^2\right)^{-\kappa}\exp\left(-\left(\frac{v}{v_0}\right)^{100}\right)$$
$$+ f_{tail}\exp\left(-\left(\frac{v_0}{v}\right)^{100}\right)\left(\frac{v}{v_0}\right)^{-5}\exp\left(-\left(\frac{v}{v_{high}}\right)^{\alpha_{high}}\right) \quad (2)$$

This single equation is used to model the phase space density of the (*a*) thermal solar wind or interstellar plasma (by setting $v_{low} = 0$ and $\alpha_{low} = 1$) with a common tail or without ($f_{tail} = 0$ and $v_0 = 10^{10}$ cm s^{-1} for no tail), (*b*) pickup ions with a common tail or without ($f_{tail} = 0$ and $v_0 = 10^{10}$ cm s^{-1}) and (*c*) common spectra above speed v_0 ($f_0 = 0$). Distribution functions observed [16] at 1 AU with *ACE*, and distribution functions of transmitted pickup ions measured [17] with *Ulysses* at ~5 AU, have been fit using equation (2).

Next, we convert the five proton distributions, $f_{1,k}(v)$, to the respective $f_{ENH,k}(v_r)$, the ENH phase space densities, using equation (3), which requires values for the neutral hydrogen gas density, n(HI),

$$f_{ENH,k}(v_r) = n(\mathrm{HI})L\sigma(v_{rel})f_{1,k}(v_r - U_r), \quad (3)$$

for L, the radial width of the respective emission regions (~92 to 122 AU for the inner heliosheath and 122 to ~152 AU for the outer heliosheath), for the H-p charge exchange cross section σ (v_{rel}) [20], which depends on v_{rel}, the relative speed between the neural gas and protons, with $\alpha = 30°$ and $\beta = 5°$,

$$v_{rel} = \sqrt{\left(V_0\cos\alpha\cos\beta - v_r - U_r\right)^2 + \left(V_0\cos\alpha\sin\beta - U_t\right)^2 + \left(V_0\sin\alpha - v_r - U_n\right)^2}, \quad (4)$$

for the direction, and $V_0 = 26$ km s^{-1}, and finally for U_r, the radial component of the bulk speed of the solar wind in the inner heliosheath, or of the plasma in the outer heliosheath. The values and errors we use for n(HI) are listed in row 3, column 2 of table 1, those for L are 30 ± 2 AU and 30 AU in the inner and outer heliosheaths, respectively, and those for U_r in the inner heliosheath are taken from row 2, column 2 of table 2. In the outer heliosheath near the heliopause $U_r \approx 0$.

The observed ENH spectrum (filled black circles of figure 1) must be matched by the sum (black solid curve) of the three spectra, curve *A* for transmitted pickup protons in the inner heliosheath, curve *B* for locally created pickup protons in the inner heliosheath and curve *E* for locally created pickup protons in the outer heliosheath near the heliopause. These three spectra make the dominant contribution to the total observed ENH spectrum above 5×10^6 cm s^{-1} in three different velocity intervals (*a*, *b*, and *e* in figure 1). The inner heliosheath solar wind spectrum (curve *C*), computed using *Voyager 2* solar wind measurements [21], makes negligible contributions to the measured ENH spectrum.

Journal of Physics: Conference Series **767** (2016) 012011 doi:10.1088/1742-6596/767/1/012011

Table 1. Values and 1-σ errors of key parameters in the outer heliosheath.

LIC parameter	Mean and standard error[a]	Value 1 and 1-σ error	Value 2 and 1-σ error	Value 3 and 1-σ error	Value 4 and 1-σ error
Flow speed, V_o (km s^{-1})	26.12 ± 0.27	25.4 ± 1.1 [24]	26.7 ± 0.5 [25]	26.08 ± 0.21 [26]	26.3 ± 0.4 [27]
Neutral Hydrogen density, n(HI) (cm^{-3})	0.125 ± 0.016	0.09 ± 0.02 [28]	0.115 ± 0.025 [23]	0.165 ± 0.035 [29]	0.13 ± 0.04 [6]
Temperature, T_o (°K)	7427 ± 423	8000 ± 1300 [24]	8150 ± 390 [25]	7260 ± 270 [26]	6300 ± 340 [27]

[a]Mean values and standard error of respective parameters of columns 3 to 6.

Table 2. Mean values and errors of solar wind parameters and magnetic field strength in the inner heliosheath (94 to 122 AU) and the outer heliosheath (122 to 152 AU) measured by *Voyager 1* and *Voyager 2*.

Parameter	~94 to ~122 AU	~122 to ~152 AU
Mean radial component of flow speed, U_r (km s^{-1})	34.0 ± 6.9 [14]	~0
Number density of protons, n_p (cm^{-3})	0.0019 ± 0.0007[a]	0.085 ± 0.01 [1]
Thermal speed, v_{th} (km s^{-1})	29.2 ± 0.07[a]	
Magnetic field strength, (nT)	0.121±0.02 [8]	0.464 ± 0.009 [7]

[a]*Voyager 2* solar wind measurements [21] in the inner heliosheath.

Next, we vary parameter values in equation (2) for each of the three populations, curves A and B in the inner heliosheath, and curve E in the outer heliosheath, to obtain the best fit (minimum reduced χ^2) to the observed ENH spectrum. The final set of parameters that gave the minimum reduced $\chi^2 = 0.955$ is listed in rows 3 to 5, columns 3 to 11 in table 3. For the Anomalous Cosmic Ray (ACR) population [3] (curve D) the parameters in equation (2) are obtained from best fits to the directly measured [3] proton spectrum above 40 keV.

Using the three parent proton velocity distribution, $f_{1,k}(v)$, in the ENA emission regions that together give the best fit to the observed ENH spectrum (filled circles in figure 1), the proton spectrum [3] of ACRs and the magnetic field strength [8] in the inner heliosheath, both directly measured by *V1*, as well as the maxwellian solar wind proton spectrum observed [21] by *Voyager 2* and assumed for the inner heliosheath, we find the total pressure in the inner heliosheath, P_{IH}, sum of magnetic pressure, $P_{IH,mag}$, and particle pressure, $P_{IH,particle}$, which in turn is the sum of thermal and ram pressures of protons, heavy particles, and electrons. In the outer heliosheath just beyond the presumed heliopause at 122 AU, the total pressure, P_{OH}, is the sum of the unknown magnetic pressure, $P_{OH,mag}$, of the draped magnetic field and the particle thermal pressures, $P_{OH,particle}$, of the interstellar plasma and of pickup ions that are produced primarily in the nose region of the outer heliosheath near the heliopause. In that region the particle ram pressures are negligible because there the particle flow is nearly parallel to the heliopause.

Table 3. Proton densities and parameter values for equation 2.

Region[a]	Type[b]	n(p) (cm^{-3}) proton density[c]	v_{low} (cm s^{-1})	α_{low}	v_{th} (km s^{-1})	κ	v_0 (cm s^{-1})	f_{tail} (s^3 km^{-6})	v_{high} (cm s^{-1})	α_{high}
					Standard Model					
INNER	Thermal	0.002	0	1	30.0	100	10^{10}	0	10^8	1.8
INNER	Local	0.008	$5\cdot10^7$	1.17	42.0	100	$1.35\cdot10^7$	$2\cdot10^4$	$5\cdot10^7$	1.8
INNER	Trans	0.000362	10^7	1.0	400	6.01	$2.2\cdot10^8$	0.03	$4\cdot10^9$	1.2
INNER	ACRs	4.1×10^{-7}	10^{30}	1	1	100	$2.2\cdot10^8$	$2.5\cdot10^{-3}$	10^{10}	1.5
					Fisk and Gloeckler Model					
HOT	Thermal	0.005	0	1	22.0	4.0	$7.5\cdot10^6$	0.003	10^7	1.50
HOT	Local	0.006	$3\cdot10^7$	1	40.0	5.0	$1.2\cdot10^7$	0.0005	$5.59\cdot10^7$	6.89
HOT	Trans	0.000351	$7\cdot10^6$	1.0	394.5	6.11	$1.8\cdot10^8$	$8.1\cdot10^{-5}$	10^9	2.00
COLD	Thermal	0.09	0	1	22.0	100	10^{10}	0	10^8	1.50
COLD	Local	0.002	$6.9\cdot10^7$	1.35	6.0	17	10^{10}	0	$5\cdot10^7$	1.20
COLD	Trans	0.000351	$7\cdot10^6$	1.0	394.5	6.11	$1.88\cdot10^8$	$7.9\cdot10^{-5}$	$2.35\cdot10^8$	8.00
OUTER	Thermal	0.09	0	1	11	10	$5\cdot10^6$	10^{-5}	10^8	1.5
OUTER	Local	$5\cdot10^{-5}$	$4\cdot10^7$	1.3	26	8	$2\cdot10^7$	$3\cdot10^{-9}$	10^8	1.2
LIC	Thermal	0.0357	0	1	8.08	10	$4\cdot10^6$	$3\cdot10^{-6}$	10^9	1.5

[a]INNER: standard inner heliosheath (94 to 122 AU); HOT: hot inner (standard) heliosheath (~90 to 122 AU); COLD: cold heliosheath (122 to ~145 AU); OUTER: outer heliosheath (~145 to ~175 AU); LIC: Unperturbed local interstellar cloud (>2000 AU).

[b]Thermal: solar wind or interstellar wind with or without tail; Local: locally created pickup ions with or without tails; Trans: transmitted pickup ions with or without tails.

[c]Densities are computed using the first equation in equation (1).

Three of the five parent proton spectra are well constrained by the measured ENH spectrum. The two that are not are (*a*) the inner heliosheath solar wind spectrum (curve *C*, previously discussed) and (*b*) the distribution of the thermal interstellar plasma in the outer heliosheath (curve *F*). We model the later distribution as a maxwellian (very large κ in equation (2)), with no suprathermal tail. This type of distribution is quite reasonable because the measured magnetic field [7] in that region is so exceptionally smooth (nonturbulent). In the absence of turbulence, maxwellian distributions without tails are generally observed [16]. For the interstellar plasma density we use the value of 0.085 ± 0.1 cm^{-3} measured [1] by the Plasma Wave System instrument on *V1* in the presumed outer heliosheath where *V1* is now located. For the interstellar protons thermal speed we use 11.0 ± 2.6 km s^{-1}, assuming that the thermal speed of the plasma is the same as the measured thermal speed of the neutral gas in that region computed from the temperature listed in row 4 of column 2 of table 1. The total particle pressures in each of the two regions is the sum of values in columns 4 and 5 of rows 3 to 5 and of rows 6 and 7, respectively, of table 1. The ACR pressure, computed using the directly measured [3] proton spectrum above 40 keV and corrected for galactic cosmic ray contributions above ~5 MeV.

We used the standard propagation of errors procedure to compute 1-σ errors of the total particle pressures, taking into account that some of the uncertainties are correlated.

Pressure balance across the heliopause requires that $P_{OH} = P_{IH}$. To achieve this pressure balance requires that $P_{OH,mag} = P_{IH} - P_{OH,particle} = 3.59\times10^{-12} - (0.20\times10^{-12} + 0.34\times10^{-12}) = 3.05\times10^{-12}$ dyne cm^{-2}. From the fourth equation in equation (1), the strength of the draped magnetic field in the region where *V1* currently resides is 0.876 ± 0.088 nT. The 4-σ lower limit is 0.521 nT, which is larger than 0.464 ± 0.009 nT, the field strength currently measured [7] by *V1*.

Journal of Physics: Conference Series **767** (2016) 012011 doi:10.1088/1742-6596/767/1/012011

Table 4. Pressures of thermal ions and electrons, locally created pickup ions, transmitted pickup ions and magnetic field in various regions of the outer heliosphere, the outer heliosheath, and the LIC.

Region[a]	Type[b]	Protons Thermal Pressure[d]	All ions and electrons[c] Thermal Pressure[d]	Ram Pressure[d]	Total Pressure[d]	Magnetic Pressure[d]	Magnetic Field (nT)
			Standard Model				
Inner	Thermal	1.54×10^{-14}	$(3.49\pm0.61)\times10^{-14}$	$(5.58\pm1.87)\times10^{-14}$	$(9.07\pm1.97)\times10^{-14}$	$(5.83\pm1.93)\times10^{-14}$	0.121 ± 0.020[e]
Inner	Local	8.94×10^{-13}	$(1.38\pm0.24)\times10^{-12}$	$(1.75\pm0.59)\times10^{-13}$	$(1.56\pm0.25)\times10^{-12}$		
Inner	Trans	1.06×10^{-12}	$(1.64\pm0.28)\times10^{-12}$	$(7.93\pm0.26)\times10^{-15}$	$(1.65\pm0.28)\times10^{-12}$		
Outer	Thermal	8.66×10^{-14}	$(1.96\pm0.34)\times10^{-13}$	~0	$(1.96\pm0.34)\times10^{-13}$	$(3.05\pm0.62)\times10^{-12}$	0.876 ± 0.088[f]
Outer	Local	2.19×10^{-13}	$(3.39\pm0.59)\times10^{-13}$	~0	$(3.39\pm0.59)\times10^{-13}$		
			Fisk and Gloeckler Model				
HOT	Thermal	4.34×10^{-14}	4.34×10^{-14}	1.40×10^{-13}	1.83×10^{-13}	5.83×10^{-14}	0.121[e]
HOT	Local	8.32×10^{-13}	1.29×10^{-12}	1.31×10^{-13}	1.42×10^{-12}		
HOT	Trans	9.27×10^{-13}	1.43×10^{-12}	7.69×10^{-15}	1.44×10^{-12}		
HOT	ACRs	1.62×10^{-13}	2.51×10^{-13}	9×10^{-18}	2.51×10^{-13}		
COLD	Thermal	8.23×10^{-13}	1.86×10^{-12}	7.84×10^{-14}	1.86×10^{-12}	8.57×10^{-13}	0.464[e]
COLD	Local	6.59×10^{-14}	1.02×10^{-13}	1.37×10^{-15}	1.03×10^{-13}		
COLD	Trans	3.73×10^{-13}	5.78×10^{-13}	2.40×10^{-16}	5.78×10^{-13}		
Outer	Thermal	4.36×10^{-15}	9.87×10^{-15}	~0	9.87×10^{-15}	3.24×10^{-12}	0.902[f]
Outer	Local	1.27×10^{-13}	1.96×10^{-13}	~0	1.96×10^{-13}		
LIC	Thermal	2.71×10^{-14}	6.12×10^{-14}	4.42×10^{-13}	5.03×10^{-13}	2.94×10^{-12}	0.859[f]
CLOUD	Thermal				$\sim3\times10^{-12}$	2.94×10^{-12}	0.859[f]

[a]INNER: standard inner heliosheath (94 to 122 AU); HOT: hot inner (standard) heliosheath (~90 to 122 AU); COLD: cold heliosheath (122 to ~145 AU); OUTER: outer heliosheath (~145 to ~175 AU); LIC: Unperturbed local interstellar cloud (>2000 AU); CLOUD: hot cloud surrounding the LIC.

[b]Thermal: solar wind or interstellar wind with or without tail; Local: locally created pickup ions with or without tails; Trans: transmitted pickup ions with or without tails.

[c]Includes contributions from protons, electrons, and heavy ions for the solar wind and interstellar thermal ion populations and protons as well as heavy ions for local and transmitted pickup ions and ACRs. Since the composition of the solar wind [30] is well known, as is the composition of transmitted pickup ions [17] and ACRs [3], the total particle pressures are readily computed. For locally accelerated pickup ions we use solar wind abundances [30].

[d]All pressures in units of dyne cm^{-2}. Thermal pressures, ram pressures, and magnetic pressures are computed using the second, third, and fourth equations, respectively, in equation (1).

[e]Average magnetic field strength measured by *V1* in the hot [8] and cold [7] inner heliosheaths.

[f]Value and 1-σ error of magnetic pressure or field strength in the outer heliosheath derived from pressure balance.

2.2. Fisk and Gloeckler Model of the Nose of the Heliosphere

The observations [7,8] that the magnetic field direction did not change to the much different direction of the interstellar field, and that the magnetic field was so incredibly smooth [7], unlike anything observed at magnetopauses of planets, already cast doubt that *Voyager 1* entered interstellar space in

Journal of Physics: Conference Series **767** (2016) 012011 doi:10.1088/1742-6596/767/1/012011

late August 2012. Now we find that there is a 99.99% certainty that the weak magnetic field strength currently measured by *V1* is not the interstellar field in the outer heliosheath. It must therefore be concluded that *Voyager 1* is currently not in the outer heliosheath, nor in the local interstellar medium. It is, however, certainly in a most unusual region of the heliosphere, a region never before explored. The Fisk and Gloeckler model [12,13] of the nose of the heliosphere describes the properties of this unusual region they called the cold inner heliosheath: The magnetic field is incredibly smooth [7] and its direction the same as it was in the hot or standard inner heliosheath [7,8]. Particles above 40 keV nuc^{-1} are no longer observed above background [2,3], they disappeared. The solar wind in the cold inner heliosheath is highly compressed. Its density is, in fact, the ~0.085 cm^{-3} high density measured by *V1* in that region [1].

We now use pressure balance (*1*) along the radial direction of the *V1* trajectory and (*2*) between our LIC and the larger hot cloud that surrounds it to determine best values of key parameters in the unperturbed (by the heliosphere) LIC. Constrained by observations (the ENH differential intensities, the >40 keV energetic particle spectra and magnetic field by *V1*, solar wind bulk velocity by *V1* and *V2* and its density by *V2*), we find the best mean values and 1-σ error limits for B, the strength of the magnetic field, for n(HI), the neutral hydrogen density, and for n(HII), the proton density.

Pressure balance along the radial direction of the *V1* trajectory requires that the average pressure in the hot heliosheath is the same as the average pressure in the cold heliosheath, is the same as the pressure in the outer heliosheath near the heliopause, and finally, is the same as the pressure in the unperturbed LIC, thousands of AU from the sun. The values of average particle pressures in the hot and cold inner heliosheath are constrained by the measured averaged ENH spectrum and the measured average energetic particle differential intensities. Since in these two regions of the inner heliosheath the magnetic field is also measured, the total average pressure in each of the two regions is also known. In figure 2 we show the derived ENH velocity distributions from (*a*) transmitted pickup protons and their tails, (*b*) locally accelerated protons and their tails, and (*c*) thermal (solar wind) protons and their tails. Parameter values in equations (2) and (3) were adjusted to (*1*) make the pressures in the hot and the cold inner heliosheath the same, and (*2*) obtain best fit (minimum reduced χ^2) of the sum (black curve) of the individual spectra to the measured spectrum (black circles). The final and observationally most constrained set of values for B, n(HI), and n(HII) are 0.86 ± 0.11 nT, 0.10 ± 0.02 cm^{-3}, and 0.036 ± 0.012 cm^{-3}, respectively. Our best mean values and 1-σ uncertainties for the LIC densities of neutral hydrogen and protons and for the magnetic field strength are listed in row 2, columns 2, 3, and 5, respectively, along with other estimates of these quantities.

Table 5. Densities of neutral hydrogen, protons and electrons magnetic field strength in the unperturbed local interstellar cloud.

Source	n(HI) neutral hydrogen density (cm^{-3})	n(HII) proton density (cm^{-3})	$n(e^-)$ electron density (cm^{-3})	B Magnetic field strength (nT)
Present work	0.10 ± 0.02	0.036 ± 0.012	–	0.86 ± 0.11
Reference [28]	0.09 ± 0.02			
Reference [23]	0.115 ± 0.025	0.040 ± 0.017	0.049 ± 0.016	< 0.43[a]
Reference [6]	0.13 ± 0.04			
Reference [29]	0.165 ± 0.035			

[a]The sizeable pressure due to transmitted pickup ions was not included.

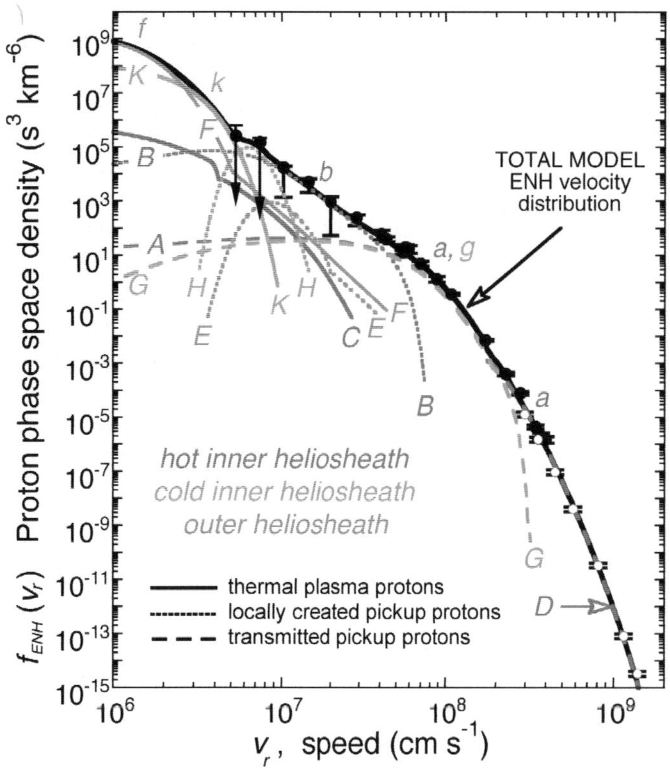

Figure 2. Same as figure 1 but here applied to the *Fisk & Gloeckler* model [12,13] for the nose of the heliosphere, which has the cold inner heliosheath. The sum of the eight ENH phase space density spectra of particle populations in the hot inner heliosheath (red curves *A*, *B* and *C*), the cold inner heliosheath (green curves *G*, *H* and *K*), and the outer heliosheath near the heliopause (~145 to ~175 AU, blue curves *E* and *F*) yield the total spectrum (bold curve) of ENH below ~4.4×10^8 cm s^{-1} from the emission region (hot inner, cold inner and outer heliosheaths) where these ENH are created. Curve *D* is the low end of the ACR ENH.

3. Concluding Remarks

The most far-reaching conclusion of this paper is the strength of the interstellar magnetic field. The ~0.9 nT field, required for pressure balance between our local interstellar cloud and the hot cloud surrounding it, we now confirmed using pressure balance between the LIC and the heliosphere moving relative to it. Because magnetic pressure exceeds dynamic pressure by almost an order of magnitude, the shape of the heliosphere is not comet-like as all of us believed, but of the type to be shaped in a strong surrounding magnetic field as shown by Parker [22]: essentially a sphere draped by the interstellar field. The motion of the heliosphere distorts the polar field lines, dragging them tailward into two lobes, as imaged by *IBEX*. As *IBEX* observations clearly show, the reason for the existence of the high LIC field is to balance the high pressure (~3.5×10^{-12} dyne cm^{-2}) of the inner heliosphere. The reason why this pressure is a factor of two higher than what we previously believed (e.g. *Gloeckler et al.* [23]) is because we grossly underestimated the thermal pressure from pickup ions. Pickup ions play a dominant role in the dynamics of the outer heliosphere and must be included along with the solar wind in any credible numerical model of the heliosphere, even though this presents an enormous challenge. These model must also use the realistic value of ~0.9 nT for the LIC magnetic field strength, neutral hydrogen density (~0.1 cm^{-3}), and proton density (~0.035 cm^{-3}).

The implications of the existence of this new region are also far-reaching and profound, and will require rethinking by our scientific community of the physics that applies to this highly unusual region. Our analysis of the measured ENH differential intensity shows that pickup ions, both transmitted and locally created, and their suprathermal tails (ACRs) account for 96% of the total inner heliosheath pressure and thus dominate the dynamics of the heliosphere. Unfortunately, current numerical models of the heliosphere ignore this fact and will thus require major revisions.

The near certainty that *Voyager 1* has not yet reached interstellar space implies that the exciting exploration of the outer heliosheath is still ahead after *V1* crosses the heliopause several years from now. But even before this historic crossing, it is very likely that the predicted [12,13] field polarity reversal will be observed. Then we will know precisely just how slow the solar wind flows in the cold inner heliosheath of the heliosphere.

Acknowledgments
We thank E. Möbius and G. Zank for their useful comments. This work was supported in part by NASA Grant NNX10AF23G.

References

[1] Gurnett D A, Kurth W S, Burlaga L F and Ness N F 2013 *Sci.* **341** 1481
[2] Stone E C, Cummings A C, McDonald F B, Heikkila B C, Lal N and Webber W R 2013 *Sci.* **341** 150
[3] Krimigis S M, Decker R M, Roelof E C, Hill M E, Armstrong T P, Gloeckler G, Hamilton D C and L. J. Lanzerotti L J 2013 *Sci.* **341** 144 doi: 10.1126/science.1235721
[4] Galli A, Wurz P, Schwadron N A, Kucharek H, Möbius E, Bzowski M, Sokół J M, Kubiak M A, Funsten H O, Fuselier S A and McComas D J 2016 *Astrophys. J.* **821** (2) 107
[5] Dialynas K, Krimigis S M, Mitchell D G, Roelof E C and Decker R B 2013 *ApJ* **778** (1), 40 doi: 10.1088/0004-637X/778/1/40
[6] Czechowski A, Hilchenbach M, Hsieh K C, Grzedzielski S and Kóta J 2008 *Astron. & Astrophys.* **487** (1) 329 doi:10.1051/0004-6361:200809555
[7] Burlaga L F and Ness N F 2014 *ApJL* **795** L19
[8] Burlaga L F, Ness N F and Stone E C 2013 *Sci.* **341** 147
[9] McComas D J, *et al.* 2009 *Sci.* **326** 959
[10] McComas D J, *et al.* 2012 *Astrophys. J. Supp.* **203** doi:10.1088/0067-0049/203/1/1
[11] Frisch P C 2011 *Physics of the Heliosphere: A 10 Year Retrospective*, *AIP Conf. Proc.*, vol. 1436, ed J Heerikhuisen et al. (Melville, NY: AIP) pp. 295
[12] Fisk L A and Gloeckler G 2013 *Astrophys. J.* **776** doi:10.1088/0004-637X/776/2/79
[13] Fisk L A and Gloeckler G 2014 *Astrophys. J.* **789** doi: 10.1088/0004-637X/789/1/41
[14] Decker R B, Krimigis S M, Roelof E C and Hill M E 2012 *Nature* **489** 124
[15] Stone E C and Cummings A C 2011 *Proc. of the 32nd Intl. Cosmic Ray Conference* vol 12 (Beijing: IUPAP) p 29
[16] Fisk L A and Gloeckler G 2014 *JGR: Space Phys.* **119** (11), 8733
[17] Gloeckler G, Geiss J and L. A. Fisk L A 2001 *The Heliosphere near Solar Minimum: the Ulysses Perspectives* eds A Balogh, E J Smith and R G Marsden (London: Springer-Praxis) p 287
[18] Bowyer S, Lieu R, Sidher S D, Lampton M and Knude J 1995 *Nature* **375** 212
[19] Cox D 1995 *Nature* **375** 185
[20] Lindsay B G and R. F. Stebbings R F 2005 *JGR* **110** A12213 doi:10.1029/2005JA011298
[21] Richardson J D 2016 MIT website for *Voyager 2* Solar Wind data: http://web.mit.edu/afs/athena/org/s/space/www/voyager/voyager_data/voyager_data.html
[22] Parker E N 1963 *Interplanetary Dynamical Processes* (New York: Interscience Publishers)
[23] Gloeckler G, Fisk L A and Geiss J 1997 *Nature* **386** 374
[24] Schwadron N A, *et al.* 2015 *ApJ Supp. Series* **220** 25
[25] Bzowski M P *et al.* 2015 *ApJ Supp. Series* **220** 28
[26] Wood B E, Müller H-R, Bzowski M, Sokól J M Möbius E, Witte M and McComas D J 2015 *ApJ Supp. Series* **220** 31
[27] Witte M 2004 *Astron. & Astrophys.* **426** 835
[28] Richardson J D, Liu Y, Wang C and McComas D J 2008 *Astron. & Astrophys.* **491** 1
[29] Quemerais E, Bertaux J-L, Sandel B R and R. Lallement R 1994 *Astron. & Astrophys.* **290** 941
[30] Ogilvie K W, Coplan M A, Bochsler P and Geiss J 1989 *Solar Phys.* **124** (1) 167

Journal of Physics: Conference Series **767** (2016) 012012 doi:10.1088/1742-6596/767/1/012012

Ionization of Interstellar Hydrogen Beyond the Termination Shock

Mike Gruntman

Department of Astronautical Engineering, University of Southern California, Los Angeles, CA 90089-1192

mikeg@usc.edu

Abstract. Models of solar wind interaction with the surrounding interstellar medium usually disregard ionization of interstellar hydrogen atoms beyond the solar wind termination shock. If and when included, the effects of ionization in the heliospheric interface region are often obscured by complexities of the interaction. This work assesses the importance of interstellar hydrogen ionization in the heliosheath. Photoionization could be accounted for in a straightforward way. In contrast, electron impact ionization is largely unknown because of poorly understood energy transfer to electrons at the termination shock and beyond. We first estimate the effect of photoionization and then use it as a yardstick to assess the role of electron impact ionization. The physical estimates show that ionization of interstellar hydrogen may lead to significant mass loading in the inner heliosheath which would slow down plasma flowing toward the heliotail and deplete populations of nonthermal protons, with the corresponding effect on heliospheric fluxes of energetic neutral atoms.

1. Introduction

The supersonic solar wind plasma undergoes a shock transition at large heliocentric distances and fills the region between the termination shock and the heliopause, the inner heliosheath. (In the article, the word heliosheath is used in the sense of the inner heliosheath.) There, the plasma flows into the heliospheric tail where it eventually mixes with the surrounding local interstellar medium (LISM). Charge exchange of energetic protons on interstellar hydrogen atoms in the heliosheath produces heliospheric energetic neutral atom (ENA) fluxes (e.g., [1] and references therein).

Two Voyager spacecraft crossed the termination shock at 94 AU and 84 AU from the Sun in 2004 and 2007, respectively, and entered the heliosheath (e.g., [2,3,4,5,6]). In 2009, the Interstellar Boundary Explorer (IBEX) mission obtained the first maps of the heliospheric interface in ENA fluxes [7] which revealed many details of the global solar wind interaction with the surrounding LISM. Cassini spacecraft also imaged the heliosphere in ENA fluxes at somewhat higher energies of atoms [8]. In-situ measurements by Voyagers and remote observations by IBEX and Cassini found a number of unexpected phenomena at the interstellar boundary of the solar system (e.g., [9]). In particular, definitive explanation of a discovered by IBEX unanticipated standing-out band of enhanced ENA emissions across the sky, the ribbon [7,10], remains challenging. A recent review [30] lists multiple possible competing concepts of this phenomena. Failure of predicting and difficulties in reproducing the ribbon in simulations of heliospheric interactions suggest that some important physical phenomena may not be identified yet and/or properly included into models.

Understanding properties of the shocked solar wind plasma between the termination shock and heliopause is indispensable for explaining observed heliospheric ENA fluxes. Even most sophisticated global models introduce important simplifications. Most common assumptions include multi-fluid and magneto-hydrodynamic descriptions of flows of gases and plasmas and Maxwellian or other postulated velocity distributions of plasma protons. Such physical simplifications and assumptions lead to uncertainties in comparing results of model simulations with observations which are hard to quantify and they are rarely, if ever, pointed out and discussed in publications.

Available Voyager in-situ measurements of the bulk solar wind plasma and higher energy (>20 keV) protons [2,4,5] incompletely characterize the termination shock. Downstream of the shock, the heliosheath plasma is not in equilibrium, with nonthermal protons accounting for most energy density while most mass and inertia being in relatively cold protons [8,10,11]. The plasma properties may not even be stationary due to solar cycle variations of the global solar wind properties. Coulomb collisions are so rare in the heliosheath, even among electrons, that they could usually be disregarded.

The complexity of the solar wind interaction with the LISM obscures and makes it difficult to assess relative importance of particular physical processes. For example, many otherwise sophisticated and complex models of the interaction disregard, without sound justification, solar photoionization and electron impact ionization in the heliosheath [9]. Photoionization is often considered to be negligible while electron impact ionization tacitly assumed either inconsequential or too complex and uncertain for inclusion into models.

Recent physical estimates show that ionization of interstellar atomic hydrogen in the heliosheath could significantly modify plasma flow patterns beyond the termination shock and heliospheric ENA fluxes [9]. This work explores effects of ionization on the plasma properties in the heliosheath. Investigation of particular contributions of other physical processes such as charge exchange is outside the scope of the article.

2. Solar photoionization

At large distances from the Sun both solar photoionization and the neutral component of the solar wind disturb the partially ionized interstellar gas, the interstellar wind, approaching the solar system. The effect of photoionization on the LISM beyond the heliopause is relatively small. In contrast, the neutral solar wind flux could significantly change the interstellar wind properties ([9], [12] and references therein).

Absorption of the solar extreme ultraviolet radiation on the heliospheric scales is negligible, and the photoionization rate of hydrogen atoms is thus inversely proportional to the square of the heliocentric distance. One can calculate the photoionization rate from solar radiation spectra.

Consider, for example, interstellar hydrogen atoms approaching the solar system with the velocity 25 km s^{-1}. A small fraction, about 0.5%, of interstellar atoms would be photoionized when they reach the heliocentric distance 130 AU under moderate solar conditions [9]. For a LISM with neutral hydrogen atoms twice more abundant than protons, photoionization would increase interstellar plasma effective temperature (by adding energetic photoelectrons) by about 100 K.

Clearly, solar photoionization is not particularly important for modifying properties of interstellar gas entering the heliosheath. However, even its small photoionization could produce a significant effect on the rarefied heliosheath plasma because the latter is two orders of magnitude less dense than interstellar neutrals. The heliosheath plasma flow picks up photoionization-produced protons and electrons, resulting in its mass loading. Electron impact ionization of atoms also adds to mass loading. While one can include photoionization into modeling in a straightforward way, accounting for electron impact ionization presents major difficulties.

To estimate the effect of photoionization, let us consider a typical simplified axisymmetric heliospheric interaction [13] shown in figure 1 (left) and calculated for realistic parameters of the LISM and the solar wind. For the upwind (interstellar wind) direction, the distances to the termination shock, heliopause, and bow shock are 89 AU, 152 AU, and 270 AU, respectively. The polar angle θ is counted from the upwind direction, and the angle-dependent heliocentric distances to the termination shock and heliopause are $R_{TS}(\theta)$ and $R_{HP}(\theta)$, respectively. The heliosheath plasma (gray region in the figure) flows across (not necessarily normally) a side surface of a cone with the half-angle θ between heliocentric distances $R_{TS}(\theta)$ and $R_{HP}(\theta)$. All the solar wind flow that crosses the termination shock at angles $< \theta$ would be evacuated to the heliospheric tail through this side surface of the cone.

Consider the isotropic supersonic solar wind consisting of only protons and electrons with the constant velocity and the flux number density inversely proportional to the square of the heliocentric distance. Let us assume that at the distance $R_0 = 1$ AU from the Sun the solar wind flux density is $f_0 = 2.2\times10^8$ cm^{-2}s^{-1}; the solar photoionization rate is $\beta_{S,0} = 1.0\times10^{-7}$ s^{-1} (moderate solar activity); and that the LISM atomic hydrogen uniformly fills space with the number density $n_0 = 0.18$ cm^{-3}.

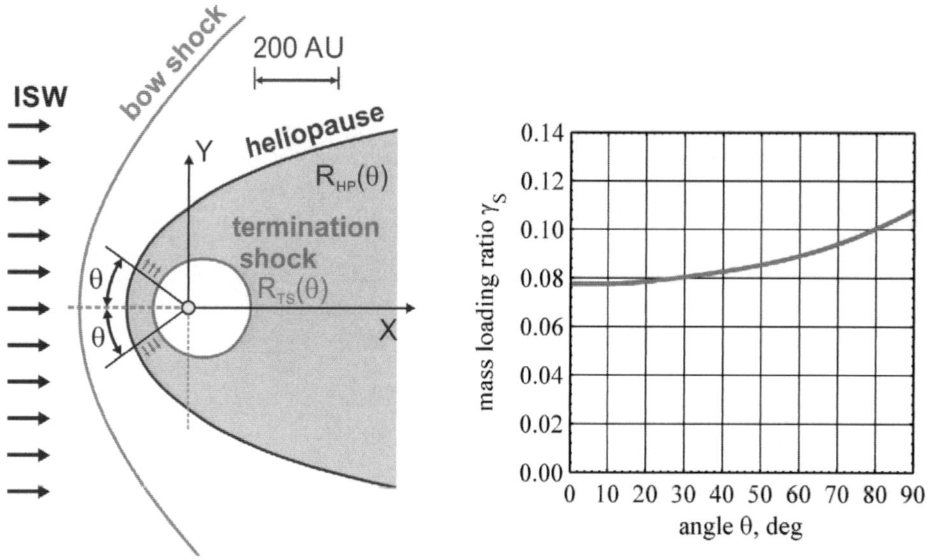

Figure 1. Left: schematic of a typical simplified axisymmetric heliospheric interaction with the bow shock, heliopause, and termination shock calculated in [13]. The angle θ is counted from the upwind (interstellar wind) direction. Small arrows show the heliosheath plasma (gray area) flow across the side surface of a cone with a half-angle θ. Right: angular dependence of the mass loading coefficient $\gamma_S(\theta)$ due to photoionization.

One can define the mass loading coefficient (ratio) of the plasma flow, $\gamma_S(\theta) = \alpha_S(\theta)/\Phi(\theta)$, as the ratio of the additional photoionization-produced plasma mass flow $\alpha_S(\theta)$ to the solar wind mass flow $\Phi(\theta)$ across a side cone surface of the heliosheath region restricted by the half-angle θ. Then,

$$\gamma_S(\theta) = g_0 \frac{\Gamma(\theta)}{1-\cos\theta},$$

where one independent of the interaction geometry dimensionless coefficient,

$$g_0 = \frac{n_0\, \beta_{S,0}\, R_0}{f_0},$$

describes solar and interstellar conditions while the other dimensionless parameter,

$$\Gamma(\theta) = \frac{1}{R_0}\int_0^\theta \left[R_{HP}(\theta) - R_{TS}(\theta)\right]\sin\theta\, d\theta,$$

characterizes the geometric properties of the heliosheath [9].

Figure 1 (right) shows the angular dependence of the mass loading coefficient $\gamma_S(\theta)$ that increases from $\gamma_S = 0.078$ at small angles θ to $\gamma_S = 0.108$ at $\theta = 90°$. So, photoionization of interstellar atomic hydrogen would increase the heliosheath plasma mass flow by approximately one tenth when it reaches and crosses the annulus at $\theta = 90°$. The mass loading coefficient increases to 0.15 and 0.18 in the heliospheric tail at distances of 200 AU and 300 AU, respectively [9]. So, photoionization would increase the mass flow of the solar wind by roughly one-fifths in the distant heliotail. For the purposes of this work, exact parameters of the interstellar medium and the solar wind are not particularly important as we focus on nonstrictly self-consistent physical estimates of ionization effects for a typical geometry.

3. Electron impact ionization

First attempts to account for electron impact ionization of interstellar hydrogen atoms beyond the termination shock in global heliospheric models date back to mid-1990s [14]. The process remains often disregarded to this date which is sometimes explained by complexity of its inclusion into models (e.g., [15]). When this ionization is accounted for, models usually assume Maxwellian distributions of electrons at plasma temperatures [13,16,17,18].

The energy-dependent cross-section of electron impact ionization of hydrogen atoms peaks in the 30–130 eV energy range. Let us assume the uniform electron (and proton) number density $n_e = 10^{-3}$ cm^{-3} in the heliosheath, consistent with the Voyager 2 observations [19]. One can calculate the electron impact ionization rate $\beta_e(T_e)$ of atoms by averaging over electron velocities. Figure 2 shows the temperature dependence of the rate $\beta_e(T_e)$ for electrons with a Maxwellian velocity distribution function at temperature T_e and negligible velocity of interstellar hydrogen atoms; two horizontal lines at 1.0×10^{-11} s^{-1} and 0.44×10^{-11} s^{-1} are the photoionization rates of atomic hydrogen at distances 100 AU and 150 AU from the Sun, respectively.

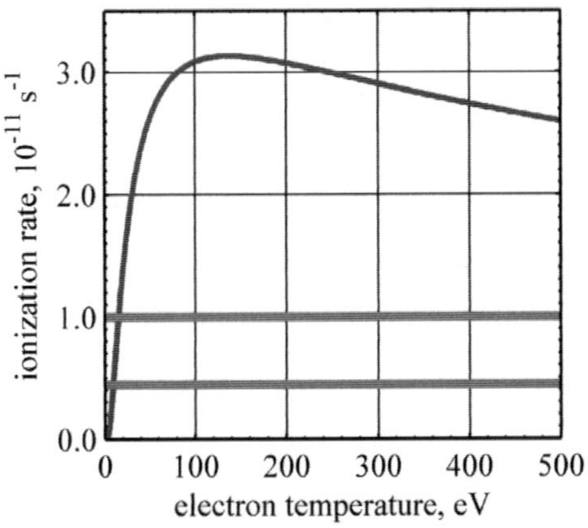

Figure 2. Temperature dependence of the ionization rate (blue curve) of hydrogen atoms in the heliosheath by electrons with a Maxwellian distribution and number density $n_e = 10^{-3}$ cm^{-3}. Two solid horizontal straight red lines correspond to the photoionization rates (moderate solar activity) 1.0×10^{-11} s^{-1} and 0.44×10^{-11} s^{-1} of interstellar hydrogen atoms at representative for the inner heliosheath distances 100 AU and 150 AU from the sun, respectively.

One can see that electron impact ionization may be significantly higher than photoionization in the heliosheath. At the same time the properties of electrons in this region remain unknown. The originally formulated concept of heliosphere ENA imaging that led to the IBEX mission had specifically pointed out lack of knowledge of "how the solar wind energy is redistributed between the proton and electron components of the postshock plasma" [20] downstream of the termination shock. There is no definitive and validated by measurements answer to this question yet, which is fundamentally important for understanding ionization in the heliosheath.

The only existing in-situ plasma measurements in the heliosheath did not probe properties of electrons in detail. One Faraday cup (side sensor D) of the Voyager 2 plasma instrument is sensitive to electrons with 10–5950 eV energies. Electron fluxes are too low, however, for reliable measurements in the distant solar wind and in the heliosheath (Richardson, 2015, private communications). Moreover, it is unclear how exactly absence of measurable electron fluxes constrains effective electron temperatures that could possibly range, as discussed below, from a few eV up to a few hundred eV. Therefore, electron impact ionization could be significantly larger than photoionization everywhere in the heliosheath, with an important effect on the plasma flow especially if electron temperatures exceed 20–30 eV.

Ionization of interstellar hydrogen increases the number density of the heliosheath plasma. It also changes effective electron temperatures in the region. Consequently, local electron impact ionization rates also change as they depend on both the number density and temperature. The average energy of photoelectrons produced by photoionization under moderate solar activity conditions is approximately 4.2 eV. Injection of such electrons into the heliosheath plasma would heat or cool its electron component depending on whether the initial temperature is lower than a few eV or higher than several eV, respectively. In addition, each act of electron impact ionization "removes" the hydrogen atom ionization energy, 13.6 eV, from the electron population. If one assumes instantaneous accommodation and Maxwellization of electrons, then ionization would increase electron temperatures in one year by 10% and 20% for the heliosheath plasma with initial electron temperatures 45 eV and 92 eV, respectively [9]. Note that it takes several years for the solar wind plasma to convect through the heliosheath and reach the distant heliotail. Consequently, if the initial effective electron temperature is $T_e > 50$ eV , then the already dominating electron impact ionization rate may increase substantially, for example double, in a control volume of the heliosheath plasma as it flows to the tail. Such an increase would be especially pronounced if initial electron temperatures are 100 eV or higher.

The concept of heating or cooling of the bulk plasma electrons is applicable only if photoelectrons reach a local equilibrium with a Maxwellian velocity distribution. Such Maxwellization not necessarily happens in the plasma of the heliosheath where Coulomb collisions are rare and negligible. (In contrast, the electron-electron collision mean free path is about 1 AU beyond the heliopause in the LISM.) So, it is not excluded that ionization-created electrons preserve their energies to some extent. Then, photoionization, for example, could produce a population of electrons with energies 0–50 eV in the heliosheath plasma, with abundances ~10% in sidewind directions at $\theta = 90°$ and ~20% in the heliotail.

4. Discussion

Photoionization alone could mass-load plasma flows by 10-20% in the heliosheath. The magnitude of this effect in some areas near the heliopause where the plasma flow slows down would be significantly larger. Figure 2 shows that electron impact ionization could be much more efficient than photoionization. At the same time, the former process is entirely ignored in the models because very little is known about electron velocity distributions in the heliosheath.

The plasma instrument on Voyager 2 directly measured velocities of the bulk solar wind plasma across the termination shock and in the heliosheath [5,19,21]. About 200 days before the termination shock crossing, the solar wind velocity was 380 km s^{-1} which corresponded to proton kinetic energy 750 eV. The flow then slowed down to about 310 km s^{-1}, or 500 eV per proton, immediately upstream of the termination shock. Downstream of the shock, the velocity dropped to 120 km s^{-1} and the corresponding 80-eV energy per proton. Voyager 2 also observed increase of the proton temperature across the shock to 10^5 K, with the corresponding average energy of proton thermal motion 13 eV.

So, about 650 eV energy per each original solar wind (not pickup) proton is transferred in the shock transition to the nonthermal proton population, solar wind electrons, plasma waves, and acceleration of higher energy ions. How exactly this transferred, or "lost," energy is partitioned among these possibilities is an important key to understanding the heliospheric interaction. For higher solar wind velocities characteristic for ecliptic polar regions during solar minima, the lost energy may be as large as 2 keV per solar wind proton, significantly larger than the conditions encountered by Voyager 2.

Observed by IBEX heliospheric ENA fluxes in the 0.5-6.0 keV energy range [10] do not seem to require transfer of energy from the bulk solar wind protons to ENA-producing nonthermal proton populations [9]. The abundance of the latter is also consistent with that of the pickup protons in the solar wind upstream of the termination shock. The energy balance then suggests that the measured 650-eV energy loss by each solar wind proton is transferred to electrons and to protons with energies > 6 keV in the termination shock transition.

Therefore, one arrives at two contrasting possibilities based on energy conservation in the shock transition, one with the negligibly small and the other with the non-negligible fractions of the bulk solar wind energy imparted to electrons in the heliosheath. In the former case, effective electron temperature would be only a few eV or smaller with the resulting electron impact ionization of interstellar hydrogen smaller than photoionization. In the latter case, electron impact ionization dominates and it could be an order of magnitude higher than photoionization if electron temperatures are 50 eV or higher (figure 2). Electron impact ionization becomes prominent in the heliosheath if only a small fraction, just one-twentieth or one-tenth of the energy lost by the bulk solar wind plasma upstream of the termination shock, is transferred to electrons. Even as much transferred energy as 100 eV or 200 eV per electron would not contradict current interpretations of experimental results or theoretical considerations regarding acceleration of higher energy particles at and beyond the termination shock.

The plasma instrument on Voyager 2 seems to be the only source of possible in-situ observational constraints on electron temperatures in the heliosheath. However, it has not detected any electron signal [5]. The situation is further complicated as the heliosheath plasma is essentially not Coulomb collisional and electron energy distributions could thus deviate from Maxwellian. Interplanetary shocks, for example, produce copious populations of 30-150 eV electrons [22]. It was recently suggested that higher energy electrons in the heliosheath may charge the Voyager spacecraft [23]. As a result, such an effect may lead to a significant increase of the possible upper limit on electron temperatures implied by absence of the measured electron signal.

Very few publications tried to theoretically consider properties of electrons in the termination shock transition and beyond. They all pointed to fundamental unresolved difficulties in understanding and describing such processes [23,24,25,26]. It was also noted that higher electron temperatures could lead to significant thermal conductivity in the heliosheath which would further modify the structure of the heliospheric interface [26].

The effect of interstellar hydrogen ionization is essentially nonlinear as the resulting mass loading slows down the plasma flow, causes compression, and further increases importance of continuing ionization. In addition, the heliosheath plasma continuously emits ENAs and thus loses its momentum and energy. The rate of electron impact ionization also increases as electrons are added to the slowing heliosheath plasma. As a result, one could expect a significant ionization effect on the heliosheath plasma in the upwind hemisphere. The distant heliospheric tail is another region where accumulated mass loading would strongly modify plasma flow properties and ENA emissions.

It is possible that temporal solar-cycle variations of the global solar wind might create conditions where a plasma flow could come practically to rest at some regions of the heliosheath, especially in the upwind direction near the heliopause. For example, unexplained properties of the so called "knot" in IBEX ENA images [10,27,30] may be a manifestation of such a phenomenon driven in part by ionization. A volume of plasma with near-zero flow velocities could first produce enhanced ENA emissions as it piles up and becomes denser and then its intensity decreases as the nonthermal proton population is depleted, the latter process sometimes described as plasma "aging" [9].

For example, a population of protons with 3-keV energy would decrease by a factor of 50 in 7 years and ENA fluxes produced by such protons practically vanish [9]. Note that a volume of plasma with typical flow velocities 150 km s^{-1}, as those observed by Voyager 2 in the heliosheath [19], would move by 220 AU in 7 years which is about the characteristic size of the heliosphere. Therefore, the solar wind plasma initially flowing into the upwind hemisphere contributes very little to ENA fluxes at such energies when it reaches the heliospheric tail at distances of 150 AU from the sun and beyond.

Depletion of ENA-producing proton populations is highly sensitive to slowing down of the plasma flow in the heliosheath and thus strongly depends on mass loading and, in turn, on the ionization rate of interstellar hydrogen. For example, just a one-year increase of the effective age of protons, from 4 to 5 years, at 3 keV would decrease ENA intensities at this energy by a factor of 2 [9]. Mass loading due to photoionization alone could produce such an effect in parts of the heliosheath. The existing observations and conservation of energy do not seem to preclude transfer of a fraction of bulk solar wind energy to the heliosheath electrons and hence causing efficient electron impact ionization that could alter plasma flows even to a greater extent. Therefore, inclusion of interstellar hydrogen ionization into heliospheric models and computer simulations is indispensable for understanding the heliosheath. Disregarding this effect could make comparison of model predictions with the in-situ and ENA observations difficult and uncertain.

Global heliospheric models are already highly complex which obscures relative importance of various physical processes involved. Dedicated parametric studies of the effects of ionization on observable ENA fluxes could advance understanding of the interaction. Present descriptions of the heliospheric interaction almost exclusively rely on the Euler variables for flow fields. Following evolution of plasma parameters in the Lagrange variables for control mass along flow lines could provide important insight to the heliosheath plasma and determine contributions of various physical processes to ENA intensity maps.

Advancing theoretical understanding of energy transfer to electrons in the termination shock transition is essential for accurate accounting of ionization of interstellar hydrogen and its effect on the heliosheath plasma. No reliable in-situ measurements of electrons at the termination shock and in the heliosheath exist or are expected in the foreseeable future. Therefore, in the near term, it is important to quantitatively determine how absence of measured electron fluxes by the Voyager 2 plasma instrument beyond the termination shock constrains possible effective temperatures and velocity distributions of electrons. In the longer term, in-situ measurement of electron properties should be among top priorities for the future Interstellar Probe mission [28, 29 and references therein].

Acknowledgments
This work is supported, in part, by NASA's IBEX program.

5. References

[1] Gruntman M 1997 Imaging of space plasmas in energetic neutral atom fluxes *Rev. Sci. Instrum.* **68** 3617–56

[2] Decker R B, Krimigis S M, Roelof E C, Hill M E, Armstrong T P, Gloeckler G, Hamilton D C, and Lanzerotti L J 2005 Voyager 1 in the foreshock, termination shock, and heliosheath *Science* **309** 2020–24

[3] Stone E C, Cummings A C, McDonald F B, Heikkila B C, Lal N, and Webber W R 2005 Voyager 1 explores the termination shock region and the heliosheath beyond *Science* **309** 2017–20

[4] Decker R B, Krimigis S M, Roelof E C, Hill M E, Armstrong T P, Gloeckler G, Hamilton D C, and Lanzerotti L J 2008 Mediation of the solar wind termination shock by non-thermal ions *Nature* **454** 67–70

[5] Richardson J D, Kasper J C, Wang C, Belcher J W, and Lazarus A J 2008 Cool heliosheath plasma and deceleration of the upstream solar wind at the termination shock *Nature* **454** 71–74

[6] Stone E C, Cummings A C, McDonald F B, Heikkila B C, Lal N, and Webber W R 2008 An asymmetric solar wind termination shock *Nature* **454** 71–74

[7] McComas D J et al 2009 First global observations of the interstellar interaction from the Interstellar Boundary Explorer *Science* **326** , 959–62

[8] Krimigis S M, Mitchell D G, Roelof E C, Hsieh K C, and McComas D J 2009 Imaging the interaction of the heliosphere with the interstellar medium from Saturn with Cassini *Science* **326** 971–3

[9] Gruntman M 2015 Interstellar hydrogen ionization in the heliosheath *J. Geophys. Res.* **120** 6119–32

[10] Funsten H O et al 2009 Structures and spectral variations of the outer heliosphere in IBEX energetic neutral atom maps *Science* **326** 964–6

[11] Roelof E C; Krimigis S M, Mitchell D G, Decker R B, Richardson J D, Gruntman M, and Funsten H O 2010 Implications of generalized Rankine-Hugoniot conditions for the PUI population at the Voyager 2 termination shock *Proc. 9th Ann. Intern. Astrophys. Conf., AIP Conf. Proc.* **1302** 133–41

[12] Gruntman M 1982 Effect of neutral component of solar wind on the interaction of the Solar system with the interstellar gas flow *Sov. Astron. Lett.* **8(1)** 24–6

[13] Izmodenov V, Malama Y G, Gloeckler G, and Geiss J 2003 Effects of interstellar and solar wind ionized helium on the interaction of the solar wind with the local interstellar medium *Astroph. J. Lett.* **594** L59–62

[14] Baranov V B and Malama Y G 1996 Axisymmetric self-consistent model of the solar wind interaction with the LISM: basic results and possible ways of development *Space Sci. Rev.* **78** 305–16

[15] Alouani-Bibi F, Opher M, Alexashov D, Izmodenov V, and Toth G 2011 Kinetic versus multi-fluid approach for interstellar neutrals in the heliosphere: exploration of the interstellar magnetic field effects *Astrophys. J.* **734** 45

[16] Izmodenov V V and Baranov V B 2006 Modern multi-component models of the heliospheric interface *The Physics of the Heliospheric Boundaries, V. Izmodenov and R. Kallenbach (eds.), ESA-ESTEC, Paris* **ISSI Scientific Report No. 5** 67–135

[17] Malama Y G, Izmodenov V V and Chalov S V 2006 Modeling of the heliospheric interface: multi-component nature of the heliospheric plasma *Astron. Astrophys.* **445** 693–701

[18] Scherer K, Fichtner H, Fahr H-J, Bzowski M, and Ferreira S E S 2014 Ionization rates in the heliosheath and in astrosheaths: spatial dependence and dynamical relevance *Astron. Astrophys.* **563** A69

[19] Richardson J D and Decker R B 2015 Plasma and flows in the heliosheath *J. Phys.: Conf. Ser.,* **577** 012021

[20] Gruntman M, Roelof E C, Mitchell D G, Fahr H J, Funsten H O, and McComas D J 2001 Energetic neutral atom imaging of the heliospheric boundary region *J. Geophys. Res.* **106** 15767–81

[21] Richardson J D and Wang C 2012 Voyager 2 observes a large density increase in the heliosheath *Astrophys. J. Lett.* **759** L19

[22] Isenberg P.A. and Feldman W C 1997 Electron-impact ionization of interstellar hydrogen and helium at interplanetary shocks *Geophys. Res. Lett.* **22** 873–5

[23] Fahr H J, Richardson J D, and Verscharen D 2015 The electron distribution function downstream of the solar-wind termination shock: Where are the hot electrons? *Astron. Astrophys.* **579** A18

[24] Chalov S V and Fahr H J 2013 The role of solar wind electrons at the solar wind termination shock *Month. Not. Roy. Astron. Soc.* **433** L40–3

[25] Chashei I V and Fahr H J 2013 On the electron temperature downstream of the solar wind termination shock *Ann. Geophys.* **31** 1205–12

[26] Izmodenov V V, Alexashov D B, and Ruderman M S 2014 Electron thermal conduction as a possible physical mechanism to make the inner heliosphere thinner *Astrophys. J. Lett.* **795** L7

[27] McComas D J. et al. 2010 Evolving outer heliosphere: large-scale stability and time variations observed by the Interstellar Boundary Explorer *J. Geophys. Res.* **115** A09113

[28] Gruntman M 2004 Instrumentation for interstellar exploration *Adv. Space Res.* **34(1)** 204–12

[29] McNutt Jr. R L, Gruntman M, Krimigis S M, Roelof E C, Wimmer-Schweingruber R F 2011 Interstellar Probe: Impact of the Voyager and IBEX results on science and strategy *Acta Astronautica* **69** 767–76

[30] McComas D.J, Lewis W S, and Schwadron N.A. 2014 IBEX's Enigmatic Ribbon in the sky and its many possible sources *Rev. Geophys.* **52** 118-155

The Sun's Dynamic Influence on the Outer Heliosphere, the Heliosheath, and the Local Interstellar Medium

D S Intriligator[1], W Sun[1,2], T Detman[1], W D Miller[1], J Intriligator[1,3], M Dryer[1] W Webber[4], C Deehr[1,2] and G Gloeckler[5]

[1]Carmel Research Center, Inc., Santa Monica, CA
[2]University of Alaska, Fairbanks, AK
[3]Bangor University, Bangor, Wales, UK
[4]New Mexico State University, Las Cruces, NM
[5]University of Michigan, Ann Arbor, MI

E-mail: devriei@aol.com

Abstract. The Sun has been observed for many years to be a dynamic influence in the heliosphere, and as the Voyager missions have continued long after achieving their original goals of observing the major planets they have provided the first *in situ* observations of the effects of solar activity in the heliosheath (HS), and the nearest portions of the local Interstellar Medium (LISM). Comparing these observations with models provides key insights. We employ two three-dimensional (3D) time-dependent models that simulate the propagation of shocks, other specific features, and the background solar wind throughout the heliosphere, starting with the solar background and solar event boundary conditions near the Sun at 2.5 Rs. The Hybrid Heliospheric Modeling System with Pickup Protons (HHMS-PI) is a 3D time-dependent Magnetohydrodynamic (MHD) simulation. HAFSS (HAF Solar Surface) is a 3D time-dependent kinematic simulation. Comparing our models with the observations indicates that solar effects are seen in the heliosphere, the HS, and the LISM in in-situ spacecraft measurements of plasma, magnetic field, energetic particles, cosmic rays, and plasma waves. There is quantitative agreement (at ACE, Ulysses, V1, V2) with data (e.g., solar wind, IMF, Ulysses SWICS pickup protons (PUPs)). Propagating shocks are slowed due to PUPs. The 3D locations of solar events and of various spacecraft are key to understanding the 3D propagation and timing of shocks, other specific features, and gradients throughout the heliosphere, HS, and LISM.

1. Introduction

It is a pleasure to recognize and honor Ed Stone and to acknowledge his many important contributions to space physics and astrophysics. Figure 1 shows three photos of Ed. ACE data and Voyager data are key for many aspects of our modeling and empirical analyses at Carmel Research Center, Inc.

Figure 1. We are indebted to Ed Stone for his many important and significant contributions including those associated with ACE and Voyager. His tireless efforts have helped make available vast arrays of data that serve to extend our knowledge of the heliosphere and of the universe.

Journal of Physics: Conference Series **767** (2016) 012013 doi:10.1088/1742-6596/767/1/012013

Figure 2a (the figure on the left) shows the two basic paths or components that are used to initiate both our HHMS-PI (Hybrid Heliospheric Modeling System with Pickup Protons), which is a three-dimensional (3D) time-dependent Magnetohydrodynamic (MHD) simulation; and HAFSS (HAF Solar Surface), which is a 3D time-dependent kinematic simulation, each of which combines ambient conditions with the effects of solar activity [1-5]. The solar magnetograms and the global photospheric

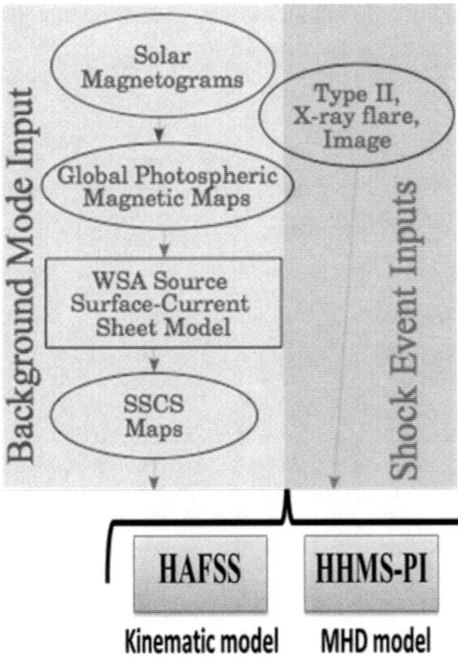

Figure 2a.

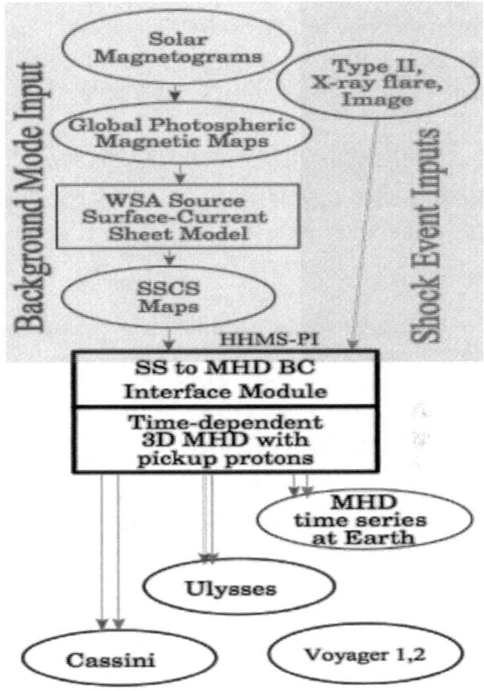

Figure 2b.

Figure 2. Descriptions of our two three-dimensional (3D) time-dependent HAFSS and HHMS-PI simulations shown in Figure 2a. A more detailed description of our magnetohydrodynamic HHMS-PI simulation is shown in Figure 2b.

magnetic maps are obtained from observations at the Wilcox Solar Observatory (WSO). The blue paths on the left (Figures 2a and 2b) indicate the background or ambient solar and solar wind conditions. The red paths on the right (Figures 2a and 2b) indicate the solar events. It is not appropriate or sufficient to launch a solar event(s) (on the right in Figure 2) without putting it in the context of the ambient or background solar and solar wind conditions (on the left in Figures 2a or 2b).

Figure 2b (the figure on the right) indicates some of the details associated with HHMS-PI, our 3D time-dependent MHD model. For example, HHMS-PI includes the effects of the Local Interstellar Medium (LISM) neutral hydrogen (N_H) flowing into the heliosphere and creating pickup protons (PUPs). This can give rise to solar wind heating, and/or slowing. A number of models for heating, turbulence, and other phenomena can be implemented and evaluated through model inter-comparisons and model comparisons with data. In their current forms both HHMS-PI and HAFSS propagate the ambient solar wind and shocks and other features out into an unbounded heliosphere. That is, there is no allowance made for the effects of the termination shock (TS) or for propagation, etc. beyond the TS in the heliosheath (HS). Similarly there are no allowances made for propagation beyond the heliopause (HP) into the LISM. Based on these omissions we assumed that in their present form our two models would not be appropriate for predicting phenomena beyond the TS. To our surprise, we found that beyond the TS the time series simulations predicted by our models were predicting an earlier arrival of

Journal of Physics: Conference Series **767** (2016) 012013 doi:10.1088/1742-6596/767/1/012013

an event or phenomenon compared to that measured in the Voyager 1 (V1) or Voyager 2 (V2) data. This early onset was expected since beyond the TS one assumes that there is a slowing (and heating) of the solar wind.

In using our models and comparing their simulations with spacecraft data we emphasize the importance of quantitative comparisons, correlation coefficients, and/or skill scores, etc. between our simulations and the data rather than only using graphs or cartoons to show the propagation of events. Often the time profile in our simulations agrees well (e.g., highly correlates) with the measured data if one time shifts the measurements to an earlier time. The magnitude of the time shift often is useful in analyzing the underlying physical processes.

2. HHMS-PI Simulation of Halloween 2003 Events

Table 1 shows the HHMS-PI shock event inputs from October and November 2003. These inputs are the values that were used in our HHMS-PI simulations as described above in the Introduction in connection with the right sides of Figures 2a and 2b.

Table 1. Shock parameters after adjusting for time of arrival at both ACE and Ulysses

date	time	doy	lat	lon	rad	shSpd	tau	FF
2003-10-19	16:50	(292.701)	5	-56.0	102.0	519.6	1.33	507
2003-10-21	03:47	(294.158)	-10	-90.0	100.0	517.0	0.67	508
2003-10-22	09:38	(295.401)	-2	-22.0	100.0	781.5	3.00	509
2003-10-23	08:27	(296.352)	-21	-88.0	108.0	1276.0	1.50	510
2003-10-25	04:15	(298.177)	-15	-43.0	120.0	530.0	2.00	511
2003-10-26	06:17	(299.262)	-18	-43.0	120.0	574.1	3.00	512
2003-10-26	17:35	(299.733)	5	32.6	70.2	1027.0	3.50	513
2003-10-28	11:02	(301.460)	-16	-8.0	120.0	1951.0	3.00	514
2003-10-29	20:44	(302.864)	-14	1.0	123.0	1612.4	1.50	515
2003-11-01	22:34	(305.940)	-10	61.0	120.0	820.9	1.00	516
2003-11-02	17:14	(306.718)	-14	82.5	158.0	1791.4	1.00	517
2003-11-03	01:24	(307.058)	10	85.0	100.0	725.0	1.75	518
2003-11-03	09:56	(307.414)	8	77.0	120.0	1131.3	1.50	519
2003-11-04	19:43	(308.822)	-19	78.8	102.0	1580.7	1.50	520
2003-11-07	15:54	(311.663)	-18	120.0	100.0	1682.7	2.00	520.2
2003-11-11	13:35	(315.566)	-3	88.8	93.7	807.0	3.00	521
2003-11-13	09:24	(317.392)	1	-90.0	101.0	718.5	3.00	522
2003-11-17	09:17	(321.387)	-1	-33.0	100.0	547.0	2.00	523
2003-11-18	07:47	(322.324)	0	-18.0	221.1	918.4	3.00	524
2003-11-20	07:47	(324.324)	1	14.0	104.9	997.9	0.75	525

Table 1. Date, Time, Day of Year: start time of metric Type II radio burst. Latitude (degrees), Longitude (degrees), Rad: width of shock (degrees). shSpd is Vs (km/s): shock speed input at the Sun from real-time radio and halo/partial halo CME plane-of-sky speed estimates. Tau (hrs): coronal shocks piston driving time above flare site. FF#: real-time "fearless forecast" events. The content of Table 1 is essentially the same as its components in Table 3 [2] and Table 1 [3].

Journal of Physics: Conference Series **767** (2016) 012013 doi:10.1088/1742-6596/767/1/012013

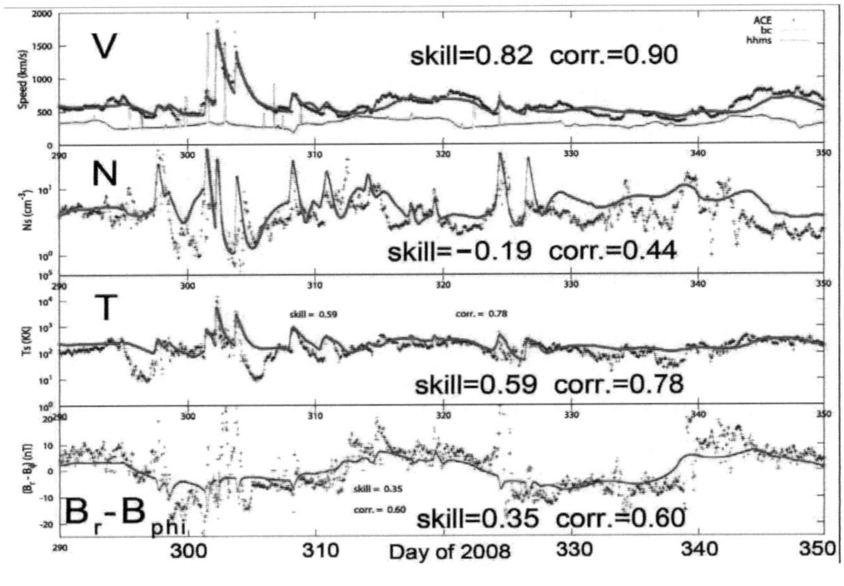

Figure 3a. Shows the comparisons [adapted from 2] of the HHMS-PI simulation (red lines) of the Halloween 2003 event with the ACE data [black lines] for the solar wind SWOOPS plasma parameters and magnetic field.

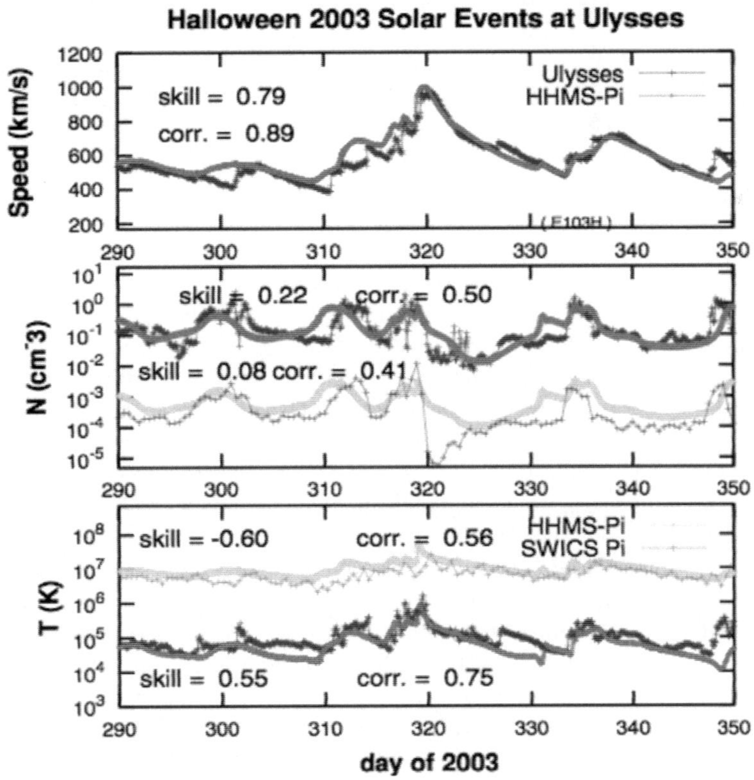

Figure 3b. Showing at Ulysses the comparisons [adapted from 3a] for the solar wind plasma parameters with the HHMS-PI simulation (red line). Figure 3b also shows at Ulysses the comparisons of the HHMS-PI simulation (green lines) of the SWICS Pick Up Proton (PUP) density and temperature with these measured PUP parameters (grey lines). These HHMS-PI simulated PUP parameters were predicted by us before the SWICS PUP parameters were available.

Figure 3b is a comparison of the HHMS-PI simulations with Ulysses measurements of the solar wind and PUPs for 60 days in 2003 including quiescent "quiet" undisturbed times and the "disturbed" times associated with the Halloween 2003 events. Shown are the HHMS-PI simulation comparisons at Ulysses with SWOOPS solar wind speed, proton density, and temperature; and with SWICS PUP density and temperature for the "quiet" solar wind both before and after the Halloween events and for the Halloween 2003 events. Note the generally good agreement between the HHMS-PI simulations (in red for the SWOOPS solar wind proton parameters and in green for the SWICS PUP parameters) and the measured plasma speed, densities, and temperatures for both the SWOOPS solar wind protons (in blue) and the measured SWICS PUP density and temperature parameters (in grey) even though at times during the Halloween 2003 event the shape of the plasma distributions is varying greatly as illustrated in [3] in the SWICS spectra (e.g., Days 319.5–320.5) in our Figure 3b.

3. HHMS-PI Simulation of Halloween 2003 Events Shows PUPs Slow Down Shock Propagation

The presence of PUPs slows down the propagation of shocks in the solar wind. This is observable even at distances as close to the Sun as ~5 AU. The presence of PUPs can have a significant impact on the propagation of both fast and slower shocks even to these distances. Table 2 and Table 3 summarize some of these results for the Halloween events [1] propagating to Ulysses at ~5.2 AU.

Table 2. HHMS-PI Halloween 2003 Transit Speeds at Ulysses Due to Pickup Protons (PUPs).

Solar Event	Distance (AU)	OBSULY speed km/s	DNH0.1 speed km/s	DNH0.0 speed km/s
514	**5.24**	**877.4**	**812.1**	**871.0**
517	**5.25**	**894.2**	**814.7**	**884.1**
520	**5.25**	**1030.8**	**969.9**	**1048.6**
520.2	**5.25**	**1239.1**	**1211.6**	**1310.6**
521	**5.26**	**807.1**	**796.3**	**864.2**
524	**5.27**	**794.4**	**711.7**	**747.3**

Table 2 lists from left to right the solar event (Table 1); Distance to Ulysses (AU) (5.24 AU = 783,904,000 km); OBSULY Speed (km/s): Transit speed to Ulysses based on Ulysses data; DNH0.1 Speed (km/s): HHMS-PI Model speed to Ulysses assuming interstellar H density = 0.1 cm^{-3}; DNH0.0 Speed (km/s): HHMS-PI Model speed to Ulysses assuming interstellar H density = 0 cm^{-3}.

Focusing on Solar Events 520.2 and 521 we can see that the first column on the left in Table 2 shows the numbering of the Solar Events. For Solar Event 520.2 the third column shows that the average speed of the event reaching Ulysses was 1239.1 km/s. The next column (fourth) shows the speed of 1211.6 km/s predicted by the HHMS-PI model resulting from deceleration by pickup from inflowing neutral hydrogen, assuming that the neutral H density was 0.1 cm^{-3}. The fifth column shows the predicted speed of 1310.6 km/s from the HHMS-PI Model with INFLOWING NEUTRAL H DENSITY = 0 cm^{-3} (i.e., no deceleration by pickup). Comparing the Ulysses measured shock propagation speed (column 3) with the HHMS-PI predicted speeds from columns 4 and 5 clearly shows that the 1211.6 km/s estimate from the 0.1 cm^{-3} assumption is more appropriate than the estimate from the 0.0 cm^{-3} assumption.

Similarly, for the next Solar Event (521 in column 1), the Ulysses data in column 3 indicates that the shock propagation speed from the Sun to Ulysses was 807.1 km/s, while the 0.1 cm^{-3} assumption provides a speed of 796.3 km/s, which is a little low. However, the 0.0 cm^{-3} assumption predicted a speed of 864.2 km/s, which is quite high compared with the ULYSSES measured speed of 807.1 km/s. Thus for solar event 521, the measured observed Ulysses speed of 807.1 km/s from the Sun to Ulysses, agrees better with the predicted slowing of the solar wind propagation and shock speed by PUPs.

In Table 2 our identifying the solar events of 520.2 and 521 clearly shows the important slowing effects of the PUPs in decelerating the ambient solar wind and shock propagation. This point also is

Journal of Physics: Conference Series **767** (2016) 012013 doi:10.1088/1742-6596/767/1/012013

shown again in Table 3. Looking at Table 3 we find correspondingly that for these same two Solar Events (520.2 and 521) the shifts in the day times agree with the shift in the speeds for these events in Table 2, so that the time to propagate from the Sun to Ulysses was greatly influenced by the PUPs slowing down the plasma and propagation speeds. As shown in the right-hand column, the predicted and actual arrival times for these events agree within about one sixth of a day, or four hours.

Thus, our Halloween 2003 simulations in Tables 2 and 3 show that the PUPs slow large shocks propagating in the solar wind from the Sun to Ulysses at 5.2 AU near the ecliptic plane. We note that others have recognized the importance of how PUPs formed by ionization of the neutral inflowing LISM hydrogen slow large propagating shocks. For example, in 2001, Zank et al. [6] used a 1-D MHD model that included PUP phenomena and interactions of large propagating shocks to predict a qualitatively similar deceleration for the propagation of the July 2000 Bastille Day solar events.

NASA has recently released some press announcements showing some cartoon graphics of the solar wind and shocks flowing from the Sun toward Pluto and the New Horizons spacecraft. We and others have noticed the multiple errors in these NASA cartoons since they exhibit implausibly early arrivals of these solar events at New Horizons because they are based on ENLIL computations that do not account for the PUPs slowing down the outward propagating disturbances from the Sun. Hence, the NASA cartoons omit these effects on the outward flow of the solar events toward New Horizons near Pluto. These erroneous associations in the NASA cartoons showing excessively early arrivals of solar events at New Horizons provide evidence of the importance of PUPs, since PUPs were not included in the 3D time-dependent MHD ENLIL model that attempted to predict the propagation of solar event shocks from the sun to New Horizons.

TABLE 3. HHMS-PI Shock Arrival Times (and Delays) at Ulysses Due to Pickup Protons (PUPs).

Solar Event	Solar Event Day	OBSULY Arrival Day	DNH0.1 Arrival Day	DNH0.0 Arrival Day	DNH0.1-DNH0.0 (Days)	OBSULY-Best Pred (Days)
514	301.460	311.801	312.632	311.877*	0.755	-0.076
517	306.718	316.884	317.876	317.000*	0.876	-0.116
520	308.822	317.641	318.194	317.491*	0.703	0.150
520.2	311.664	319.000	319.167*	318.600	0.567	-0.167
521	315.566	326.851	327.004*	326.105	0.899	-0.153
524	322.324	333.810	335.145	334.534*	0.611	-0.724

Table 3 lists from left to right the solar event (Table 1); Distance to Ulysses (AU) (5.24 AU = 783,904,000 km); OBSULY Arrival Day: Transit Times to Ulysses based on Ulysses data; DNH0.1 Arrival Day: Transit Times to Ulysses based on HHMS-PI Model with INFLOWING NEUTRAL H DENSITY = 0.1 cm^{-3}; DNH0.0 Arrival Day: Transit Times to Ulysses based on HHMS-PI Model with INFLOWING NEUTRAL H DENSITY = 0 cm^{-3}; asterisks in the DNH0.1 and DNH0.0 columns indicate for each event which model gave the best prediction, which is used in the right-hand column.

4. HHMS-PI Simulations of Halloween 2003 Events Show They Can't Unambiguously Help Us Determine the HCS Tilt Orientations at V1 and V2 in 2004

The orange lines in Figure 4 show the vertical displacement of the Heliospheric Current Sheet (HCS) for V1 and V2 in 2004, as estimated by taking data from the Wilcox Solar Observatory (WSO) and allowing for possible propagation delays to the spacecraft by shifting the WSO data 270 days (WSO + 270 days: Figure 4a) and 360 days (WSO + 360 days: Figure 4b). In these panels the V1 Trajectory is shown as a red horizontal line at 31 degrees North in heliographic inertial (HGI) coordinates, meaning that the spacecraft remained at this latitude without significant change as it travelled outward a distance of about 3.5 AU during this year, since the latitudinal component of its velocity vector was

Journal of Physics: Conference Series **767** (2016) 012013 doi:10.1088/1742-6596/767/1/012013

close to 31 degrees North. Similarly, the V2 trajectory is shown as an almost exactly horizontal blue line at ~ 26 degrees South HGI, since it travelled outward about 3 AU during this year, but its latitude changed by only about 0.25 degrees, which is too small to be visible on this scale.

In each panel, the asterisks (in colours matching the corresponding trajectory lines) show the sector boundaries found in the magnetometer data from the respective spacecraft during the times that each was observing phenomena that appear to be the shock and other disturbances generated by the Halloween 2003 flares [4,5] (and during some of the previous period in the case of V2, since the plasma instrument on V2 is still working, while the V1 plasma instrument failed in 1980; on the other hand, currently the V2 magnetometer data are not publicly available from June 18, 2004 (day 170) to the end of the year, though the available data resume at the beginning of 2005). At the bottom of the figures the red vertical arrows sticking up on each figure indicate the times of the two forward shocks at V2 (Day 119, 2004) and V1 (Day 209, 2004).

As can be seen from these panels, these magnetometer data do not provide strong evidence for identifying the propagation delay more definitely than the previously estimated range of 270 to 360 days. Thus, until the V2 magnetometer data for the last half of 2004 become available, we are unable to determine what delay of the HCS results from the WSO data (which, as noted in the Introduction, are the fundamental inputs to our modelling computations) may be better suited for analyses of the effects of the Halloween 2003 events and the periods before and after, as observed at the Voyagers.

We should note that when additional V2 magnetometer data become available either in 2004 or later (such as in 2013, 2014, etc.) we also should be able to make some additional comparable plots to those in Figures 4a and 4b that may allow us to distinguish the various sector boundaries in the V1 and V2 magnetometer data and their respective relations to the WSO HCS tilt angles. This may provide a great deal of help not only for understanding how well the estimated WSO tilt angles predict when the spacecraft should be crossing the current sheet and hence observing sector boundaries, but it may provide useful comparisons with the models described in the previous sections and also may supply important information about the transport of the magnetic field within the heliosheath and perhaps eventually in the LISM.

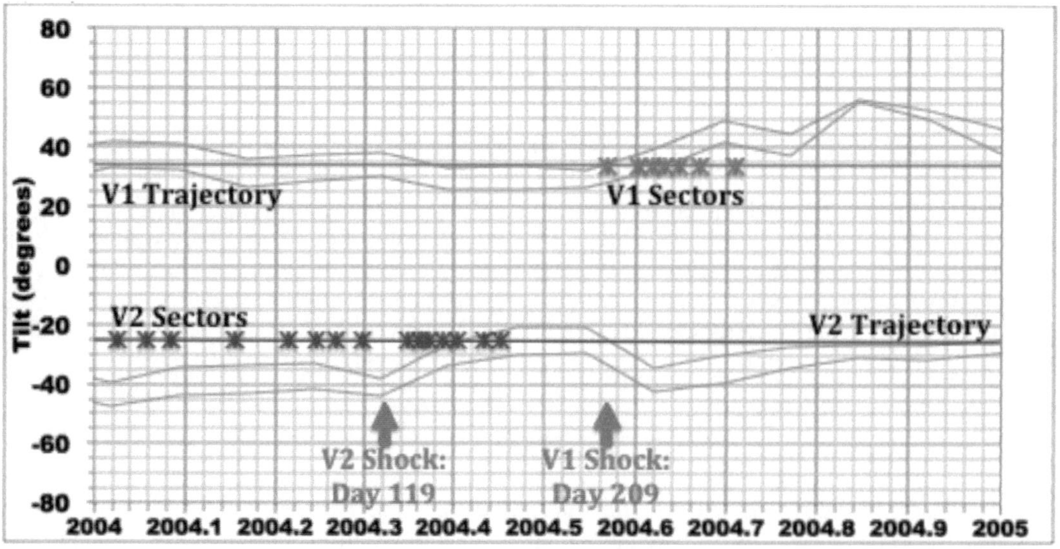

Figure 4a. Halloween 2003 WSO HCS Tilt Angles + 270 days with V2 & V1 Sector Boundaries.

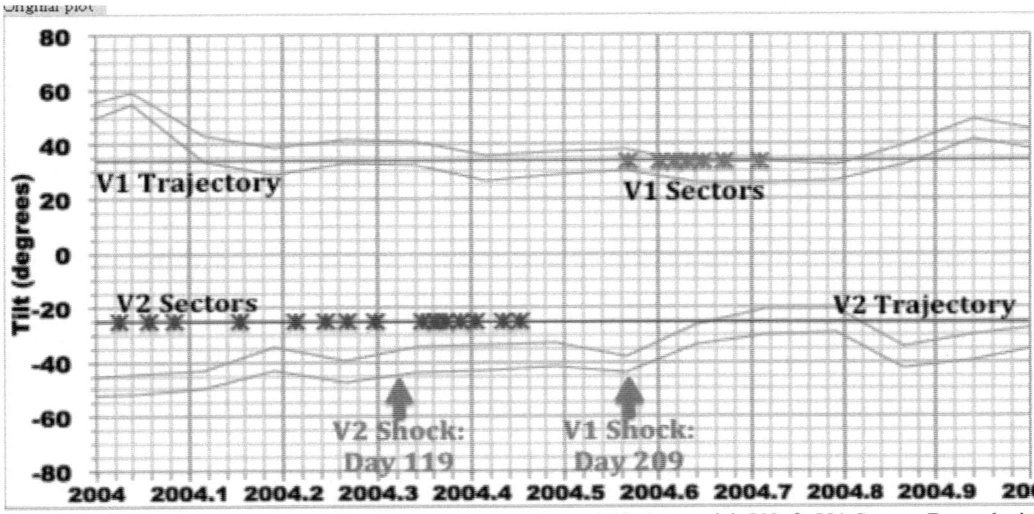

Figure 4b. Halloween 2003 WSO HCS Tilt Angles + 360 days with V2 & V1 Sector Boundaries.

5. The Importance of Geometry in the Simulations: the Locations of Solar Events and the Spacecraft

Figures 5a, 5b, and 5c indicate the locations of the flares at the Sun and the positions of the various spacecraft for, respectively, the Oct./Nov. 2003 solar events, the March 2012 solar events, and the July 2012 solar events, as if viewed from a point far above the solar north pole. In each of these plots the orange round dot in the center represents the Sun and the adjacent red asterisk shows the location of the flare that occurred at the time given at the top of the plot. As indicated by the labels and abbreviations, the positions of the various spacecraft are shown in each plot with a logarithmic scale of radial distance from the Sun. Thus, the blue round dots at 1 AU represent the positions of Earth (with ACE at the L_1 Lagrangian point about 1.5 million km radially sunward of the Earth at all times), and for the 2012 events STEREO A and B (SA and SB) also are located at 1 AU with ~120 degrees of longitude separating Earth, SA, and SB. For the Oct/Nov 2003 events the Ulysses spacecraft are at the green squares about 5.2 AU from the Sun; and the much more distant labelled green squares show the respective V1 and V2 spacecraft locations for all three series of these solar events. The projections onto the ecliptic plane in these figures do not indicate that in these years V1 was about 30 degrees North of the ecliptic plane, while V2 was about 26 degrees South, with trajectories taking them respectively to gradually increasing North and South latitudes.

In each figure the various spacecraft are shown at the positions where they were at the times of the flares. The shocks, CMEs, and energetic particle effects of the flares reached the spacecraft after propagation delays that ranged from hours or days for the spacecraft at 1 AU to a year or more for V1 and V2 far beyond the outer radius of the Kuiper Belt. However, in each case the velocities of the spacecraft were small enough, relative to their distances from the Sun, that they did not move significantly on the scales of these figures during their respective propagation delays, so that it is possible in each figure to extrapolate from the location of each flare to the location of each spacecraft in interpreting the observations, even though the spacecraft closer to the Sun moved significantly during the propagation delays to the more distant spacecraft. Thus, in each figure, when they detected the shocks the spacecraft near the Earth, the STEREO spacecraft, and Ulysses were respectively where they are shown, even though they all moved significantly during the subsequent months that the shocks took to reach the Voyagers, and because of the logarithmic distance scale the Voyagers are in nearly the same positions in all the figures, despite the years that elapsed from late 2003 to the middle of 2012.

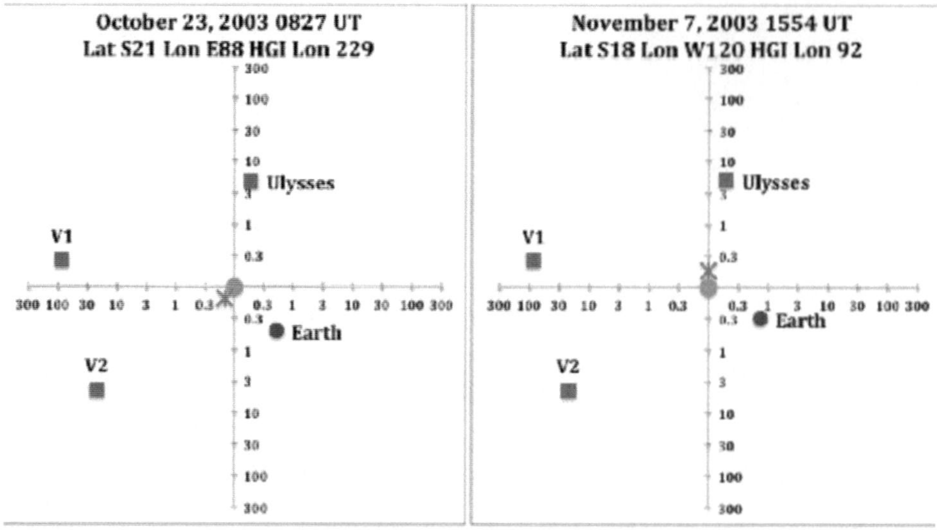

Figure 5a. October-November (Halloween) 2003 Solar Events.

The left panel of Figure 5a shows that the Oct. 23, 2003 solar event was located on the side of the Sun facing V2 at ~East 88 degrees as viewed from Earth, or a HGI longitude of 229 degrees. Note that Ulysses at ~5.2 AU was about 140 degrees from this flare. By contrast, the right panel of Figure 5a indicates that the Nov. 7 flare was located about 120 degrees west of Earth at HGI longitude 92 degrees, in the direction of the Ulysses spacecraft.

Thus, the Oct. 23 flare and several that occurred in the same active region in the next few days were all well observed from the Earth, but since the Nov. 7 flare occurred after the active region had rotated over the west limb of the Sun as viewed from the Earth, its importance was not recognized until the Ulysses experimenters sought an explanation for the unusually high solar wind speed of 993 km/s observed on Nov. 15, after which evidence of the flare was found in coronagraph observations. (This is why this event had to be added *post facto* as event 520.2 to the master list of events used to prepare Tables 1, 2, and 3.)

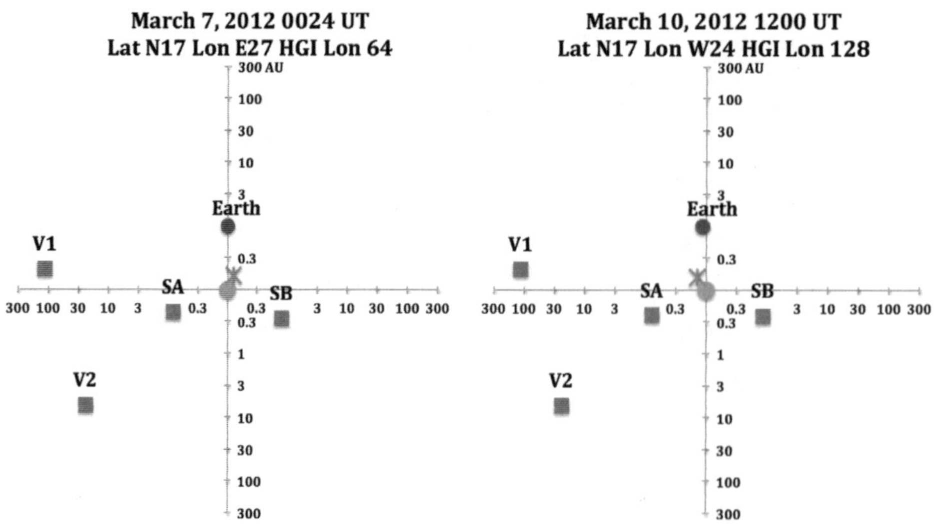

Figure 5b. March 2012 Solar Events.

The figures also show that both the Oct. 23 and the Nov. 7 flares were well located for their shocks to be observed by the Voyagers, although because of the rotation of the active region some of the flares during intermediate days, though well located for observations from Earth, were at unfavourable locations to be detected after propagation by the Voyagers. As described in Table 4 below, events associated with this period of major flares were detected at both V1 and V2 in 2004. However, owing to interaction region merging over the long distance to the Voyagers (we have previously reported [4] an analysis of the merged interaction regions (MIRs) observed in the V2 data in April and May, 2004), it is not possible to identify the contributions of the individual flares to the phenomena observed at the Voyagers.

As Figure 5b shows, the March 7, 2012 flare was located in a good position to be observed at the Earth and STEREO B, but was nearly on the opposite side of the Sun from STEREO A and V2, while the March 10, 2012 flare was nearly on the opposite side of the Sun from STEREO B, but was well positioned for the Earth, STEREO A, and the Voyagers, especially V1, since the flare was about 50 degrees east of V1.

This event may be responsible for important phenomena observed by Voyager 1 in April 2013. As these two March 2012 flares were both located on the Sun at a North latitude of 17 degrees and V1 was located near North 30 degrees, V1 was in a particularly favourable position to observe the effects of the March 10 flare, with possible contributions from the March 7 flare and lesser ones in the previous few days. In particular, the shocks and other phenomena generated by this group of flares are

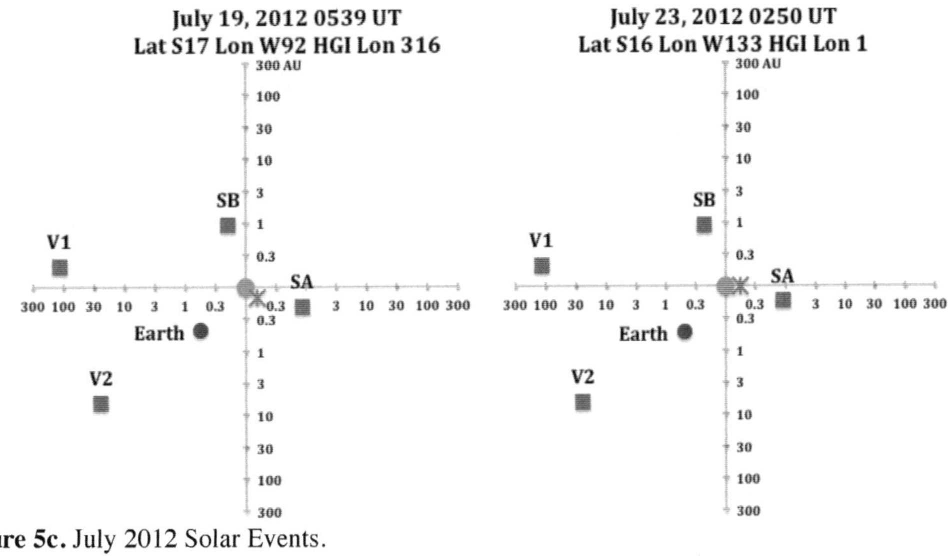

Figure 5c. July 2012 Solar Events.

believed to be the source of the 3 kHz signals detected by the V1 plasma wave system (PWS) that provided conclusive evidence (by corresponding to a plasma electron density of about 0.08 cm^{-3}) that the jump in the cosmic ray count rate observed by the Voyager instruments in August 2012 did in fact mark the spacecraft's transit through the heliopause (HP) and entry into the local interstellar medium (LISM).

Many colleagues likened the July 23, 2012 solar event (on the right side of Figure 5c) to the Carrington 1859 solar event. The July 23, flare occurred at a latitude of south ~ 26 degrees and at a HGI longitude of 1 degree, so V2 was more favourably located in latitude than V1 to observe the effects of this flare, but the large longitude separation diminished the effect.

As the right panel of Figure 5c shows [5], the July 23 flare occurred when its active region had rotated well beyond the west limb of the Sun, as viewed from the Earth, so the world escaped what might have been an international technological catastrophe if the shock from this flare had caused a Carrington-class geomagnetic storm. As it was, the effects at the Earth were small, while STEREO A, being almost radially aligned with the flare, was subjected to solar wind speeds that exceeded the measurement range of its instruments. Elsewhere [5] we have estimated the peak speed at STEREO A of the plasma in the CME from this event at nearly 3000 km/s, but this is based on the HAFSS kinematic simulation, for a period during which the STEREO experimenters have no usable data from their solar wind instrument.

However, in view of the extraordinary power of this event, it is plausible to associate it with phenomena observed at V1 in 2014, even though, as the figure shows, V1 was nearly on the opposite side of the Sun from this flare. In particular, on May 20, 2014, the V1 PWS began to detect another 3 kHz plasma wave event similar to the one in April and May, 2013, that we attributed above to a shock associated with the March 2012 flare events, so it seems likely that this plasma wave event also resulted from a flare shock, for which the July flare events, especially the July 23 flare, are the most likely source. We attribute the low average propagation speed to V1 to the unfavourable location of the spacecraft, while a higher average propagation speed to V2 (both shown below in Table 4) is probably due to the more favourable location with respect to this spacecraft of both this flare and the July 19 flare.

Figures 5a, 5b, and 5c all illustrate the advantages of looking at pairs of plots for the periods of powerful flares to see how the geometric relationships between the flare locations and the spacecraft locations aid in understanding the propagation of the shocks and CMEs generated by these solar events. Table 4 in the next section presents quantitative ranges for the delay times and average transit speeds of the shocks from the flares that produced them to when the Voyagers observed them.

6. Shock Delays and Average Transit Speeds

Table 4 summarizes the shock transits from the Sun to V2 and to V1 for the solar flare events of Oct./Nov. (Halloween) 2003; March 2012; and July 2012. All of the dates, delays, and transit speeds in Table 4 are based strictly on the times of the events at the Sun and the times of the measurements of the events at the spacecraft that have been identified as corresponding to the respective solar flares, without referring to any outside models not connected with these measurements.

Halloween 2003 Flare Events	**V2 observations**	**V1 observations**
Delay (days) Apr. 28, 2004 at V2 Jul. 27, 2004 at V1	173-188	263-278
Average transit speed (km/s)	674-732	577-610
March 2012 Flare Events		
Delay (days) Apr. 18, 2013 at V2 Apr. 25, 2013 at V1	404-407	411-414
Average transit speed (km/s)	432-435	518-522
July 2012 Flare Events		
Delay (days) Aug. 30, 2013 at V2 May 20, 2014 at V1	403-407	666-670
Average transit speed (km/s)	436-441	330-332

Table 4. Shock Summaries and Comparisons for Three Solar Events: October/November (Halloween) 2003 solar events; the March 2012 solar events; and the July 2012 solar events.

This point is important. We decided to use only dates, measurements, etc. that were obtained from solar observations and in spacecraft data when those numbers are beyond reproach. Thus, if the measured shock speed or its related values can't be verified, we don't want to consider using them. Thus, any *a priori* methods or numbers based on other types of analyses are not accepted.

In Table 4 the first (top) columns indicate for the Halloween (Oct./Nov) 2003 solar events that the start date of the first event reaching Voyager 2 (V2) was on April 28 (Day 119), 2004. This corresponds to a delay time from the Sun to V2 of 173-188 days. The second top column indicates that these Halloween solar events reached V1 on July 27 (Day 209), 2004. This corresponds to a delay time from the Sun to V1 of 263-278 days. The Average Transit Speed delay time (see bottom row for the Halloween events in Table 4) for these events were for V2: 264-732 km/s; and for V1: 577-610 km/s. Similar numbers are shown in Table 4 for the March 2012 and July 2012 solar events and for their corresponding measured phenomena at V2 and V1.

For this Halloween 2003 solar event, since all of the final shocks reached V2 and V1 when these spacecraft were both still in the heliosphere, this does not imply any simplified imaging or backward projections of these events. However, the shocks from the March 2012 and the July 2012 solar events reached V2 and V1 after these two spacecraft had travelled beyond the termination shock (TS) and into the heliosheath (HS). In addition, the solar shocks from the July 2012 solar event appear to have reached V1 after V1 had left the HS and had entered the Local Interstellar Medium (LISM), since V1 crossed the heliopause (HP) on August 25, 2012. Thus, any correct modeling of these March 2012 and July 2012 solar events would have to take account of the propagation of these events in the heliosheath. Furthermore, for the July 2012 solar events of which the effects appear to have reached V1 in 2014, since the spacecraft had already crossed the HP in August of 2012 the modeling would also have to estimate propagation in the LISM. It is not obvious now what would be the correct modeling formulas for these events when the spacecraft are in the HS and/or in the LISM. Thus, we decided to base our current estimates only on the data presently available. In the future when more V2 and V1 data are available and when V2 crosses the HP, we will plan to employ these newly available data for future estimates and calculations.

7. Summary

We presented some examples that show the Sun is a dynamic influence in the heliosphere, the HS, and the LISM. Our models and data comparisons provide key insights. We prefer to use measured quantitative data at the Sun and at the spacecraft rather than relying on inferred numerical values at these various locations. We employ our two 3D time-dependent models that start near the Sun (2.5 Rs) where they input solar background and solar event boundary conditions derived from solar observatory magnetograms. These two 3D time-dependent models simulate the propagation of shocks, other specific features, and the background solar wind throughout the heliosphere. The Hybrid Heliospheric Modeling System with Pickup Protons (HHMS-PI) is a 3D time-dependent Magnetohydrodynamic (MHD) simulation. HAFSS (HAF Solar Surface) is a 3D time-dependent kinematic simulation. Our models and data comparisons indicate that quantitative solar effects are seen in the heliosphere, the HS, and the LISM in in-situ spacecraft measurements of plasma, magnetic field, energetic particles, cosmic rays, and plasma waves. There is quantitative agreement (at ACE, Ulysses, V1, V2) with data (e.g., solar wind, IMF, Ulysses SWICS pickup protons (PUPs)). Quantitatively we showed examples of propagating shocks slowed due to PUPs. The 3D locations and measurements of solar events and of their corresponding features at various spacecraft are key to understanding the 3D propagation and timing of shocks, other specific features, and gradients throughout the heliosphere, HS, and LISM.

8. References

[1] Intriligator D S, Sun W, et al., 2005, *J. Geophys. Res.*, **110**, A09S10, doi:10.1029/2004JA010939.
[2] Detman T M, Intriligator D S, et al., 2011, *J. Geophys. Res.*, **116**. A03105, doi:10. 1029/2010JA015803.
[3] Intriligator D S, Detman T M, Gloeckler G, et al., 2012, *J. Geophys. Res.*, **117**, A06104, doi:10.1029/2011JA017424.
[4] Intriligator D S, et al., AGU presentation, 2011.
[5] Intriligator D S, et al., 2015, *J. Geophys. Res.*, **120**, 8267–8280, doi:10.1002/2015JA021406.
[6] Zank G P, et al., 2001, *J. Geophys. Res.*, **106**, A12, 29363-29372, doi: 10.1029/2000JA000401.

Acknowledgments

We gratefully thank Professor Edward C. Stone, and the ACE and Voyager projects for the vast information we have received. We acknowledge the support of Carmel Research Center, Inc.

Journal of Physics: Conference Series **767** (2016) 012014 doi:10.1088/1742-6596/767/1/012014

The Role of Particle Drifts in the Heliosphere

József Kóta

The University of Arizona, Lunar and Planetary Laboratory, Tucson, AZ 8521-0092, USA

E-mail: kota@lpl.arizona.edu

Abstract. Some of the basic effects of particle drifts on cosmic rays in the inner and outer Heliosphere are discussed. We bring arguments supporting that particle drifts, convection and adiabatic cooling can be treated by the same matching condition at discontinuities, the apparent infinite value of drift speed is consistent with the diffusive description. Simple illustrative examples are presented to highlight the differences between drift and no-drift solutions of the Parker equation. A simplistic model simulation is presented to demonstrate the role of drifts in the 22-year solar modulation cycle. The nature and possible effects of drift in the heliosheath are briefly addressed.

1. Introduction

It was 40 years ago that the two Voyager spacecraft started their amazing journey through our Heliosphere and beyond. It is quite a remarkable coincidence that two major developments in the theory of energetic charged particles were launched almost the same time. Diffusive shock acceleration (DSA) was proposed in 1977 (Axford et al., 1977; Bell et al., 1978; Blandford and Ostriker, 1978; Krymskii 1977) and it was in the same year that Jokipii et al. (1977) called attention to the profound role particle drifts play in the transport of cosmic rays. The 2×40=80 years of Voyager mission, with the leadership of Edward Stone, brought many fundamental observations inspiring further research, and hopefully this shall continue for many more years to come.

While the idea of DSA was immediately embraced, the concept of particle drifts met some resistance in the cosmic ray community. A beautiful observation of Voyagers bringing compelling evidence for the action of drifts meant a milestone in 1987. Comparing V1, V2 and Pioneer 10 measurements, Cummings et al. (1987) found that, the intensity of anomalous oxygen was decreasing away from the flat heliospheric current sheet (HCS). This occurred during the so called $A < 0$ solar minimum, in agreement with the prediction of drift models (Jokipii, 1986) This observation demonstrated the effects of both the DSA and particle drifts at the same time.

The present work is devoted to cosmic-ray effects connected with particle drift. Drifts are important both in the acceleration and spatial transport of energetic particles. We shall focus on the solar modulation of anomalous and galactic cosmic rays (ACRs and GCRs), examples will be brought for GCRs. Sections 2 and 3 reiterate some general features of drift effects and illustrative examples are presented for the simplest 2-dimensional (2D)convection-diffusion model. Section 4 presents a simplistic simulation model of the 22-year solar cycle focusing on drift effects and omitting other important mechanisms of GCR modulation. Finally the nature and possible role of particle drifts in the heliosheath is briefly addressed in Section 5.

2. Drift effects on cosmic rays in the Heliosphere

The most widely used tool describing the diffusive transport of energetic charged particles in the heliospheric magnetic field (HMF) is the equation of Parker (1965). This robust equation assumes that scattering is frequent enough to maintain quasi-isotropic particle distribution. Then the omnidirectional phase-space density $f(x_i, p, t)$ at position, x_i, momentum, p, and time, t obeys

$$\frac{\partial f}{\partial t} = \frac{\partial}{\partial x_i}\left(\kappa_{ij}\frac{\partial f}{\partial x_j}\right) - V_i\frac{\partial f}{\partial x_i} + \frac{1}{3}\frac{\partial V_i}{\partial x_i}\frac{\partial f}{\partial ln\, p} + Q \tag{1}$$

where the consecutive terms represent diffusion, advection by the solar wind moving at velocity, V_i, and adiabatic energy change due to the expansion or compression of the fluid. Energy and momentum are measured in the frame of the fluid. $Q(x_i, p, t)$ on the RHS accounts for possible sources, or sinks. Double indices refer to summation. The full 3D diffusion tensor, κ_{ij} contains an anti-symmetric component, κ^a_{ij}, which accounts for regular particle motion in the large-scale HMF, and whose divergence yields the drift velocity, V^{dr}_i, in the non-uniform HMF.

$$\kappa^a_{ij} = \kappa_A \epsilon_{ijk} b_k \qquad V^{dr}_i = \frac{\partial \kappa^a_{ij}}{\partial x_j} \tag{2}$$

For weak scattering, $\kappa_A = r_g v/3 = Pv/3B$ where r_g stands for the gyroradius of a particle of speed, v, and rigidity, P. $b_i = B_i/B$ is a unit vector along the magnetic field, B_i. The drift motion, V^{dr}_i, is frequently treated as a separate term, then only the symmetric part of κ_{ij} is kept in Eq. (1). For an excellent review on the Parker equation see Moraal (2013).

Modulation, in most cases, is connected with energy change. In a static magnetic field, drift would be present but, in itself, would not cause density gradients, anisotropy or energy change. In a moving fluid, on the other hand, drifts do affect energy change, by virtue of controlling the path and dwelling time of GCRs in the heliosphere. A conceptual problem is posed by particle drifts at current sheets, shocks, or other discontinuities, where drift derived form Eq. (2) can be infinitely fast. A great deal of work has been done on this subject. Isenberg and Jokipii (1979) pointed out that the standard form of drifts applies under broad conditions. Other efforts substitute the step-like discontinuity by a sharp but continuous change of κ_A, hence distribute the drift over a finite thickness (Burger et al., 1985).

We note that the Parker equation can also be re-written in a more compact form as

$$\frac{\partial f}{\partial t} = \frac{1}{p^3}\frac{\partial}{\partial \xi_a}\left(p^3 K_{ab}\frac{\partial f}{\partial \xi_b}\right) + Q \qquad a, b = 1, ..., 4 \tag{3}$$

where ξ_a is a 4D vector comprising of the three spatial coordinates, x_i, with $\xi_4 = lnp^3 = 3\,lnp$ added as the 4th component. The corresponding 4D diffusion tensor is

$$K_{ab} = \begin{pmatrix} \kappa_{ij} & V_j \\ -V_i & 0 \end{pmatrix} \qquad i, j = 1, 2, 3 \tag{4}$$

This implies that, at least technically, there is no essential difference between advection, energy loss, and drift, all being derived from the anti-symmetric part of the 4×4 diffusion tensor. In general, regular reversible motion appears in the antisymmetric part, while diffusive irreversible motion is in the symmetric part of the 4D diffusion tensor. A purely anti-symmetric tensor has imaginary eigenvalues corresponding to reversible motion. The form of Eq. (3) suggests that drifts, advection and energy change are best treated in the same way at discontinuities. The normal component of $K_{ab}\,\partial f/\partial \xi_b$ should match at the two side of the discontinuity:

$$n_i\left(\kappa_{ij}\frac{\partial f}{\partial x_j} + V_i\frac{\partial f}{\partial ln\, p^3}\right)\Big|_1^2 = 0 \tag{5}$$

which indicates that the use of matching condition (5) is sufficient and there is no compelling reason to distribute the drift over a finite thickness. These two approaches may differ in subtleties but they should give very much the same results.

Finally we note that the infinite value of drift velocity is not an inconsistency in a diffusive description. Diffusion equation is a robust approximation which takes the limit of infinitely fast particle speed and infinitely small diffusion mean free path, keeping the product finite. This shows up in a number of features of the diffusive equation. It is known that, in diffusive solutions, particles appear instantly (though with little chance) at large distances. Similarly, the frequently used $f = 0$ condition at absorbing boundaries implies an infinite particle speed. One further example is DSA. The diffusion solution implies an infinite number of shock crossings with infinitesimal energy gain at each crossing. All these give good approximation as long as the conditions of diffusive transport hold, and break down where the diffusive description is no longer valid.

3. Simple illustrative examples

To illustrate the basic drift effects imagine a 2D (x,z) slab model with the fluid velocity, V_i, pointing in the x direction, and the magnetic field pointing in the ignorable y direction. In an analogy to GCR modulation, x and z may play the role of radius and latitude, respectively. For brevity, assume that both V_i and κ_{ij} are uniform in space. This represents a case when modulation takes place without energy change, and Parker's equation reduces to a convection-diffusion equation which has the obvious solution

$$f_0(x_i) \propto exp\Big(Vx/\kappa_{xx}\Big) \qquad g_x = \frac{\partial ln(f)}{\partial x} = V/\kappa_{xx} \tag{6}$$

There are other, z dependent solutions, which may appear if we impose a z-dependent boundary condition. These solutions do, however decay faster and f_0 dominates at large distances from outer boundary.

The z-independent solution, f_0, involves a unobstructed streaming in the positive or negative z direction depending on the sign of B_y. This changes if we place reflecting boundaries or periodic currents sheets (their effect are the same for this case) at say $z = 0$ and $z = z_0$. Then particles drift along the current sheets (or reflecting boundaries) and the solution changes to (Kóta, 1981)

$$f_0^\star(x_i) \propto exp\Big(x_i\kappa_{ij}^{-1}V_j\Big) \qquad g_x^\star = V/(\kappa_{xx} + \kappa_A^2/\kappa_{zz}) = V/\kappa^\star \tag{7}$$

where $f_0^\star$ is a completely isotropic solution. Eq. (7) indicates that drifts will reduce the gradients in the x direction, and this reduction depends on the value of the latitudinal diffusion coefficient κ_{zz}. Roughly speaking, drift assists GCRs to find a faster way into the heliosphere, hence reduce modulation. GCRs follow the fast drift path as long as κ_{zz} is small, but larger values of κ_{zz} destroys the coherent drift pattern (Kóta, 1989; 2013).

We note that, results were similar for an oblique magnetic field, the effective $\kappa^\star$ would be somewhat modified, and an anisotropy, analogous to the co-rotational anisotropy, could also appear in the y direction.

Figure 1 shows numerical solutions with keeping $\kappa_{xx} = \kappa_{xz} = -\kappa_{zx} = \kappa_A$ constant. The important dimensionless parameter is $\eta = \kappa_A^2/(\kappa_{xx}\kappa_{zz})$. We find that, after some transient regime near the outer boundary, the solutions settle to a fully isotropic solution given by Eq. (7). A larger value of η implying smaller 'latitudinal' diffusion gives milder modulation relative to the 1D 'no-drift' solution of (6).

To demonstrate the effect of a wavy HCS, we present a simulation carried out with a sectored field placed between two reflecting boundaries at $z = -z_0$ and $z = +z_0$. A simple step-function is used, the sectored field extends from $-0.5z_0$ to $+0.5z_0$. The simulation presented in Figures

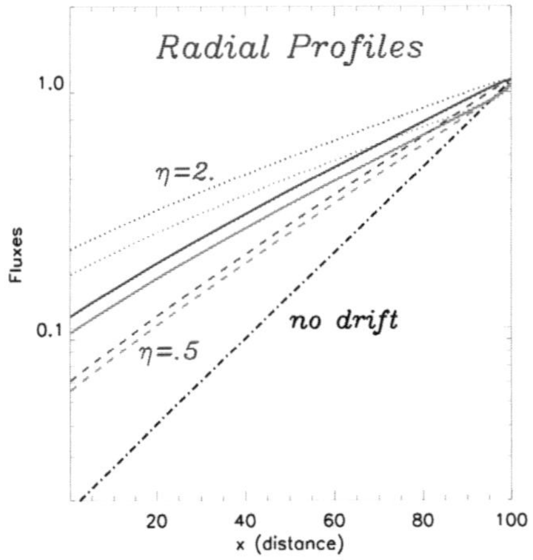

Figure 1. Solution of the 2D convection-diffusion equation. Shown are drift solutions with $\eta = 0.5$ (dashed), $\eta = 1.0$ (solid), and $\eta = 2.0$ (dotted). Blue lines refer to positions at $z = 0$, magenta lines refer to $z = z_0$. Smaller κ_{zz} (larger η) values results in less modulation. Drift solutions always result in less modulation than the no-drift solution (black dashed-dotted line).

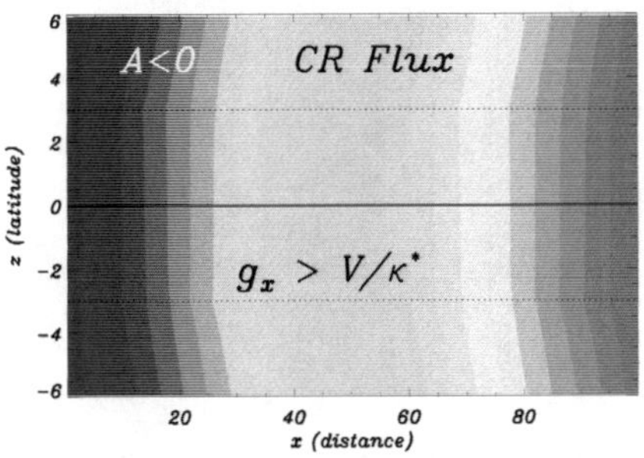

Figure 2. Color plot of GCR density (red high, blue low) obtained with sectored HCS extending between the dotted lines. Note that the latitudinal (z) gradients essentially vanish in the sectored area. Radial x) gradients and modulation increase with the increase of the latitudinal extent of the HCS.

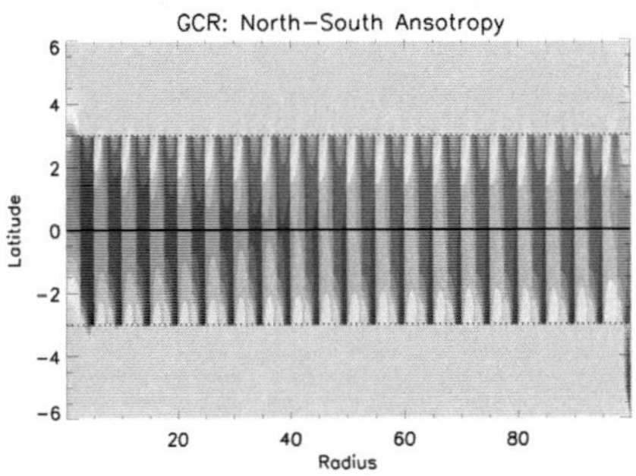

Figure 3. Color plot of cosmic-ray streaming in the z (latitudinal) direction in a sectored field. 'N-S' anisotropies arise, which alternate between upward (red) and downward (blue) directions according to the polarity of the respective sector. Anisotropies are small at higher latitudes in the unipolar region (green).

2 and 3 use $\eta = 1.0$ and the same other parameters used previously (Fig. 1), the only difference is the inclusion of a sectored field.

Inspection of Figure 2 showing the predicted spatial distribution of GCRs, indicates that the 'latitudinal' (z) gradient does essentially vanish in the sectored area (between the dotted lines). The radial gradient increases, the radial variation moves toward the 'no-drift' solution. What happens is that drift along the HCS allows a fast latitudinal transport, the effect of which is the same as that of a larger κ_{zz}.

An important consequence of the sectored field is the rise of N-S streaming in the z or $-z$ direction depending on the polarity of the respective sector. This is seen in Figure 3. The presence of this sector-dependent N-S anisotropy has been long known and been widely used to infer the local radial gradient. Studies of the GCR anisotropies offer a unique tool to obtain constraints on the local transport parameters (see for instance Bieber and Chen, 1991; Chen and Bieber, 1993; Gil et al., 2012; Munakata et al., 2014). The modeling of GCR anisotropies requires fully 3D simulations including the proper κ_A parameter. The presence of the N-S anisotropy seems to indicate that drift is present even during solar maxima, but global drift effects are reduced due to the fast latitudinal transport. In other words, drift effects do diminish without scaling down the anti-symmetric component of the full diffusion tensor.

4. 22-year solar modulation cycle of GCRs

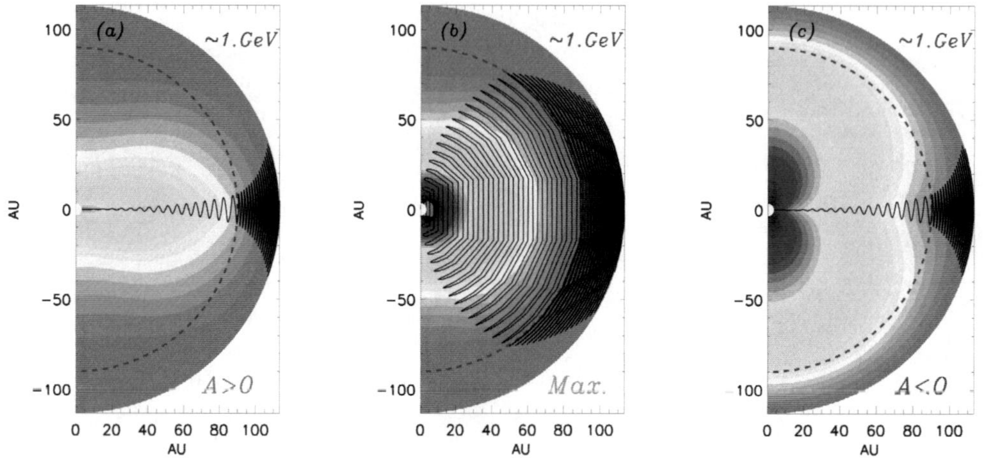

Figure 4. Contour-plots of GCR intensity at $A > 0$ solar minimum (a), near solar maximum (b), and at $A < 0$ solar minimum (c). Red means high, blue means low intensities. The TS is shown by the dashed circle, the wavy black lines indicate the evolving HCS. Note the different latitudinal variation during the two solar minima of opposite polarities.

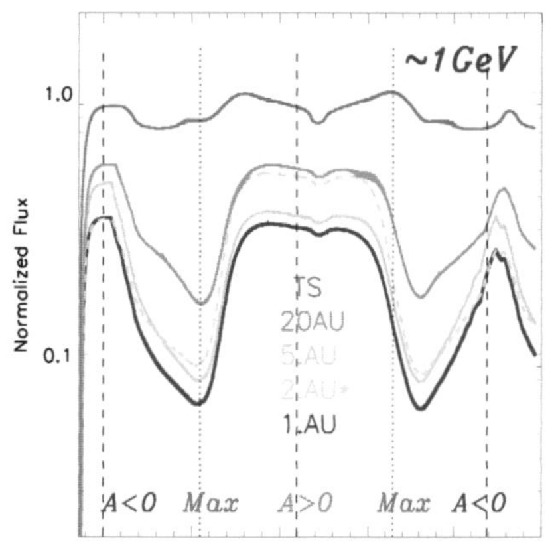

Figure 5. 22-year solar-cycle variation of ~1GeV GCR intensity at different heliocentric radii in the Heliosphere. All curves refer to the equatorial plane except for the 2AU curve, which refers to high latitude. Vertical lines mark times when the tilt angle at the Sun is 0° (dashed) or 90° (dotted). GCR fluxes follow the changing tilt with some delay.

In this section we present a simplistic simulation using an axially symmetric model to highlight the role of particle drifts in the 22-year cycle solar modulation of GCRs. We focus on the effects connected with the polarity reversal of the solar magnetic field and the changes in the tilt angle of the heliospheric current sheet (HCS) dividing the two hemishperes of opposite polarities. Other important aspects of the modulation process, such as the cyclic variation of solar wind speed, magnetic field strength and diffusion coefficients are not included here, so results are illustrative and are not intended to fit observations.

We consider a spherical supersonic solar wind with a uniform radial speed of $400km/s$, which undergoes a shock transition at the termination shock (TS) placed at 90AU heliographic radius. The solar wind speed is reduced at the TS by a factor of 3, and is kept to flow radially as an incompressible fluid ($V \propto 1/r^2$). An artificial outer boundary is imposed at ≈115 AU, where the GCR flux is set equal to the presumed interstellar spectrum of Webber and Higbie (2009).

We note that, under broad conditions, a stable corotation yields a HMF

$$B_i = \Big(\frac{B_0}{n_0 V_0}\Big)\, n\, (V_i - \epsilon_{ijk}\Omega_j x_k) \tag{8}$$

where Ω_j stands for the rotation of the Sun, and the factor $(B_0/n_0 v_0)$ remains constant along each magnetic field line carried by the solar wind. A temporal change in the HCS introduces a more complex HMF, and an additional meridional component appears inevitably. Modeling the effects of sectored HMF requires 3D simulations (Kóta and Jokipii, 1983, 1991; Hattingh and Burger, 1995). Potgieter and Moraal (1985) developed a 2D model that captured the modulating effects of the wavy HCS. This method has been successfully applied in a sequence of work to reconstruct the variation spatial and temporal variation of GCRs (see Burger and Potgieter, 1989; Burger and Hattingh 1995, Potgieter, 2013).

Here we take the simplistic approach: in calculating drifts we omit the radial component of the HMF, which is important in the inner heliosphere but its relative role decreases fast toward larger distances from the Sun. Magnetic field lines shall be assumed to be hoops that are passively carried by the wind:

$$B_i = -\Big(\frac{B_0 n}{n_0 V_0}\Big)\epsilon_{ijk}\Omega_j x_k \tag{9}$$

For a uniform radial wind speed the azimuthal field decreases as $\propto n * \rho$, where $\rho = rcos(\Theta)$ is the distance from the Sun's rotation axis at latitude, Θ. The polarity is preserved from the time $t - r/V$, when the ring originated from the Sun. The virtue of this field model is that it is less restrictive than the full spiral, and allows the continuous change of the tilt angle without changing the geometry of the field.

The parallel mean free path, $\lambda_{||}$, is conveniently assumed to be proportional to particle rigidity, P, and inversely proportional to the local magnetic field so that 1 GV particles in a 6 nT field have $\lambda_{||} = 0.075$AU. A relatively low value of $\kappa_\perp/\kappa_{||} = 0.02$ is chosen to highlight drift effects, a larger value of $\kappa_\perp$ would reduce these to some extent.

The spiral field (8) becomes weak near the poles, so the polar region is dominated by randomly oriented transverse fields (Jokipii and Kóta, 1989). We add a random transverse component to the field at all latitudes with $\delta B_t = \delta B_0(n\,r)/(n_0 r_0)$, so that $(\delta B/B) \approx 0.2$ at the Earth. In calculating the field strength and the diffusion coefficients we include the full spiral field together with the contribution of the random component. The random dB_t component tends to reduce the radial and enhance the latitudinal diffusion.

Figure 4 shows three phases of the 22-year solar cycle. The successive panels depict the spatial distribution of $\sim 1GV$ GCRs around $A > 0$ solar minimum (a), near solar maximum (b) and $A < 0$ solar minimum (c). The robust drift features are clearly visible, the latitudinal dependence of GCRs is markedly different at consecutive solar minima. GCRs find easier to penetrate into the inner heliosphere through the polar region in the $A > 0$ solar minimum, and along the near flattened equatorial HCS in the $A < 0$ minimum. The distribution is largely spherical with small latitudinal gradients when the HCS is highly tilted around solar maximum.

Figure 5 shows the 22-year cycle caused solely by the changing tilt of the HCS, as predicted for different locations in the Heliosphere. All curves refer to equatorial locations except for the dashed $2AU^{\star}$ curve, which refers to high latitude and predicts a distinctly different profile. The familiar peaked and plateau GCR maxima are clearly visible. GCRs follow the change of tilt with some delay since the flattening of the HCS takes time to propagate to the outer heliosphere. Inspection of Fig. 5 also indicates that the momentary tilt angle, in itself, cannot be used as a single parameter to determine the level of modulation. Modulation is weakest and GCR fluxes are the highest when the HCS is flat in the whole computational domain. This is apparent in the asymmetry between year 0 and year 22. The momentary tilt is 0° in both cases, but a flat HCS was assumed for the whole period between years 0 and $= 1.5$ (when numerical simulation was started with an empty heliosphere).

No drift solutions would be horizontal lines through the GCR fluxes at solar maxima. The calculations presented here were using a relatively small value of $\kappa_\perp$ in order to emphasize drift effects. The low value of $\kappa_\perp$ enhances the GCR flux at the $A < 0$ solar minimum, while has less effect during the $A > 0$ cycle. Global drift effects diminished during solar maxima without any change in the value of κ_A.

5. Drifts in the Heliosheath

The diffusion and drift motion of cosmic rays in the heliosheath critically depends on the structure of the HMF beyond the TS. Currently there are two different views on the global structure of the heliosheath. The traditional view has been that a long heliotail is formed in the direction of the interstellar wind. Because of the slow radial speed of the subsonic wind in the nose section, magnetic sectors in this model become closely stacked and may become essentially transparent for GCR whose gyro-radii are comparable or larger than the width of the magnetic sectors (Florinski, 2011). Further away from the TS, the HCS becomes completely distorted and disintegrates due to the shear of the flow (Pogorelov et al., 2015 and references therein).

Recently Opher et al. (2015) proposed an alternative 'croissant ' model suggesting that, the magnetic tension causes the solar fluid to be lifted southward and northward forming a pair of jets, and this avoids the further stretching of field lines. Drake et al. (2015) gave a simple model to illustrate the physics involved in the jet model. The omission of interstellar wind and interstellar magnetic field allowed these authors to construct an axially symmetric model. Furthermore, the omission of the radial component of the HMF, which has negligible role in the dynamics, led to a HMF where field lines are hoops carried by the solar wind.

Figure 6 depicts a schematic picture of the HMF in this simplified jet model. The left panel (a) shows the hoop-like field lines directed oppositely on the two sides of the flat HCS. The magnetic field for this case is given by Eq (9). We note that retaining the radial field component Eq. (8) would hold, the omitted field component would remain directed along the stream lines, and be proportional to $n\,V_i$.

The right panel (b) shows a meridional cut allowing a wavy HCS. The structure of the HCS is distorted due to the shear of the fluid but to a lesser extent than the distortion in the 'classic' long-heliotail case. The width and placement of the wavy HCS are not in scale, the HCS is drawn largely stretched in Fig. 6 for brevity.

The particle drift, V_i^{dr} can be readily obtained. Taking the weak scattering limit, the resulting drift in this model of magnetic hoops is

$$V_i^{dr} = \epsilon_{ijk}\frac{vP\,B_k}{3B^2}\frac{\partial}{\partial x_j}\Big(ln(\frac{B_0}{n_0 V_0}) + ln\,n\Big) \tag{10}$$

In the simplest approximation the factor $B_0/(n_0 V_0)$ can be taken as constant except for its sign which changes at the HCS. Then, the first term in Eq. (10) gives the the inward drifts along the HP and the HCS, which is balanced by an outward drift concentrated at the polar line.

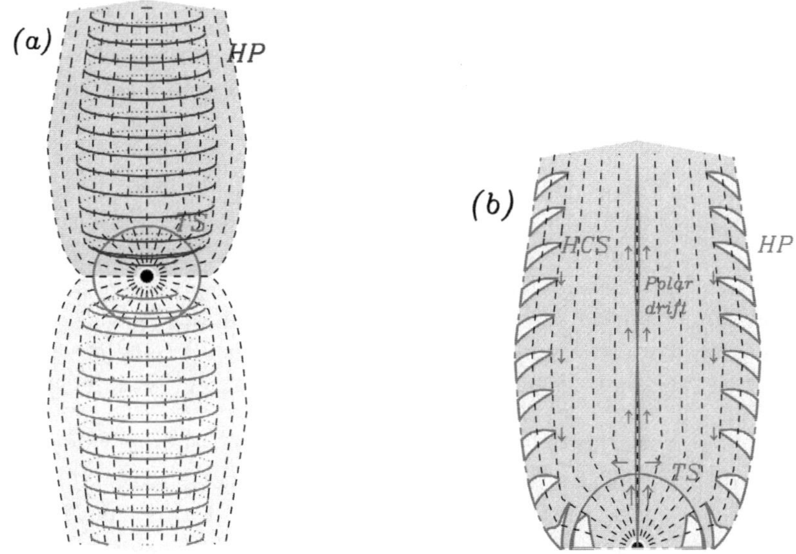

Figure 6. Schematic view of an axially symmetric toy-model of the Heliosphere similar to that of Drake et al. (2015). The left panel (a) shows that the solar wind is deflected to form a pair of jets. Magnetic field lines are oppositely directed hoops below and above the helioequatorial plane. The right panel (b) depicts a schematic view with a tilted HCS in the Northern meridional section. Red arrows indicate the direction of particle drifts for the $A < 0$ polarity state. The extension of the stretched HCS is not in scale. (see text).

The second term of Eq. (10) has no contribution in the the heliosheath where the density, n, is assumed to be constant. This term yields an equatorward drift along the TS where the density, n, has a jump. These directions refer to the $A < 0$ polarity state, and would reverse for the opposite $A > 0$ state.

The inclusion of random transverse field component, δB_t, is complicated by the velocity shear in the heliosheath, which will stretch and increase this component beyond the TS (in addition to the increase by the compression at the TS). The magnitude of this increase requires further study.

Finally we note that drift along the wavy HCS can provide fast particle transport normal to the stream lines. This can potentially be important for the $A > 0$ polarity case when GCR ions drift inward at the poles. Most of the early models of GCR modulation imposed an artificial boundary some distance from the Sun. The simulation presented in the previous section also suffers from this inherent problem: the polar line offers an easy access for GCRS from the heliopause to the inner heliosphere. In fact, the polar field lines do not to connect the HP, so the question remains how GCR reach the polar regions (Kóta, 2013). We suggest that this could be assisted by drifts along the HCS in a sectored heliosheath. To answer this question requires advanced simulations incorporating the full heliosphere (see Florinski and Pogorelov, 2009; Florinski et al. 2013; Lou et al., 2013; Strauss et al., 2013) and further theoretical research is needed to explore how GCR enter the heliosphere. A surprising finding of the HP crossing of Voyager-1 was the ≈ 20 percent increase seen is the $\sim 200 GeV$ GCR flux (Stone et al., 2013; Krimigis et al., 2013). It will be exciting to see if the same intensity jump is repeated at the HP crossing of Voyager-2.

6. Conclusions

Particle drifts are important in both the propagation and acceleration of energetic particles. This work focused on some aspects of drifts in the solar modulation of cosmic rays. Simple numerical simulations were presented to illustrate drift effects, our examples served to highlight the role of drifts and were not intended to fit observations.

Drifts can be derived from the anti-symmetric term of the full 3D diffusion tensor, κ_{ij}^a, hence we may obtain infinitely fast drift speed at discontinuities, such as the HCS or shocks. We argue that this apparently infinite speed does not pose a conceptual problem, the diffusive approach does already represent a limit of infinite particle speed. The Parker equation (1) can be re-cast into a compact form, where drifts, advection, and adiabatic energy change are all derived from the anti-symmetric component of a 4×4 diffusion tensor, which makes it plausible to use the same mathematical treatment and same matching condition for all these quantities. The frequently used method of spreading drift over a finite area is another possible way, the two approaches should give very much the same results.

We re-iterate that global drift effects critically depend on the latitudinal diffusion (Kóta, 1989). Well organized drift motion can occur as long as $\kappa_A \geq \kappa_\perp$. Large $\kappa_\perp$, on the other hand, destroys the coherent drift motion and results in larger radial gradients, i.e. stronger modulation. This will happen without any change in the value of either κ_A or $\kappa_\perp$. Drift along a highly tilted HCS allows fast latitudinal spread of GCRs, which acts in the same way as a larger $\kappa_\perp$ does.

In turbulent magnetic fields, we expect some reduction of κ_A relative to its weak scattering limit (Giacalone et al.; 1999, Minnie et al., 2007). Anisotropy measurements do, however, indicate that this reduction cannot be very strong: a polarity dependent N-S anisotropy connected with radial gradient and anti-symmetric diffusion is clearly seen in neutron monitor data even during solar maxima (Bieber and Chen, 1991). The presence of this anisotropy component would be hard to explain without a substantial value of κ_A. Simulation models should explain both GCR intensities and anisotropies at the same time. This is a challenging task, requiring advanced 3D simulations.

A simplistic model was presented to simulate the 22-year solar modulation cycle of GCRs solely due to the changing tilt of the HCS. We tried to capture the effect of the HCS in an axially symmetric model, where the tilt of the evolving HCS was the sole parameter to vary in time. All other parameters were kept unchanged. The robust features of drift solution, such as the peaked/plateau intensity maxima of GCRs during consecutive 11-year cycles and low fluxes during solar maxima were qualitatively reproduced. This again suggests that global drift effects tend to diminish even without scaling down κ_A during solar maxima.

The possible role of drifts in the heliosheath was briefly touched. While the traditional elongated heliosphere with along heliotail is likely to have a fragmented sector structure, the jet-model of Opher et al (2015) may retain a more coherent magnetic structure where large scale drifts could play a role. Particle drift can certainly be expected along the surface of the HP. We suggest that current sheet drift might play a role in transporting GCRs toward the polar regions during $A > 0$ polarity cycles. It is not inconceivable that HCS reaching the HP or close to the HP can act as cracks where GCRs find easier to penetrate and fill up the heliosphere. The anticipated HP crossing of Voyager-2 may shed light on this question.

Acknowledgments

The author benefited from discussions with J.R. Jokipii and M. Krainev. This work was partially supported by NASA under Grants NNX15AJ72G and NNX16AB77G. The help from the Theoretical Astrophysics Program of the University of Arizona is gratefully acknowledged.

Journal of Physics: Conference Series **767** (2016) 012014 doi:10.1088/1742-6596/767/1/012014

References

Axford W I, Leer E and Skadron G 1977 *Proc. 15th Int. Cosmic Ray Conf.* (Plovdiv, Bulgaria, 1977), **11** 132-7
Bell A R 1978 *MNRAS* **182** 147
Blandford A D and Ostriker J P 1978 *Astrophys. J.* **221L** 29
Bieber J W and Chen J 1991 *Astrophys. J.* **372** 301
Burger R A, Moraal H and Webb G M 1985 *Astrophys. Space Sci.* **116** 107
Burger R A and Potgieter M S 1989 *Astrophys. J.* **339** 501
Burger R A and Hattingh M 1995 *Astrophys. J.* **230** 375
Chen J and Bieber J W 1993 *Astrophys. J.* **405** 375
Cummings A C, Stone E C and Webber W R 1987 *Geophys. Res. Lett.* **14** 174
Drake J F, Swisdak M and Opher M 2015 *Astrophys. J.* **808** 44
Florinski V and Pogorelov N V 2009 *Astropys. J.* **701** 642
Florinski V 2011 *Adv. Space Res.* **48** 308
Florinski V, Ferreira S E S and Pogorelov N V 2013 *Space Sci. Res.* **176** 147
Giacalone J, Jokipii J R and Kóta J 1999 *Proc. 26th Int. Cosmic Ray Conf.* (Salt Lake City, Utah, 1999), **7** 37
Gil A, Modzelewska R and Alania M V 2012 *Adv. Space Res.*, **50** 712
Hattingh M and Burger R A 1995 *Adv. Space Res.* **16** 9
Isenberg P A and Jokipii J R 1979 *Astrophys. J.* **234** 7431
Jokipii J R 1986 *J. Geophys. Res.* **91** 2929
Jokipii J R, Levy E H and Hubbard W B 1977 *Astrophys. J.* **213** 861
Jokipii J R and Kóta J 1989 *Geophys. Res. Lett.* **16** 1
Kóta J 1981 *Adv. Space Res.* **176** 391
Kóta J 1989 in *Physics of the Outer Heliosphere* Eds: S Grzedzielski and D E Page, Pergamon Press, COSPAR Colloquia Series **1** 119 - 31
Kóta J 2013 *Space Sci. Rev.* **176** 391
Kóta J and Jokipii J R 1983 *Astrophys. J.* **265** 573
Krimigis S M, Decker R B, Roelof E C, Hill M E, Armstrong T P, Gloeckler G, Hamilton D C and Lanzerotti L J 2013 *Science* **341** 144
Krymskii G F 1977 *Dokl. Akad. Nauk SSSR* **234** 1306
Luo Xi, Zhang M, Potgieter Marius, Feng Xueshang and Pogolerov N V. 2013 *Astrophys. J.* **808** 82L
Minnie J, Bieber J W, Matthaeus W H and Burger, R A 2007 *Astrophys. J.* **670** 1149
Moraal H 2013 *Space Sci. Rev.* **176** 299
Munakata K, Kozai M, Kato C and Kóta J 2014 *Astrophys. J.* **791** 22
Opher M, Drake J F, Zieger B and Gombosi T I 2015 *Astrophys J* **800** 280
Parker E N 1965 *Planet. Space Sci.* **13** 9
Pogorelov N V, Borovikov S N, Heerikhuisen J and Zhang M 2015 *Astrophys. J.* **812** 6
Potgieter M S 2013 *Space Sci. Rev.* **176** 165
Potgieter M S and Moraaal H 1985 *Space Sci. Rev.* **294** 425
Stone E C, Cummings A C, McDonald F B, Heikkila B C, Lal N and Webber W R 2013 *Science* **341** 150
Strauss R D, Potgieter M S, Ferreira S E S, Fichtner J and Scherer K 2013 *Astrophys. J.* **765** 91
Webber W R and Higbie P R 2009 *J. Geophys. Res.* DOI: 1029/2008JA013689

Journal of Physics: Conference Series **767** (2016) 012015 doi:10.1088/1742-6596/767/1/012015

The Weibull functional form for the energetic particle spectrum at interplanetary shock waves

M Laurenza[1], G Consolini[1], M Storini[1], G Pallocchia[1], A Damiani[2]

[1] INAF-Istituto di Astrofisica e Planetologia Spaziali, Roma, Italy

[2] Japan Agency for Marine-Earth Science and Technology, Yokohama, Japan

E-mail: monica.laurenza@iaps.inaf.it

Abstract. Transient interplanetary shock waves are often associated with high energy particle enhancements, which are called energetic storm particle (ESP) events. Here we present a case study of an ESP event, recorded by the SEPT, LET and HET instruments onboard the STEREO B spacecraft, on 3 October 2011, in a wide energy range from 0.1 MeV to $\sim$ 30 MeV. The obtained particle spectrum is found to be reproduced by a Weibull like shape. Moreover, we show that the Weibull spectrum can be theoretically derived as the asymptotic steady state solution of the diffusion loss equation by assuming anomalous diffusion for particle velocity. The evaluation of Weibull's parameters obtained from particle observations and the power spectral density of the turbulent fluctations in the shock region, support this scenario and suggest that stochastic acceleration can contribute significantly to the acceleration of high energetic particles at collisioness shock waves.

1. Introduction

It is generally accepted that collisionless shock waves are strong sources of high energy particles. In the heliosphere the main sources for the acceleration of energetic particles are CME driven shocks, which are thought to accelerate solar energetic particle (SEPs) during gradual SEP events [1], and more generally transient interplanetary shocks which can be associated with enhancements of energetic charged particles [2, 3, 4, 5, 6] called ESP events. They can present a wide variety of different types, ranging from large particle intensity enhancements lasting several hours to small spikes lasting only a few minutes. It is also well established that energetic particles are accelerated in association with corotating interaction regions [7, 8]; they form when a stream of fast solar wind emerging from a coronal hole, overtakes slow solar wind, emitted at an earlier time. As a consequence of solar rotation these different speed plasmas became radially aligned and interact. This interaction creates a compression region, corotating with the Sun, and can strengthen to form a forward shock at the leading edge and a reverse shock at the trailing edge.

The association of energetic charged particles with collisionless shocks suggests an acceleration mechanism that is associated with the shock itself. The mechanism of diffusive shock acceleration (DSA) [9, 10, 11] has received the most attention as it naturally predicts a power law spectrum, where the spectral index (γ) is related only to the shock compression ratio (R), regardless of species, as $\gamma = 1/2[(R+2)/(R-1)]$. Nevertheless, it has been shown that the majority of the ESP events, measured in the period 1997 - 2001 at energies >47 keV, do not follow the prediction [12], although the agreement is better at low energies (47-187 keV) at strong shocks (having Mach number Ma > 3 and R > 2.5 [13]). A variety of effects may truncate the power law behaviour, such as adiabatic deceleration in expanding blast waves, shock lifetime comparable to particle acceleration time, shock size comparable to particle diffusion lengths. Ellison and Ramaty 1985 [14] argued that the power law spectra should roll over at high energies due to increasing diffusion coefficient with energy and assumed an exponential function, modulating the power law. On the other hand, it has been shown [15] that curving proton spectra could be better fitted by the Bessell functions, predicted in the frame of the stochastic acceleration. Nevertheless, both the Ellison-Ramaty and the Bessel functions do not provide a reasonable fit up to the highest observed energies [18]. The so-called double power law shape has been used [16] (following [17] for fitting gamma-ray burst spectra), which is identical to the Ellison-Ramaty form below the a transition energy; then, there is a smooth transition at higher energies to a second power law. It was shown that all of the proton spectra during the 16 Ground Level Enhancements (GLEs) of Solar Cycle 23 exhibit spectral breaks at energies ranging from ~ 2 to ~ 46 MeV and all are well fitted by a double power-law shape up to relativistic energies. Nevertheless, there have been few attempts to explain why the spectral breaks result in double power-law spectra. Afanasiev et al 2014 [19] studied the effect of an additional process that can modify the power-law spectral form produced by the diffusive shock acceleration: the stochastic re-acceleration of energetic protons by enhanced Alfvenic turbulence in the downstream region of a shock wave. They model stochastic acceleration of protons downstream from the shock by performing Monte Carlo simulations with self-consistently treated wave-particle interactions. The modelled spectra for different values of the intensity of magnetic fluctuations have a broken power law dependence on particle rigidity. The spectra were fitted at high rigidities by power law functions, with power law index values at high rigidities from 9.8 to 4.5, which fully covers the range of values into which the majority of the observed power-law indices for GLEs falls.

Recent observations [20, 21] have shown that the observed spectra at lower energies (≤ 100 MeV) in several SEP, ESP and CIR events can be reproduced through the Weibull function [22]. Hence, they argued that the Weibull distribution (WD) can be associated with shock acceleration and explained it in terms of a stochastic multiplicative process. The WD (also known as the two-parameter stretched exponential - see [23]) is a sub-exponential distribution interpolating smoothly between the exponential distribution and power law family.

Here, we analyze the 3 October 2011 ESP event, show that the Weibull function reproduces the observed spectrum with a very good agreement and try to get insight on implications for particle acceleration. We present the ESP event as well as the solar wind conditions in section 2. A theoretical derivation of the WD is presented in section 3, and its consistency with the observed WD parameters and physical conditions of the turbulent medium is discussed in section 4. Conclusions are drawn in section 5.

2. 3 October 2011 ESP event

Transient and corotating shocks are systems where particles are assumed to be locally accelerated by the shock, as they are usually associated with energetic particle enhancements, although these might contain particles trapped around the shock that were accelerated before the shock arrival at the observing point. However, in ESP events the characteristics of both the injected particles and the medium can be directly observed. Hence it is possible in principle to assess what conditions are necessary for shock acceleration to occur and what role irregularities in the ambient solar wind and/or shock surface may play in producing and modifying shock-accelerated particle populations.

We selected a transient interplanetary shock for which particle enhancements were detected up to high energies and occurred on unperturbed background. On 3 October 2011 at 22:23 UT a strong forward shock was observed by the STEREO B spacecraft (located at 1.08 AU, -98.09° 1.08° heliographic longitude and latitude, respectively), with prominent plasma signatures, along

Journal of Physics: Conference Series **767** (2016) 012015 doi:10.1088/1742-6596/767/1/012015

with upstream and downstream waves. Figure 1 illustrates the solar wind parameters, where the shock can be identified from the abrupt changes in their time behavior.

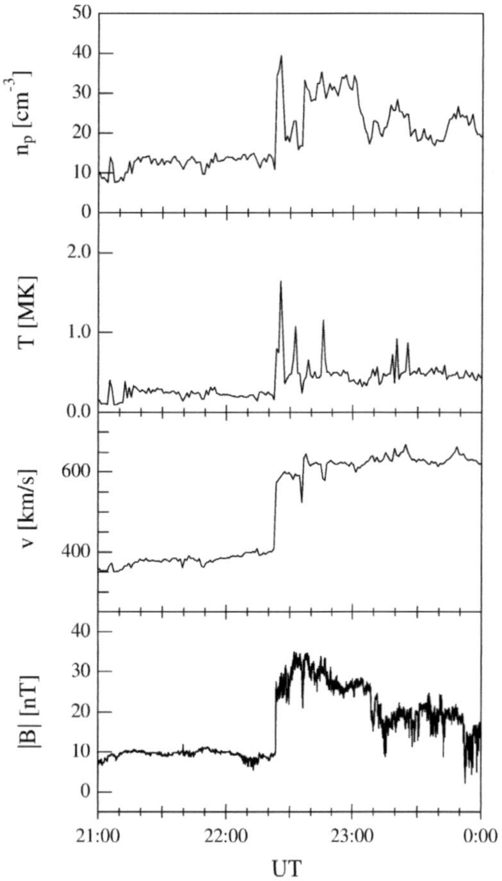

Figure 1. Time history of solar wind plasma parameters and magnetic field magnitude, as recorded on board the STEREO B spacecraft between 21:00 UT and 24:00 UT on 3 October 2011. From top to bottom, the proton density n_p, Temperature T, velocity v, the magnetic field intensity $\mid \mathbf{B} \mid$.

In concomitance, a particle enhancement was recorded by the Solar Electron Proton Telescope (SEPT [24]), the Low Energy Telescope (LET [25]) and the High Energy Telescope (HET [26]) in the energy range 0.1 − 30 MeV. Data used to study this event are 1 minute averaged proton fluxes measured by the three instruments[1] in 39 energy differential channels. Left panel of Figure 2 shows that the event occurs on a quiet background and the intensities start to rise sharply at the shock passage, the proton peak being at 22:23 UT. The energy spectrum was derived for the time interval 22:14 - 22:31 UT around the proton peak and the best fit was performed by using the function $dJ/dE = C \times N(E) \times E^{1/2}$, which takes into into account the conversion from the particle density $N(E)$ to the differential flux, where $N(E)$ is the Weibull distribution:

$$N(E) = E^{\beta-1} e^{-(\frac{E}{E_\tau})^\beta} \tag{1}$$

The obtained values for the best-fit parameters are: $C \sim 2 * 10^5 cm^{-2} s^{-1} sr^{-1} MeV^{-1}$, $\beta = [0.50 \pm 0.07]$ and $E_\tau = [95 \pm 50]$ keV. Right panel of Figure 3 shows that the Weibull like function well reproduces the observed energy spectra. A very good agreement can be observed

[1] Available at:
http://www.srl.caltech.edu/STEREO/index.html
http://www2.physik.uni-kiel.de/stereo/browseplots/

between the Weibull shape and the experimental data over the wide energy range 0.3 - 30 MeV spanning around two orders of magnitude. Nevertheless, at lower energies the fit departs from the actual spectrum. On the other hand, the best fit performed by using the Ellison-Ramaty form has the best agreement at lower energies (< 10 MeV).

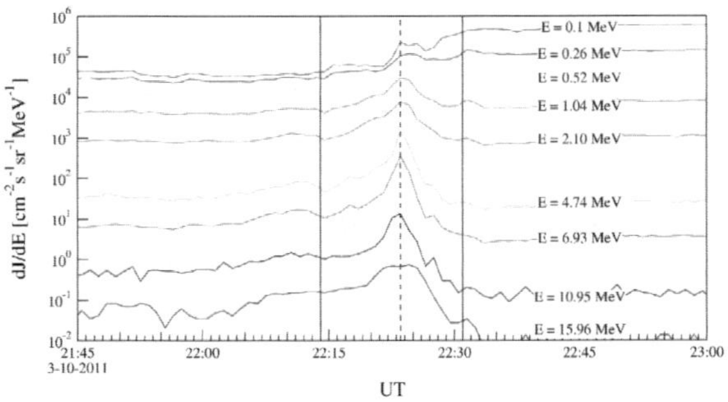

Figure 2. Time history of the proton flux for a selected number of energy channels, as recorded on board the STEREO B spacecraft on 3 October 2011. Black vertical lines delimit the time interval around the shock arrival (dashed line), in which the spectrum (shown in right panel) is obtained.

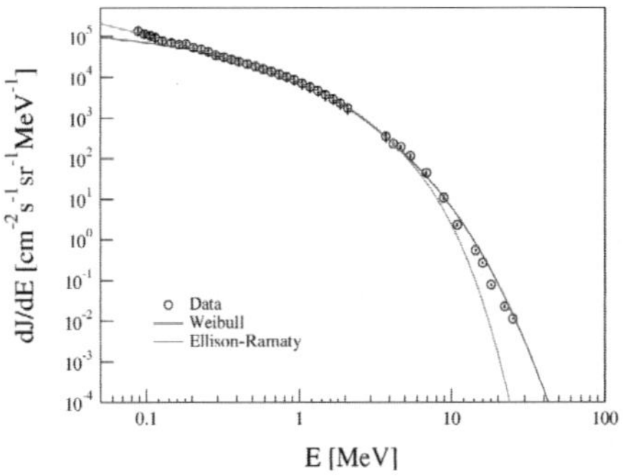

Figure 3. Time averaged spectrum over the time interval 22:14 - 22:31 UT on 3 October 2011. Blue (green) curve is the Weibull (Ellison-Ramaty) function used to fit the spectrum. Data errors are inside the marker size.

3. Derivation of the Weibull distribution

The Weibull distribution can be derived in the framework of stochastic acceleration. As in the majority of astrophysical particle acceleration schemes, we consider a particle population confined in an acceleration region for a characteristic time (leaky box model) and acquire energy through a succession of small increments and can therefore be treated as diffusing and/or being convected through momentum space, along with a spatial diffusion associated with these changes. This can be described through a diffusion loss equation in the presence of energy losses and injection of fresh particles from sources, also including the diffusion of particles in momentum or phase space as described by [27]. Let us assume an isotropic distribution of particles $N(E, \vec{x}, t)$, for which $4\pi^{-1}N(E, \vec{x}, t)\Delta E$ is the number of particles per unit volume and per unit solid angle (at given time t and point $\vec{x}$) with energies in the range E to $E+\Delta E$. The diffusion loss equation at different points in space reads:

$$\frac{\partial N}{\partial t} = D\nabla^2 N + \frac{\partial(b(E)N(E))}{\partial E} + \frac{1}{2}\frac{\partial^2(d(E)N(E))}{\partial E^2} - \frac{N}{\tau} + Q \tag{2}$$

where the five terms on right-hand side account respectively for: spatial diffusion of the particles (D is the scalar diffusion coefficient), the mean drift of the particles ($b(E) = -\frac{dE}{dt}$ represents the average acceleration rate), and the broadening of the energy distribution due to energy space diffusion ($d(E) = \frac{d\langle(\Delta E)^2\rangle}{dt}$), particles leakage from the acceleration region ($\tau(E)$ is a charateristic escape time) and supply of fresh particles from sources $Q(E, \vec{x}, t)$. We can derive a stationary solution of eq. 2 when the spatial diffusion is not considered (i.e., isotropic particle distribution), no particles sources are present and the escape time τ is indipendent from particle energy. Under these conditions, eq. 2 reduces to:

$$\frac{\partial(b(E)N(E))}{\partial E} - \frac{N}{\tau} + \frac{1}{2}\frac{\partial^2(d(E)N(E))}{\partial E^2} = 0 \tag{3}$$

This equation describes the original Fermi's acceleration mechanism (second order) when seeking solutions of power law form [28].

Here we assume that the particles energy increases according to a power law of the time t' as follows:

$$E(t') = E_\tau(\frac{t'}{\tau})^{1/\beta} \qquad E_\tau, \beta > 0 \tag{4}$$

where the constant $E_\tau = E(\tau)$ represents the energy of a particle after a typical time τ; eq. 4 can be written in terms of the dimensionless quantities $\epsilon = \frac{E}{E_\tau}$ and $t = \frac{t'}{\tau}$ as $\epsilon(t) = t^{1/\beta}$. Such assumption is quite natural within the framework of Fermi's acceleration mechanism, where the particle velocity undergoes diffusion due to the stochastic nature of the energy gain. For instance, [29] developed a minimal stochastic model for Fermi's acceleration mechanism and produced an acceleration process characterized by anomalous diffusion in the phase space, taking into account only the general features of the diffusion process in phase space, i.e. that a particle may absorb kinetic energy (accelerate) through collisions against moving scatterers. The relative importance of the broadining of the energy distribution and the mean drift energy gain can be estimated by the ratio R between the first and third term on left-hand side of eq. 3; expressing the time growth of moments as $\langle E^n \rangle \sim t^{q(n)}$, R reads:

$$R = \frac{d(E)}{b(E)} \sim \frac{\langle E^2 \rangle}{\langle E \rangle^2} \sim t^{(\xi-1)2/\beta} \sim \epsilon^{2(\xi-1)} \tag{5}$$

when $\xi = q(2) - q(1)$. In the hypothesis that $q(n)$ is a convex function ($q(2) < q(1)$, $\xi < 0$), the R time decrease is enough rapid that asymptotically ($E >> E_\tau$) the spreading of the energy distribution due to Fermi's acceleration can be neglected with respect to the mean drift term, already on timescale of the order of τ. Note that the required time evolution of the distribution moments was found for the tracer particle radial positions in an anomalous (superdiffusive) transport mechanism in a plasma turbulence [30]. Taking into account the above approximation eq. 3 reduces:

$$N = -\frac{1}{\beta}\frac{\partial(\epsilon^{1-\beta}N)}{\partial\epsilon} \tag{6}$$

A straightforward integration yields the solution:

$$N(\epsilon) = A\epsilon^{\beta-1}e^{-\epsilon^\beta} \tag{7}$$

where A is an integration constant. Therefore the particles are found to be distributed according to a Weibull statistics. Note that in turbulent plasmas the theoretical escape time from the acceleration region due to the spatial transport is a power law in energy, that is $\tau(E) \sim E^{-\gamma}$ [31]. If we assumed such $\tau(E)$, the resulting (softer) spectrum would differ from that in eq. 7 just for the replacement $(E/E_\tau)^\beta \to \beta/(\beta+\delta)(E/E_\tau)^{\beta+\delta}$ in the exponential factor. We verified that in the present case study such a correction to Weibull's spectrum ($\delta = 0$) results to be negligible for energies lower than several tens of MeV. Therefore, in spite of extreme simplicity, a constant escape time proves to be a reasonable approximation.

4. Wavelet ananlysis of the turbulent fluctuations

Let us now see if this scenario is plausible in case the acceleration region is a collissionless shock wave and if that obtained estimates of the parameters are congruent with the physical picture of the ESP event.

If the acceleration region is a collisionless shock, turbulence can provide efficient particle scattering [32, 27, 33] and it can be responsible for momentum diffusion [31, 32, 29, 34, 35]. Hence, we analyzed the variation of the magnetic fluctuations across the shock, by using the high time resolution data (0.125 s) of the magnetic field intensity recorded by the MAG experiment on board STEREO B. Top left panel of Figure 4 depicts the time profile of the interplanetary magnetic field (IMF) intensity.

In order to understand how the turbulence spectrum varies with time, we applied the wavelet analysis (WA, [36]) to the IMF intensity, as the WA decomposes a time series into the time-frequency (t - f) space. The middle panel of Figure 4 shows the Wavelet Power Spectrum $|W(f,t)|^2$ (WPS) of the observed IMF intensity, which indicates that turbulent fluctuations are especially enhanced at the shock front at all frequencies. The variance ($\sigma^2 = \int_{f_2}^{f_1} |W(f,t)|^2$) is represented in the bottom panel of Figure 4, where it can be observed an increase of the magnetic energy at the shock, followed by a second peak still in the region (delimited by the vertical lines) of the particle peak flux. Note that the σ time profile is similar to the one of the particle flux as observed for the high energy channels (see left panel of Figure 2).

If we consider the Weibull parameters obtained through the fit (β=0.5 and E_τ=0.95 keV) and assume the scheme presented in section 3 to be valid, we expect from eq. 4 that the greatest observed particle energy of 30 MeV is reached after a time T_{high}=18τ. In case of nearly local acceleration, T_{high} must be of order of the time width of particle enhancement, that, in present case, is about 20 min, implying an upper limit for $\tau \sim 1$ min. Hence, we investigate the turbulence spectrum across the shock only at small temporal scales (under $2\tau = 2$ min), which are smaller than the confinement time and thus are the important ones for particle acceleration. Therefore we detrended the total field intensity by using the Empirical Mode Decomposition (EMD) technique [37], through which a time series $X(t)$ is decomposed into a finite number m of oscillating modes, each of them having their own time-scale and represents a zero mean oscillation experiencing amplitude and frequency modulations. The signal can be partially reconstructed in a selected time range (in the present case < 2 min) through partial sums of a proper number of modes. The magnetic field intensity reconstructed at time scales < 2 min is shown in the top right panel of Figure 5.

Then we repeated the WA on the signal of magnetic field intensity reconstructed at time scales < 2 min, to determine the properties of the turbulence, such as its WPS (middle right panel of Figure 5) and its variance (bottom right panel of Figure 5), in the vicinity of the shock. An enhanced turbulent energy is observed in the downstream region. Moreover, a relationship between the variation of the energetic particle intensity (see left panel of Figure 2) and the variation of the magnetic fluctuations across the shock, is apparent at < 2 min scales.

5. Conclusions

The energy spectrum obtained for the 3 October 2011 ESP event was found to be reproduced by the Weibull functional form, as previously found [20, 21] in case of several SEP events, as well as in their ESP phase, at the Earth' s shock passage, and at forward shocks of two corotating interaction regions. This result confirms that the Weibull distribution can be associated with shock acceleration. We propose a theoretical derivation of the Weibull spectrum, involving the effect of stochastic acceleration at shock waves, in case of momentum anomalous diffusion. This scenario seems to be also consistent with the observation of enhanced magnetic fluctuations (as expected [38, 39, 40, 41]) around the shock front, which is assumed to be the particle acceleration region. First we note that magnetic irregularities (hydromagnetic waves, turbulence) in a moving fluid background, can produce isotropy of the particle distribution through the particle-wave interaction as well as momentum superdiffusion, so that the energy of the turbulent field in the downstream and/or the upstream regions could be efficiently transferred into particle energy through the second order Fermi acceleration. In addition, a second order acceleration process acting in the vicinity of the shock wave has been shown to be efficient [42] both in terms of energy budget and acceleration time scale, due to the presence of high amplitude MHD turbulence. Finally, [35] proposed that due to efficient momentum diffusion of particles in the

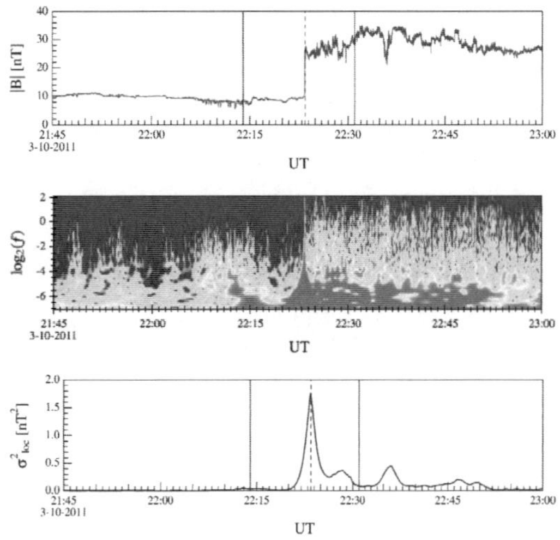

Figure 4. Time history of the magnetic field intensity (top). Wavelet power spectrum (middle) and variance (bottom) of the magnetic field intensity.

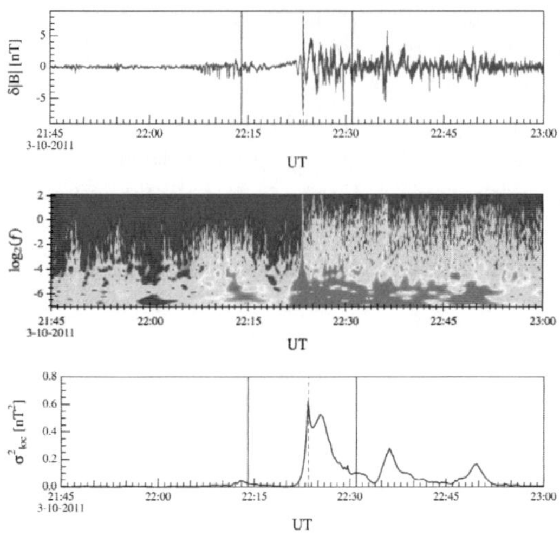

Figure 5. Time history of the magnetic field intensity at scales < 2 min (top). WPS (middle) and variance (bottom) of the magnetic field intensity at scales < 2 min.

downstream region of the shock, the acceleration is dominated by the second order acceleration mechanism and the particle spectra become flatter than in the original treatment of diffusive shock acceleration. Accordingly, the derived accelerated particle distribution can be expressed as a Weibull function with $\beta < 1$, i. e. having a slower dacay with respect to the power law modulated by an exponetial function, proposed by Ellison and Ramaty to reproduce the roll over at high energies in the frame of the diffusive shock acceleration. In particular, the considered ESP spectrum is better reproduced by the Weibull function at high energies (greater than E_T), and by the Ellison-Ramaty one at lower energies.

Both the spectrum and the flux time profile of the considered ESP event suggest that diffusive shock acceleration and stochastic acceleration should be taken into account simultaneously. First, we note that the acceleration time scale (< 2 min) derived from our interpretation of the Weibull parameters is comparable with that expected from DSA. Moreover, the energetic particle flux profiles indicate that they are diffusing across the shock. Therefore, it is natural to assume they experience energy gain by interaction with the shock as well. In particular, the particle profiles at low energy are consistent with those expected from DSA, while at high energies seem to be related to the variations of the magnetic energy, represented by the variance of the power spectrum of the magnetic field intensity. Hence, stochastic acceleration seems to give a significant contribution at high energies.

We conclude that the stochastic acceleration of protons in the downstream region of a near-perpendicular shock can be an additional mechanism for producing the observed particle energy spectra associated with interplanetary shocks. We suggest that stochastic acceleration may act as a re-acceleration of energetic protons by enhanced turbulence in the downstream region of the shock, as proposed by [19] for relativistic energies in case of GLE events. Further studies about the microphysical processes in the shock region are needed, e.g., the turbulence level, the nature and geometry of small-scale dissipative structures of turbulence, which can be complicated and possibly fractal, and how they define the particle diffusion regimes [44].

References

[1] Reames D V 1999 *Space Sci. Rev.* **90** 413
[2] Armstrong T P, Krimigis S M and Behannon K W 1970, *J. Geophys. Res.* **75**, 5980
[3] Sarris E T, Krimigis S M and Armstrong T P 1976 *Geophys. Res. Lett.* **3** 133
[4] Gosling J T, Asbridge J R, Bame S J et al. 1981 *J. Geophys. Res.* **86** 547
[5] Tsurutani B T, Smith E J, Pyle K R and Simpson J A 1982 *J. Geophys. Res.* **87** 7389
[6] Pesses M E, Decker R B and Armstrong T P 1982 *Space Sci. Rev.* **32** 185
[7] Barnes C W and Simpson J A 1976 *Astrophys. J.* **210** L91
[8] McDonald F B, Teegarden B J, Trainor J H and von Rosevinge T T 1975 *Astrophys. J.* **203** L149
[9] Krymskii G F 1977 *Akademiia Nauk SSSR Doklady* **234** 1306
[10] Blandford R D and Ostriker J P 1978 *Astrophys. J. Lett.* **221** L29
[11] Lee M A 1983 *J. Geophys. Res.* **88** 6109
[12] Lario D, Ho G C, Decker, R B, Roelof E C, Desai M I and Smith C W 2003, in American Inst. Phys. Conf. Series, Vol. 679, Solar Wind Ten, ed. M. Velli, R. Bruno *et al.*, 640–643
[13] Giacalone J 2012 *Astrophys. J.* **761**:28 (12 pp)
[14] Ellison D C and Ramaty R 1985 *Astrophys. J.* **298** 400
[15] McGuire R E and von Rosenvinge, T T 1984 *Adv. Space Res.* **4** 117
[16] Tylka A J, Cohen C M S, Dietrich W F, Lee M A, Maclennan C G, Mewaldt R A, Ng C K and Reames D V 2005 *Astrophys. J.* **625** 474
[17] Band D., Matteson J, Ford L, Schaefer B., Palmer D, Teegarden, B, Cline T, Briggs M, Paciesas W, Pendleton G, Fishman G, Kouveliotou C, Meegan C, Wilson R and Lestrade P 1993 *Astrophys. J.* **413** 281
[18] Mewaldt R A, Cohen C M S, Labrador A W, Leske R A, Mason G M, Desai M I, Looper M D, Mazur J E, Selesnick R S and Haggerty D K 2005 *J. Geophys. Res.* **110** A09S18
[19] Afanasiev A, Vainio R, and Kocharov L 2014 *Astrophys. J.* **790** 36
[20] Laurenza M, Consolini G, Storini M and Damiani A 2013, in American Inst. Phys. Conf. Series, Vol. 1539, American Inst. Phys. Conf. Series, ed. G. P. Zank, J. Borovsky, *et al.*, 219–222
[21] Laurenza M, Consolini G, Storini M and Damiani A 2015, J. Phys. Conf. Series 632 012066
[22] Weibull W 1951 *J. Appl. Mech. - Trans. ASME* **18** 293
[23] Frisch U and Sornette D 1997 *J. Phys.* I **7** 1155
[24] Müller-Mellin R et al 2008 *Space Sci. Rev.* **136** 363
[25] Mewaldt R A et al 2008 *Space Sci. Rev.* **136** 285
[26] Von Rosevinge T T et al 2008 *Space Sci. Rev.* **136** 391
[27] Blandford R and Eichler D 1987 *Phys. Rew.* **154** 1
[28] Longair M S 1994 High energy astrophysics. Vol.2: Stars, the galaxy and the interstellar medium, ed. Cambridge: Cambridge University Press —c1994, 2nd ed.
[29] Bouchet F, Cecconi F and Vulpiani A. 2004 *Phys. Rev. Lett.* **92** 040601
[30] Carreras B A, Lynch V E and Zaslavsky G M 2001 *Phys. Plasmas* **8** 5096
[31] Fedorov Y, Shakhov B and Stehlik M 2012, *J. Phys. B: At. Mol. Opt. Phys.* **45** 165702
[32] Tverskoĭ B A 1968 *JETP* **26** 821
[33] Bykov A M, Ellison D C, Osipov S M and Vladimirov A E 2014 *Astrophys. J.* **789** 137
[34] Petrosian V 2012 *Space Sci. Rev.* **173** 535
[35] Schlickeiser R and Achat U 1993 *J. Plasma Phys.* **50** 85
[36] Torrence C and G. P. Compo 1998 *Bull. Am. Meteorol. Soc.* **79(1)** 61
[37] Huang N.E., Shen Z, Long S R et al. 1998, Royal Society of London Proc. Series A, 454, 903
[38] Kennel C F, Coroniti F V, Scarf F L, Smith E J and Gurnett D A 1982 *J. Geophys. Res.* **87** 17
[39] Zank G P, Li G, Florinski V et al. 2006 *J. Geophys. Res.* A **111** 6108
[40] Hu Q, Ao X, Peltzer R and Zank G P 2012, in American Inst. Phys. Conf. Series, Vol. 1500, American Inst. Phys. Conf. Series, ed. Q. Hu, G. Li *J. Geophys. Res.*, 192–197
[41] Hu Q, Zank G P, Li G, and Ao X 2013, in American Inst. Phys. Conf. Series, Vol. 1539, American Inst.Phys. Conf. Series, ed. G. P. Zank, J. Borovsky *et al.* 175–178
[42] Ostrowski M. 1994 *Astron. Astrophys.* **283** 344
[43] Ostrowski M and Schlickeiser R 1993 *Astron. Astrophys.* **268** 812
[44] Verkhoglyadova O P and Le Roux J A 2005 *J. Geophys. Res.* A **110** 10

Visualizing the three-dimensional structure of the local interstellar medium and possible physical causes for this structure

Jeffrey L. Linsky[1], **Seth Redfield**[2], **and Max Schwarz**[3]

[1] JILA, University of Colorado and NIST, Boulder CO, USA 80309-0440

[2] Astronomy Department and Van Vleck Observatory, Wesleyan University, Middletown CT, USA 06459-0001

[3] APS Department, University of Colorado, Boulder CO, USA 80309-0391

E-mail: jlinsky@jila.colorado.edu

Abstract. Analysis of high-resolution interstellar absorption lines observed in the ultraviolet spectra of nearby stars provides the basis for identifying 15 partially ionized warm "clouds" of interstellar gas located within a few parsecs of the Sun. We show visualizations of these clouds as seen from different directions. The presence of discrete clouds implies one or more physical processes that establish these structures and the presence of gas with very different properties between these clouds. While classical models of the interstellar medium based on the assumptions of pressure and ionization equilibrium in a nonmagnetic medium are not consistent with observations, dynamical simulations without these assumptions are likely more realistic. Possible physical processes that could structure the clouds include ram pressure, magnetic fields, and photoionization by extreme-ultraviolet radiation. We describe the available evidence concerning these three processes.

1. The many connections between the LISM, the heliosphere, and the Galactic interstellar medium

The local interstellar medium (LISM) is a collection of partially ionized gas clouds located within a few parsecs of the Sun. Each of these clouds is identified by a common velocity vector inferred from interstellar absorption-line radial velocities measured in high-resolution spectra towards stars covering a broad region of the sky. Redfield & Linsky (2008) [1] used HST ultraviolet spectra of 157 stars to identify 15 such clouds, each located within 0–15 pc of the Sun. Each cloud is located either in front of or surrounding the nearest star that shows interstellar absorption lines with a radial velocity consistent with the cloud's vector velocity. Observations of interstellar absorption towards 40 additional stars [2] confirmed the existence and location of these clouds. Gry and Jenkins (2014) [3] subsequently proposed that these 15 clouds are actually a single cloud with a complex velocity structure. However, Redfield et al. (2015) [4] made the case for discrete clouds on the basis of a better fit to the observed radial velocities for the new data set of 40 observed stars and the presence of quasar-scattering sites at the edges of the clouds [5]. For comprehensive reviews of LISM science and the interaction between the solar wind and the LISM, see [6,7].

Journal of Physics: Conference Series **767** (2016) 012016 doi:10.1088/1742-6596/767/1/012016

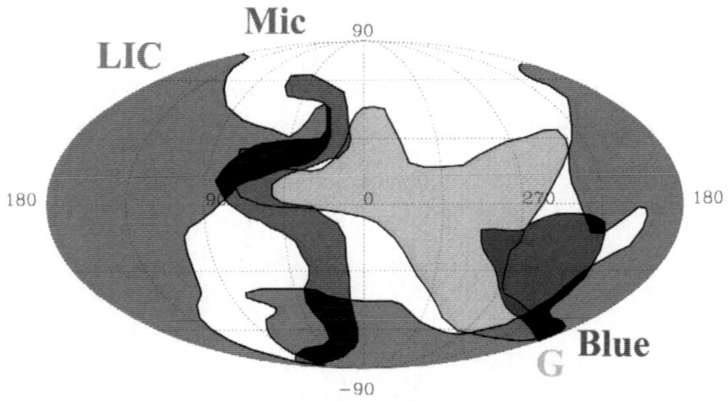

Figure 1. The four closest LISM clouds (LIC, G, Blue, and Mic) to the heliosphere plotted in Galactic coordinates. The Galactic Center is at the center of the plot. Figure from [1].

Understanding the structure and properties of the LISM clouds is important, because the LISM serves as an interface between the heliosphere and the Galactic interstellar medium, providing physical insights concerning both. The Sun is located barely inside the Local Interstellar Cloud (LIC) (see Figure 1), and the motion of the Sun through space will take the heliosphere outside of the LIC and into either: (a) the adjacent G cloud, (b) an interface between the LIC and G clouds, or (c) the highly ionized gas that may lie between the LISM clouds. This transition will occur in less than 4000 years [8], perhaps far less. Müller et al. (2006) [8] calculated the location of the termination shock, heliopause, and bow shock for a wide range of interstellar densities, temperatures, and flow velocities that the heliosphere would see during its past and future trajectory through the inhomogeneous LISM. For example, the passage of the heliosphere through a cloud with a neutral hydrogen density of 11 cm^{-3} rather than the present value of about 0.2 cm^{-3} would shrink the termination shock from about 90 AU in the upwind direction to only 14 AU. Since the LISM clouds are located nearby, our understanding of their properties will provide insight into the properties of partially ionized gas in more distant regions of the Galaxy. The ISM plays a critical role in the chemical evolution of the Galaxy with enrichment from supernova explosions and stellar mass loss and the subsequent formation of a new stars from the chemically enriched interstellar gas.

The physical properties and evolution of interstellar gas have been the subject of numerous theoretical and simulation studies. The now classic theoretical models [9], [10], and [11] postulate a steady-state interstellar medium in pressure and thermal equilibrium. These models predict that the ISM includes two or three thermal components, including a warm component with a temperature of about 8,000 K and density about 0.1 cm^{-3} that is similar to the properties of the LIC [6]. De Avillez & Breitschwerdt (2005) [12] proposed a very different model for the ISM on the basis of their three-dimensional MHD simulations. In these simulations, the energy source is random supernova events resulting in an ISM that is highly turbulent with no stable thermal phases and a wide range of plasma temperatures, densities, pressures, and magnetic field strengths. They find that, on average, magnetic pressure dominates for low temperature gas ($T < 200$ K), ram pressure dominates for plasma at temperatures 200–10^6 K, and thermal pressure dominates at $T > 10^6$ K. Do the observed physical properties of the LISM support the theoretical or simulations models? Which of these pressure terms dominates in the LISM, or is there another process that controls the properties, in particular the structure, of the LISM clouds?

2. Visualizing the LISM in three dimensions

Our analysis of interstellar absorption lines [1] provided neutral hydrogen column densities N(H I) and cloud identifications along the lines of sight to many nearby stars with known distances.

To visualize the spatial layout of the clouds in three dimensions requires several assumptions. First, we assume that the neutral hydrogen density in all of the clouds is the same as for the LIC, n(HI)=0.20 cm^{-3} [6]. Second, we assume that along each line of sight the cloud is centered at the midpoint to the star, and its thickness is $\Delta d = N(H\ I)/n(H\ I)$. We require that stars with observed astrospheres are located within one of the partially ionized clouds and that clouds do not overlap. Finally, we smooth the cloud surfaces.

Our simplifying assumptions should be tested when sufficient data become available. A major concern is that the neutral hydrogen number density could vary considerably from cloud to cloud. There is no obvious way of measuring the neutral hydrogen number density directly. Density-sensitive line ratios can measure electron densities, but these are a function of the total hydrogen number density and the ionization state in the clouds. The placement of clouds along each line of sight and the fuzziness of the cloud surfaces are interesting but not critical to LISM morphology. Progress in understanding cloud morphology would benefit from analysis of many more lines of sight and the inclusion of physical constraints such as the location of ionization zones (e.g., Stromgren spheres) set by the EUV radiation field and the locations of interstellar shock fronts and strong magnetic fields.

The video clip located at http://jila.colorado.edu/~jlinsky/Interstellar_Clouds.mp4 (or .wmv) shows the clouds in three dimensions as seen from various directions in the Galactic plane.

3. Potential causes for the structure of partially ionized clouds

We now consider three physical mechanisms that could shape the clouds in the LISM.

3.1. Ram pressure and intercloud collisions

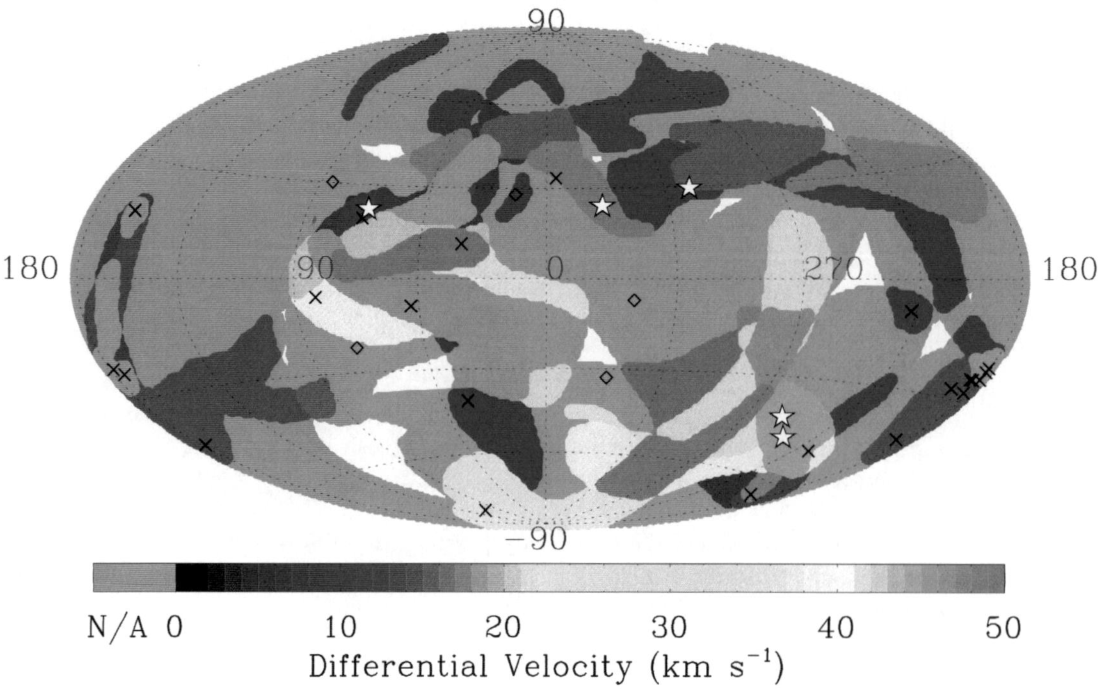

Figure 2. Zones of LISM cloud interactions. Colored regions indicate directions in which multiple LISM clouds are detected along the line of sight. Figure from [5].

The simulations [12] predict that ram pressure should be the dominant pressure term that controls the structures of warm interstellar gas such as the LISM clouds. Figure 2 shows the velocity differences between two or more clouds along the same line of sight. The velocity differences are supersonic for many lines of sight, but we do not know whether the clouds along a given line of sight are in contact, in which case velocity differences greater than about 8 km/s would indicate shocks, or whether the clouds are not in contact, and there are no shocks. In either case, the figure provides support for the presence of large velocity differences in the ISM and thus potentially large ram pressures. A second argument for the importance of ram pressure in the LISM is shown in Figure 1. A considerable portion of the filamentary Mic cloud lies in the line of sight between the LIC and G clouds. Since the LIC and G clouds are approaching each other at a speed of 5.5 km s^{-1} and the LIC has the hottest gas ($T = 9900$ K) of the 15 clouds [1], the LIC and G clouds may be compressing and thus heating the Mic cloud between them. This possible interaction supports the hypothesis that ram pressure or intercloud collisions stucture the LISM clouds. However, we have no independent evidence that the Mic cloud actually lies between the LIC and G clouds rather than being simply more distant. The Mic cloud could be anywhere along the 5.1 pc line of sight to the nearest star α Aql behind the cloud.

As a test of whether ram pressure controls the structure of the LISM clouds, we computed the cloud shapes as seen from the upwind direction of the interstellar flow vector (l=155°, b=7°). Figure 3 shows no evidence for cloud elongations along the flow direction. We then plotted the cloud shapes perpendicular to the flow to search for elongations produced by the flow. None of the clouds showed thin structures aligned perpendicular to the flow.

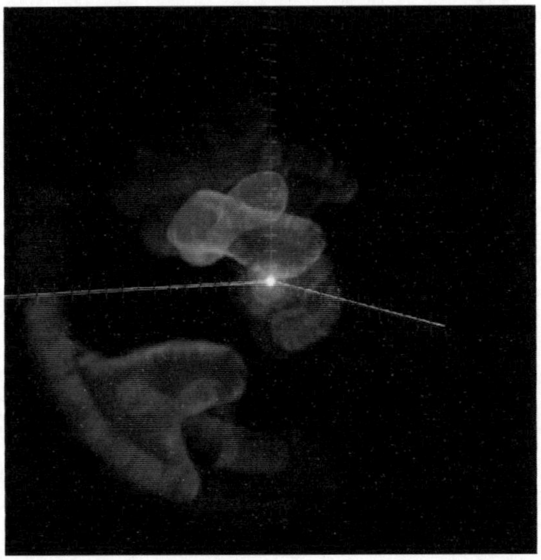

Figure 3. The LISM clouds as viewed from the upwind direction of the interstellar flow.

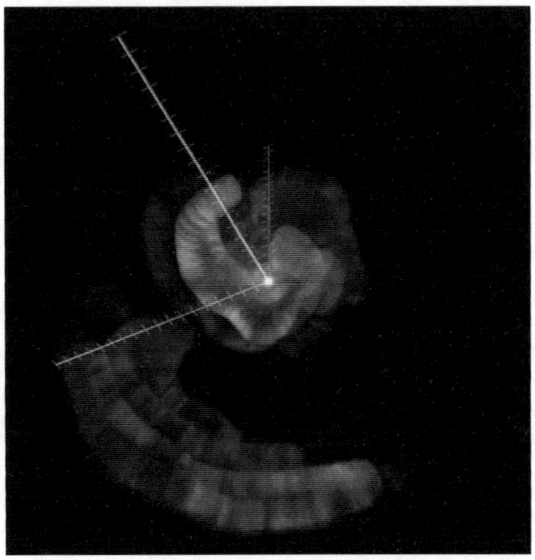

Figure 4. The LISM clouds as viewed from the magnetic field direction.

3.2. Magnetic fields

The Voyager 1 spacecraft has been measuring magnetic fields in the heliopause, and the IBEX spacecraft has observed energetic neutral atoms (ENAs) with higher fluxes along a ribbon now referred to as the "IBEX ribbon". Heerikhuisen et al. (2010) [13] proposed and Zirnstein et al. (2016) [14] refined a model for the ribbon in which the ribbon's shape as a function of ENA energy is produced by the interstellar magnetic field draped around the heliopause. In their model the extrapolated-interstellar magnetic field away from the heliosphere is 2.93 ± 0.08 μGauss. This magnetic field strength is consistent with Voyager 1 observations [15] and evidence from interstellar polarization [6]. For this value of the interstellar magnetic field, the magnetic pressure, $P_{\rm mag}/k = 2480 \pm 130$. This pressure can be compared with the LISM cloud thermal pressures, $P_{\rm th}/k = 2050 - 5030$ [6]. The plasma $\beta = P_{\rm th}/P_{\rm mag} = 0.5 - 1.2$, which suggests that magnetic fields may play a role in shaping the LISM clouds.

As a test of whether magnetic pressure is the pressure term controlling the structure of the LISM clouds, we computed the cloud shapes as seen from the direction of the interstellar magnetic field (l=206°, b=-50°) [14]. Figure 4 shows no evidence for cloud elongations along the magnetic field direction. We then plotted the cloud shapes perpendicular to the magnetic field vector to search for elongations produced by the magnetic field, but none of the clouds showed thin structures perpendicular to the magnetic field.

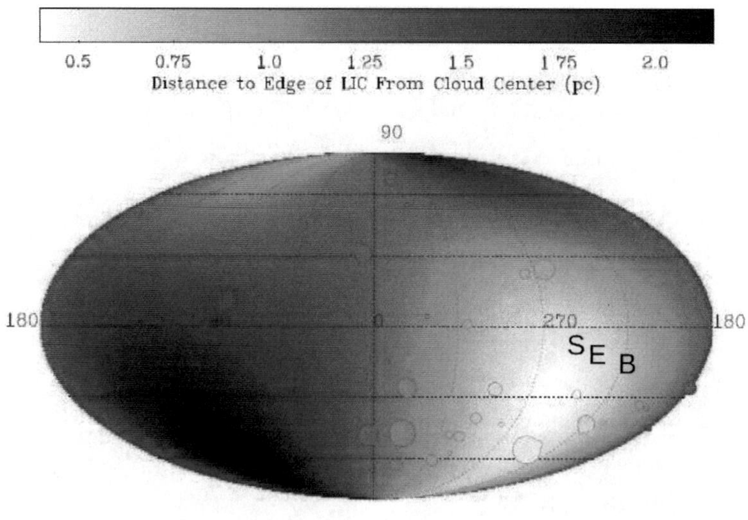

Figure 5. Distances from the geometric center of the LIC to its edge and the line of sight directions to three strong EUV radiation sources: ϵ CMa (E), β CMa (B), and Sirius B (S).

3.3. Photoionization by EUV radiation

The Extreme Ultraviolet Explorer (EUVE) spacecraft, which observed the sky in the 70–730 Å EUV spectral range, detected many stars as sources of EUV radiation. In particular, the B1 V star ϵ CMa is by far the brightest nonsolar EUV source despite its distance of 124 pc. Besides ϵ CMa, there are four other important EUV sources [16], β CMa and three hot white dwarfs

(G191-B2B, HZ 43, and Feige 24). Figure 5 shows the distances from the geometric center of the LIC to its edge in Galactic coordinates. Centered near l=270°, b=-30° is a region of minimal distance from cloud center to edge with ϵ CMa, β CMa, and the nearby white dwarf Sirius B located in this region. The apparent depletion of neutral hydrogen suggests that EUV radiation from ϵ CMa and to a lesser extent the other sources are the photoionizing sources.

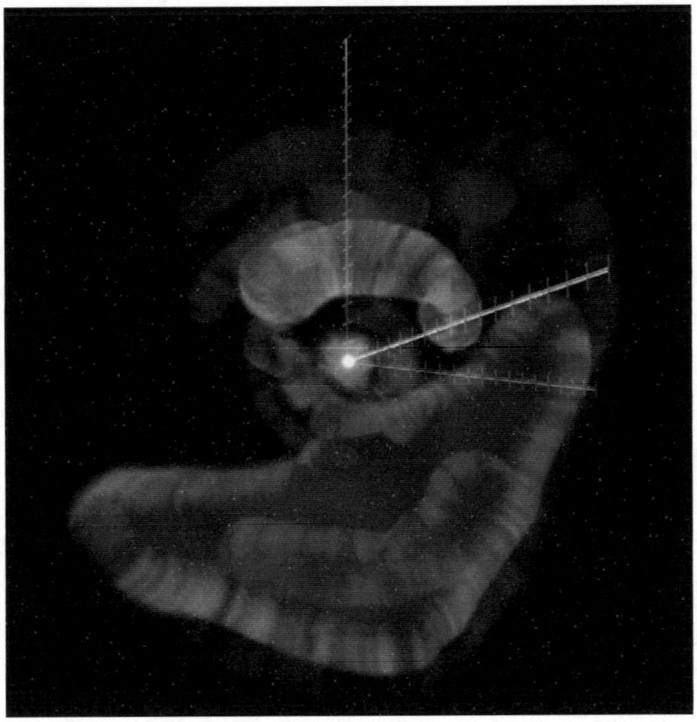

Figure 6. The LISM clouds as viewed from ϵ CMa.

EUV photons penetrate to a 1/e depth d=1/n(H I)σ, where $\sigma = 1 \times 10^{-17}$ cm^2. For a density of 0.2, hydrogen atoms per cm^{-3}, the 1/e penetration depth is d=0.2 pc and essentially all of the EUV photons will be absorbed by 0.6 pc. Since the mean radius of the LIC is about 1.5 pc, the decrease in radius from the center of the LIC shown in Figure 5 is consistent with photoionization by EUV radiation from ϵ CMa.

As a test of whether EUV photons from ϵ CMa shape other LISM clouds, we computed the cloud shapes as seen from the direction of ϵ CMa (l=59°, b=11°). We note that no other clouds lie in the path from ϵ CMa to the LIC that could shield the LIC. This is consistent with the "interstellar tunnel" proposed by [17] and [18] to explain the strong flux measured by EUVE. Figure 6 shows no evidence for cloud elongations along the ϵ CMa direction. We then plotted the cloud shapes perpendicular to the line of sight from ϵ CMa to search for elongations produced by EUV photoionization. We found four clouds with alignments that are perpendicular to the line of sight from ϵ CMa. These are shown in Figures 7 and 8 (for the Aur and Hyades clouds) and Figures 9 and 10 (for the NGP and Cet clouds).

4. Conclusions and perspectives

We have presented a three dimensional model for the location and structure of the 15 partially ionized LISM clouds [1]. Using this model, we have tested whether ram pressure, magnetic fields, or EUV radiation are the most likely physical process that shapes these clouds. We find that EUV radiation from the brightest nonsolar source, ϵ CMa, is the most likely source for shaping the LIC and four other LISM clouds.

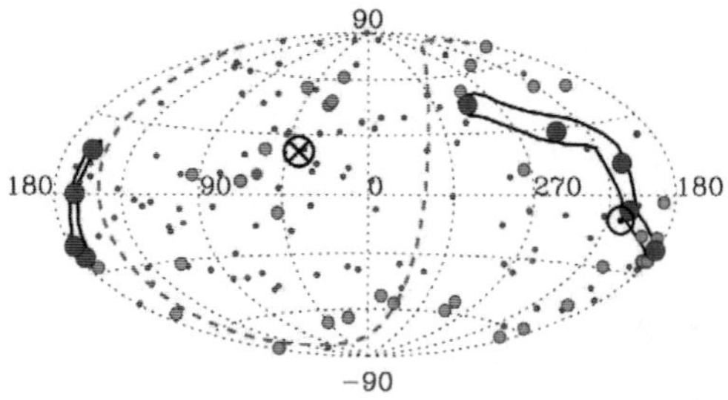

Figure 7. Boundary of the Auriga cloud (solid line) and 90° from the ϵCMa line of sight (dashed line).

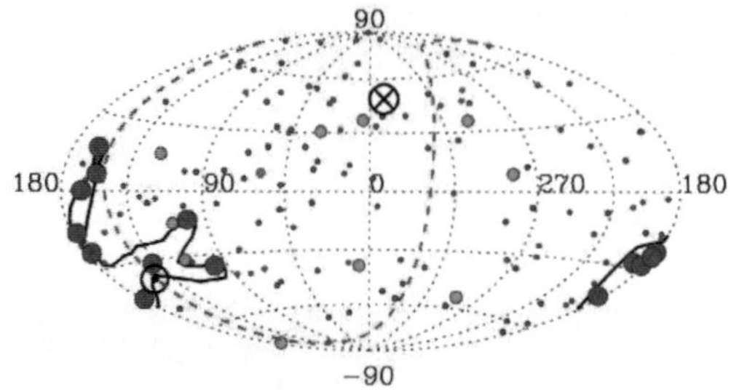

Figure 8. Boundary of the Hyades cloud (solid line) and 90° from the ϵCMa line of sight (dashed line).

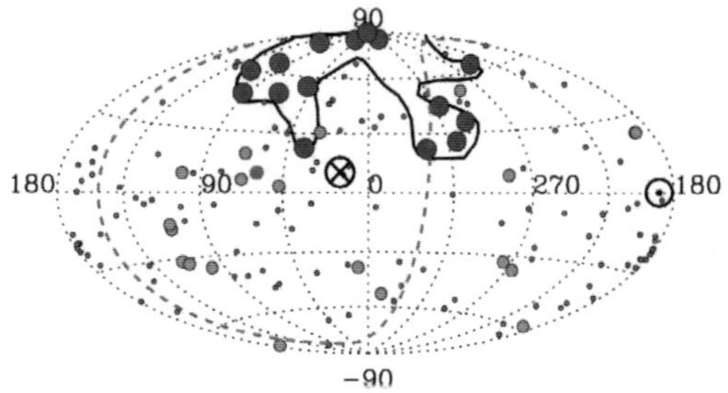

Figure 9. Boundary of the NGP cloud (solid line) and 90° from the ϵCMa line of sight (dashed line).

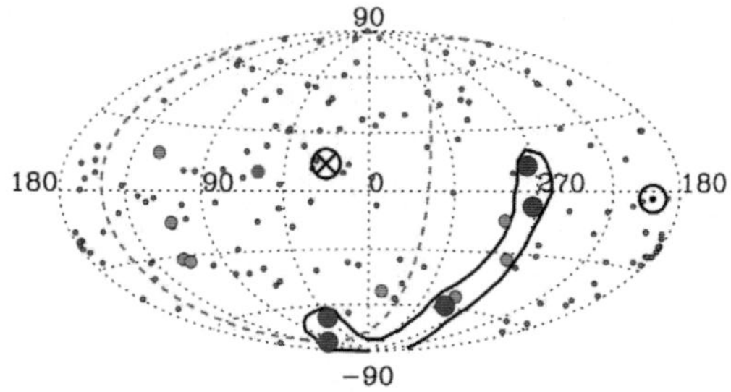

Figure 10. Boundary of the Cet cloud (solid line) and 90° from the ϵCMa line of sight (dashed line).

5. Acknowledgements

We acknowledge support through the NASA HST Grant GO-11568 from the Space Telescope Science Institute, which is operated by the Association of Universities for Research in Astronomy, Inc. for NASA, under contract NAS 5-26555. We thank Steven Burrows (JILA) for creating the visual clip and Figures 3, 4, and 6. We also thank the referee for helpful suggestions.

6. References

[1] Redfield S and Linsky J L 2008 *Astrophys. J.* **673** 283
[2] Malamut C, Redfield S, Linsky, J L, Wood B E and Ayres T R 2014 *Astrophys. J.* **787** 75
[3] Gry C and Jenkins E B 2014 *Astron. Astrophys.* **567** A58
[4] Redfield S and Linsky J L 2015 *Astrophys. J.* **822** 125
[5] Linsky J L, Rickett, B J and Redfield S 2008 *Astrophys. J.* **675** 413
[6] Frisch P C, Redfield S and Slavin J D 2011 *Ann. Rev. Astron. Astrophys.* **49** 237
[7] Zank G P 2015 *Ann. Rev. Astron. Astrophys.* **53** 449
[8] Müller H-R, Frisch P C, Florinski V, and Zank G P 2006 *Astrophys. J.* **647** 1491
[9] Field G B, Goldsmith D W and Habing H J 1969 *Astrophys. J.* **155** L149
[10] McKee C F and Ostriker J P 1977 *Astrophys. J.* **218** 148
[11] Wolfire M, McKee C, Hollenbach D and Tielens A 1995 *Astrophys. J.* **453** 673
[12] de Avillez, M A and Breitschwerdt, D 2005 *Astron. Astrophys.* **436** 585
[13] Heerikhuisen J, Pogorelov N V, Zank G P, et al. 2010 *Astrophys. J. Letters* **708** L126
[14] Zirnstein E J, Heerikhuisen J, Funstein, H O, Livadiotis, G. McComas, D J, and Pogorelov N V 2016 *Astrophys. J.* **818** L18
[15] Zank G P, Heerikhuisen J., Wood B E, et al. 2013 *Astrophys. J.* **763** 20
[16] Vallerga J V 1998 *Astrophys. J.* **497** 921

[17] Welsh B Y 1991 *Astrophys. J.* **373** 556
[18] Welsh B Y, Wheatley J, Dickinson N J and Barstow M A 2013 *Pub. Astron. Soc. Pacific* **125** 644

Journal of Physics: Conference Series **767** (2016) 012017 doi:10.1088/1742-6596/767/1/012017

Interstellar flow longitude from pickup ion cut-off observations at 1 AU with STEREO and ACE

E Möbius[1], M A Lee[1], G Gloeckler[2], C Drews[3], D Keilbach[3]

[1] Space Science Center & Department of Physics, University of New Hampshire, 8 College Road, Durham, NH 03824, U.S.A.
[2] Department of Atmospheric, Oceanic & Space Sciences, University of Michigan, 2455 Hayward Street, Ann Arbor, MI 48109, U.S.A.
[2] Institut für Experimentelle und Angewandte Physik, Christian-Albrecht-Universität, Leibnizstraße 11, 24118 Kiel, Germany

eberhard.moebius@unh.edu

Abstract. A precision determination of the interstellar neutral (ISN) flow direction is important in several ways. As a cardinal axis of the heliosphere it has strong leverage on the plane subtended by the ISN velocity and interstellar magnetic field vector, which controls the shape of the heliosphere and its interaction with the interstellar medium. Continuing observations of the ISN flow through the heliosphere for several decades allow the search for potential temporal variations of the ISN flow and comparison with astronomical observations. Recent efforts to obtain a consistent ISN vector and temperature with Ulysses and IBEX neutral gas observations point to remaining uncertainties and potential systematic effects. In particular, IBEX measurements provide a very precise relation between ISN flow longitude and speed via the hyperbolic trajectory equation, but they contain larger uncertainties along the parameter tube defined by this relation. The pickup ion (PUI) cut-off variation with ecliptic longitude at 1 AU can provide a complementary determination of the ISN flow longitude with high precision. We compare STEREO PLASTIC and ACE SWICS observations with a simple analytical model of the cut-off. We perform a Pearson correlation analysis of the cut-off as a function of ecliptic longitude with its mirrored function and obtain the symmetry axis with a statistical uncertainty <0.1°. Here we test variations of this value due to Poisson fluctuations in the original data with simulations and due to systematic effects with multi year and location ACE SWICS and STEREO PLASTIC samples.

1. Introduction

The interstellar neutral (ISN) velocity vector $\boldsymbol{V}_{ISN\infty}$ relative to the Sun is one of the key quantities that control the global interaction between the interstellar medium and the heliosphere. Together with the interstellar magnetic field $\boldsymbol{B}_{IS}$ it defines the $\boldsymbol{B}_{IS}$ - $\boldsymbol{V}_{ISN\infty}$ plane, which determines the shape of, and the ISN flow deflection in, the outer heliosheath [1, 2]. $\boldsymbol{B}_{IS}$ is deduced from the IBEX ribbon, consistent with the orientation based on the heliospheric asymmetry [3, 4, 5] and TeV cosmic ray anisotropy [6]. After a long history of studying the local interstellar parameters with UV backscatter observations [7, 8], pickup ions (PUIs) [9, 10], and finally direct observations of ISN He with Ulysses GAS [11], the Interstellar Boundary Explorer (IBEX) provides a more complete view of the H, He, O, and Ne ISN flow [12]. A precise ISN flow vector taken over an extended time of more than one decade will also enable detection of small temporal variations in the ISN flow or place tight upper limits on them, and

Journal of Physics: Conference Series **767** (2016) 012017 doi:10.1088/1742-6596/767/1/012017

thus resolve a current debate [13, 14, 15]. In addition, such local measurements over time, in comparison with astronomical absorption line observations of nearby interstellar clouds [16], will facilitate the identification of local structures, such as turbulent eddies, in the local interstellar cloud.

IBEX returns a very precise relation between ISN flow longitude $\lambda_{ISN\infty}$ and $V_{ISN\infty}$ via the hyperbolic trajectory equation, but leaves a much larger uncertainty separately for $\lambda_{ISN\infty}$ and $V_{ISN\infty}$. These results define a 4-dimensional parameter tube [17] connecting flow longitude, latitude, speed, and temperature, which have been consolidated recently [18, 19, 20, 21]. Related values and uncertainties for $\lambda_{ISN\infty}$ are 75.9±0.5° [18] and 75.5±1.4° [21], respectively, in agreement with most recent analyses of Ulysses GAS data that have led to $\lambda_{ISN\infty}$ = 75.3-1.1+1.2° [22] and 75.54±0.19° [23]. However, these results lead to a much higher ISN temperature than deduced previously in [24]. While all results now agree within their mutual uncertainties, the best-fit values still vary noticeably. An independent determination of $\lambda_{ISN\infty}$ will substantially tighten the flow vector in combination with further narrowing the IBEX parameter tube through a growing database. The analysis of the variation of the pickup ion (PUI) energy cut-off as a function of ecliptic longitude, which reflects the ISN flow pattern at 1 AU, has recently been suggested as the appropriate precision method [25].

Reporting on this new PUI cut-off method is a very appropriate contribution for the celebration of Ed Stone's 80th birthday. This topic touches upon two of his most remarkable achievements, the Voyager interstellar mission [26], which continues to pose challenges to our understanding of the connection between the interstellar flow and magnetic field at the heliospheric boundary, and the ACE mission [27], which has provided the legacy PUI data set with the SWICS sensor [28] that is being used together with continuing STEREO PLASTIC [29] observations to test for systematic effects and to evaluate long term variations in the ISN flow direction.

In this paper, we will take a first look at the effects that statistical and systematic uncertainties have on the ISN flow longitude as deduced from the PUI cut-off. We will start by briefly laying out the method in Section 2. Will evaluate in Section 3 how Poissonian noise added to an artificial STEREO PLASTIC-like dataset affects the statistical uncertainty of the result. In Section 4 we will analyze several individual annual data sets of ACE SWICS and STEREO A PLASTIC, to determine average values and standard errors, which we then compare with the results of the entire multi-year data set. Section 5 will contain a summary, a discussion of the implications, and suggested future work.

2. Determination of the ISN flow longitude from the PUI cut-off

The ISN flow longitude can be derived independently from characteristic spatial structures in the inner heliosphere, the so-called focusing cone [30, 31] and crescent, both identifiable with PUI observations at 1 AU [32]. Although a careful statistical study largely eliminated the influence of solar wind structures and potential sensor efficiency variations [32], PUI flux observations likely retain considerable variations relative to the parent neutral gas structures due to solar wind structures [33], temporal and spatial variations in the ionization rates, and PUI transport effects [34, 35]. A more robust method uses the fact that the PUI cut-off speed (the high-energy edge of the relatively flat PUI velocity distribution [9]) is a function of the radial ISN flow component V_r and the solar wind speed V_{SW} and thus varies with the observer longitude λ_{Obs} [36]. In its dimensionless form $v_r = V_r/V_E$, normalized to the Earth's speed at 1 AU, the radial ISN flow varies with λ as [25]

$$v_r^2 = 2 + v_{ISN\infty}^2 - (1-\cos\lambda) - \left\{ v_{ISN\infty}^2 \sin^2\lambda + v_{ISN\infty} \sin|\lambda| [v_{ISN\infty}^2 \sin^2\lambda + 4(1-\cos\lambda)]^{1/2} \right\} / 2 \qquad (1)$$

where $\lambda = \lambda_{Obs} - \lambda_{ISN\infty} - 180°$ and $\lambda_{ISN\infty}$ is the ISN flow direction outside the heliosphere. v_r reaches its maximum exactly upwind, i.e. for $\lambda_{Obs} = \lambda_{ISN\infty} + 180°$ where $\lambda = 0$. v_r decreases symmetrically with increasing angle $|\lambda|$ and $v_r = 0$ holds at $\pm\lambda_0$, where the ISN bulk flow reaches its perihelion at 1 AU [37, 38] with $\cos\lambda_0 = -(1 + v_{ISN\infty}^2)^{-1}$.

As discussed in [25] the PUI cut-off speed w'_{CutOff} in the solar wind frame (primed variable), normalized to the solar wind speed V_{SW}, is to a good approximation given by

$$w'_{CutOff} = (V_{SW} + V_r) / V_{SW} \qquad (2)$$

Note in equation (2) and in Figure 1 V_r is taken positive for inward ISN flow directions. Because PUI distributions are gyrotropic and mostly scatter in pitch angle it is advantageous to accumulate their spectra on phase space surfaces constant in w'. With the energy and angular resolution of STEREO PLASTIC the He^+ observations taken in the spacecraft frame have been transformed into the solar wind frame to facilitate accumulation along constant w'. ACE SWICS integrates He^+ over its entire field-of-view in constant energy per charge in the spacecraft frame, or in $w = w' + 1$. As we will see, this makes the ACE SWICS observations more sensitive to variations in the interplanetary magnetic field (IMF) direction and in the solar wind speed. For consistency in the presentation, we show all PUI cut-off results in terms of w'_{CutOff} and transform the ACE SWICS observations according to $w'_{\text{CutOff}} = w_{\text{CutOff}} - 1$. To obtain the most reliable cut-off determination, the initial torus of the PUI velocity distribution, which has been clearly identified with STEREO PLASTIC observations [39], needs to be in the sensor field-of-view. To achieve this condition, STEREO PLASTIC observations have been constrained so the angle between $\mathbf{B}_{IMF}$ and the solar wind lies between 70° and 110°. Because ACE SWICS observations have to be used in the spacecraft frame, where changes in the IMF orientation lead to variations in the cut-off, these data have been constrained to angles in the range 75° - 105°.

To obtain a proxy value for the cut-off the He^+ PUI counts accumulated in a 1° cadence in ecliptic longitude by STEREO PLASTIC are used on a fixed grid in w', normalized to the maximum count value for each 1° bin to stay as close to the observable as possible. For ACE SWICS the normalized phase space density is used to correct for differences in duty cycle and efficiencies as a function of particle energy. A hyperbolic tangent is fitted to these data sets, as described in [25], according to

$$f_n(w') = 0.5 - 0.5\tanh(A \cdot (w' - w'_{CutOff})) \tag{3}$$

where f_n represents the normalized counts or the normalized phase space density, respectively, and A is the steepness of the cut-off. How the proxy for the cut-off determined in this way is shifted by a constant relative to the actual cut-off of the PUI distribution for the two data sets has been discussed in [40]. They also showed that this shift does not influence the determination of the ecliptic longitude of the ISN flow because of the independence of the shift in value and the longitude dependence and because of the symmetry of the longitude variation.

The cut-off $w'_{CutOff}(\lambda_{Obs})$ then undergoes a correlation analysis that invokes the mirror symmetry of the longitude variation according to equation (1). The longitude with the maximum of the Pearson correlation coefficient is identified as the ISN flow direction λ_{ISN}, with a very small statistical fit uncertainty for the combined STEREO A PLASTIC 7-year data set that includes March 2007 through March 2014 [25]. It is evident that the small uncertainty reported in this first analysis may underestimate effects due to stochastic variations in the observations and that most systematic effects have not been considered yet. In the following, we will first obtain a more realistic estimate of the statistical uncertainties involved and then look into systematic effects with individual year ACE and STEREO data sets. Most of the important systematic effects that influence PUI observations, such as solar wind speed and flux variations, changes in IMF strength and orientation, as well as in wave power, are distributed stochastically over ecliptic longitude and thus vary in the same fashion from year to year. Therefore, multiple year compilations will see substantially reduced systematic effects.

3. Estimation of the statistical uncertainties

To estimate the statistical uncertainties, artificial data sets were generated and then varied in a Monte Carlo scheme with Poisson distributed counts. To construct the artificial data we followed the dependence of the observed PUI distribution as a function of w' and λ_{Obs}. For the variation in w' the decrease near the cut-off and the decrease toward lower energy were represented by different hyperbolic tangent functions. Only the hyperbolic tangent at the high-energy cut-off matters for this analysis. For the longitude variation, a cosine dependence was chosen, with the maximum in w'_{CutOff} at the presumed upwind direction of the ISN flow. The parent function of the artificial PUI distribution data sets is shown in a color-coded spectral representation in the left panel of figure 1. Instead of w' the radial component of the PUI injection velocity V_r is shown. Poissonian noise was added to each pixel in w'

and λ_{Obs} with a random generator, according to the mean total number of counts expected over the course of one year. In this way 200 trials were prepared, one of which is shown in the right panel of figure 1.

To find the cut-off w'_{CutOff} two different methods were used. In method I, the fit of a hyperbolic tangent to the decrease toward the high-energy end was applied as described above. In method II, a typical decrease function was determined from the data set averaged over 240° ≤ λ_{Obs} ≤ 270°, which was then used in a correlation analysis for each 1° bin, with w' varying to obtain the maximum correlation coefficient.

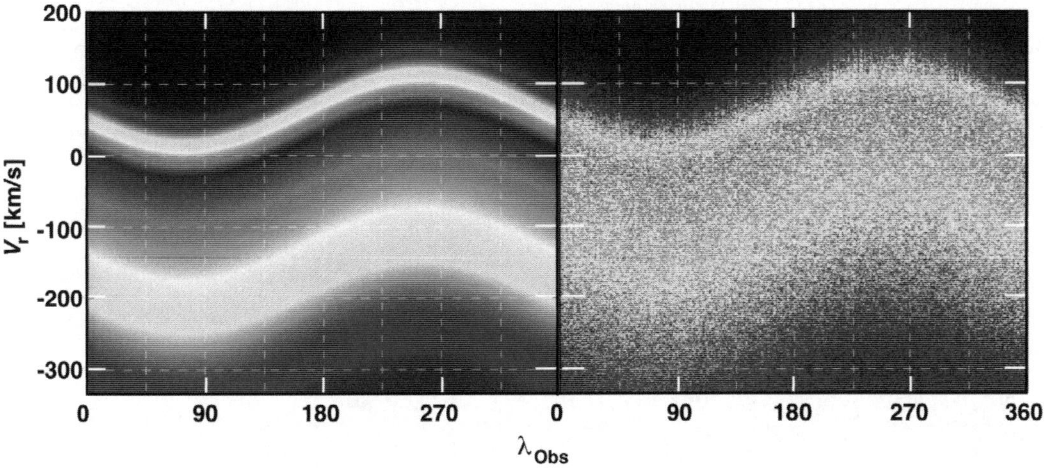

Figure 1: Artificial data for the PUI distributions as a function of λ_{Obs} and V_r in a color-coded spectral representation. Left: Parent distribution with a hyperbolic tangent variation in V_r on both sides of the maximum of the distribution and a cosine in λ_{Obs}, with the maximum in the upwind direction set at λ_{Upwind} = 255°. Right: Distribution with Poisson noise added

The results from this Monte Carlo simulation with artificial PUI distribution data, assuming λ_{ISN} = 75°, analyzed with both methods are compiled in Table 1. The presumed ISN flow direction is found very close to the input value. The statistical uncertainties are ±0.62° for method I and 0.40° for method II, both noticeably larger than the mere fit uncertainty of ±0.04° reported in [25].

Table 1: Results from the Monte Carlo simulation with artificial PUI distributions

Method	λ_{ISN} [°]	λ_{Upwind}[°]
I	75.129 ± 0.622	255.119 ± 0.623
II	75.073 ± 0.396	255.065 ± 0.396

As pointed out above, in this simulation the counting statistics for 1 year was used. Considering that [25] used a 7-year data set, the uncertainties will likely be reduced by √7, which would leave the uncertainty obtained with method II at 0.15° about a factor of 3.5 higher. Barring any systematic effects, a statistical uncertainty ≤0.15° would then be achievable. Method II appears to be superior to the fit with a hyperbolic tangent, likely because this correlation method is less sensitive to any variations in the slope or shape in the decreasing part of the PUI distribution. None of these uncertainties contain any systematic effects, the magnitude of which we tackle next.

4. Systematic effects on the ISN flow longitude in multiple year observations

To obtain a first estimate of systematic effects on the ISN flow longitude derived with the mirror correlation method described in [25], we apply the same method to several individual year data sets from STEREO A PLASTIC and ACE SWICS.

Journal of Physics: Conference Series **767** (2016) 012017 doi:10.1088/1742-6596/767/1/012017

4.1. Analysis of annual PUI data sets

Figure 2 shows the PUI cut-off values obtained with STEREO A PLASTIC for 2007 in the center panel and those with ACE SWICS for 2007 in the bottom panel, as a function of λ_{Obs}, along with the March 2007 – March 2014 STEREO data set in the top panel. 2007 has been chosen for this comparison because during that year the STEREO spacecraft were still close to the Earth and the STEREO and ACE observations are co-located.

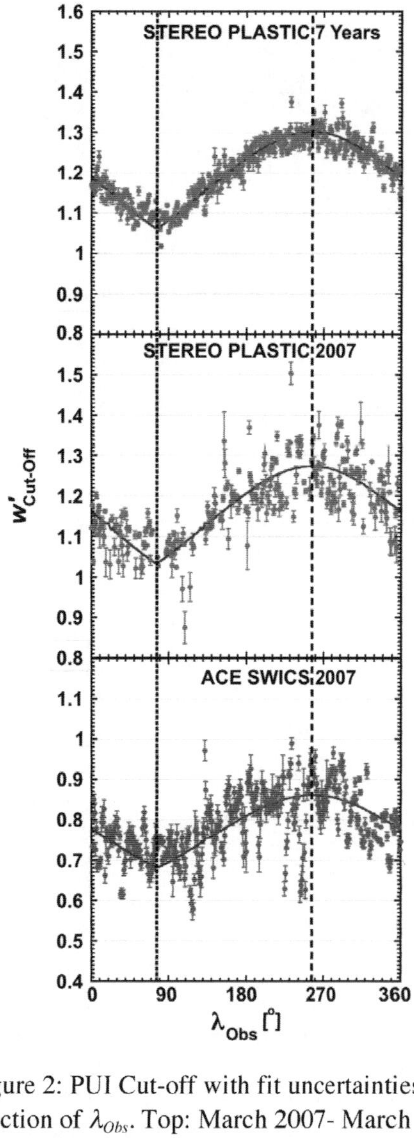

Figure 2: PUI Cut-off with fit uncertainties as a function of λ_{Obs}. Top: March 2007- March 2014 STEREO PLASTIC observations of the PUI cut-off in the solar wind frame (w'), constrained to $V_{SW} < 450$ km/s and $\boldsymbol{B}_{IMF}$ so that the torus is completely within the PLASTIC FOV. Center: 2007 STEREO PLASTIC observations. Bottom: 2007 ACE SWICS observations, constrained in the angle between $\boldsymbol{B}_{IMF}$ and the solar wind to >75°.

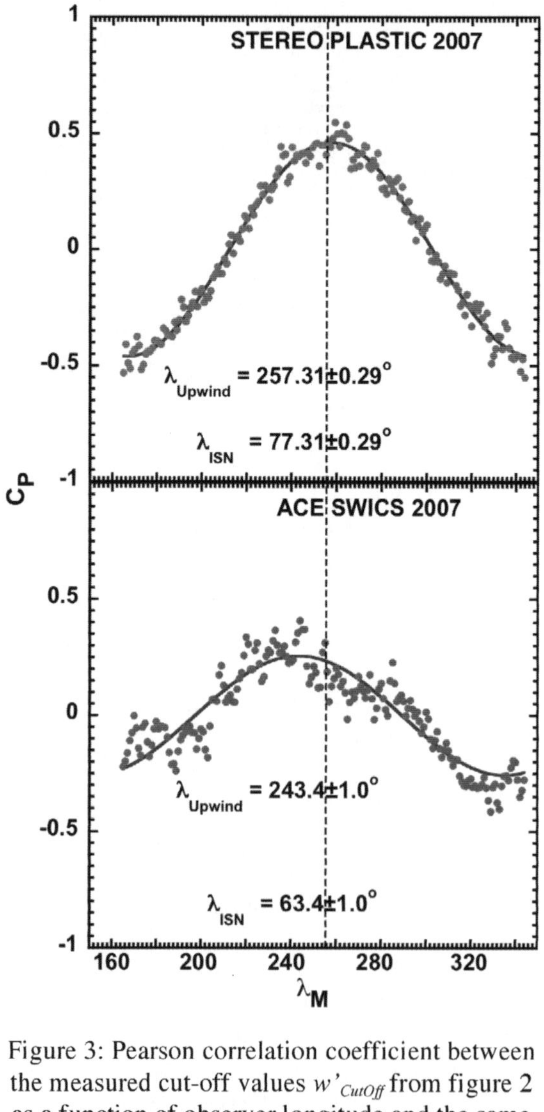

Figure 3: Pearson correlation coefficient between the measured cut-off values w'_{CutOff} from figure 2 as a function of observer longitude and the same values, mirrored about the longitude λ_M, incremented in 1° steps for ±90° about λ_{Upwind}. Also shown is the best-fit curve to a cosine function, along with fit values and uncertainty for λ_{Upwind} and $\lambda_{ISN\infty}$. The vertical dashed line indicates the upwind direction based on the entire STEREO PLASTIC data set.

It is obvious that the single year data sets show much larger variations. While this is already expected due to much reduced counting statistics, as is evident from the larger statistical error bars in the plots, there also appear to be visible systematic deviations from the rather smooth variation of the cut-off with λ_{Obs} in the STEREO A multi-year data set. Near $\lambda_{Obs} = 100^{\circ}$ as well as around 240° and 270° there appear to be visible systematic deviations to smaller w'_{CutOff} values, surrounded by somewhat larger values. These deviations are visible in both data sets, but much more pronounced in the ACE SWICS observations. We will discuss potential reasons for these variations at the end of this section.

Figure 3 shows the Pearson correlation coefficient obtained by mirroring the results about a chosen ecliptic longitude λ_M, as described in Section 2 and in [25] for the two 2007 data sets in figure 2. Both mirror correlations vary generally according to a cosine function with λ_M and show a pronounced maximum near the ISN upwind direction. However, the maximum correlation coefficient is substantially reduced compared to the result in [25], approximately 0.5 for STEREO PLASTIC and 0.3 for ACE SWICS, and the longitudes for the maximum deviate noticeably from the value found for the combine STEREO data set. Because the original data set shown in figure 2 contains systematic deviations, such a behavior is expected. Table 2 shows the ISN flow direction λ_{ISN} along with the fit error and the maximum correlation coefficient for each annual data set used from ACE SWICS and STEREO A PLASTIC. The third line from the bottom shows the results obtained with the combined ACE SWICS and STEREO A PLASTIC data set, for which figure 4 shows the longitude variation of the mirror correlation coefficient for the combined ACE SWICS (left panel) and combined STEREO A PLASTIC (right panel) observations. The bottom line contains the weighted mean values for both data sets, with the maximum correlation coefficient used as the weight, along with the standard deviation of the individual results and the standard error for the multi-year data sets.

Table 2: Results from the multi-year PUI cut-off analysis with ACE and STEREO

ACE SWICS				STEREO A PLASTIC			
Year	λ_{ISN}	$\Delta\lambda$	C_{pMax}	Year	λ_{ISN}	$\Delta\lambda$	C_{pMax}
2001	95.3°	1.2°	0.26	2007	77.3°	0.29°	0.46
2002	86.7°	1.2°	0.24	2008	75.75°	0.22°	0.52
2003	86.7°	2.4°	0.10	2009	72.49°	0.13°	0.66
2005	86.4°	1.4°	0.17	2010	76.42°	0.34°	0.44
2006	76.5°	0.6°	0.37	2011	77.68°	0.17°	0.57
2007	63.4°	1.0°	0.26	2012	78.87°	0.18°	0.58
2009	74.5°	0.3°	0.53				
2010	79.3°	0.5°	0.42				
Combined	74.0°	0.3°	0.61	Combined	76.69°	0.04°	0.86
	Weighted Mean	Std Dev	Std Error		Weighted Mean	Std Dev	Std Error
	81.6°	9.5°	3.4°		76.30°	2.20°	0.90°

The year-to-year results of STEREO PLASTIC show a 2.2° standard deviation, but a weighted mean that falls close to the value obtained with the combined data set and a standard error of 0.9° that signals that the two different results are in agreement within the stated error bars. The annual ACE SWICS results show a much larger standard deviation of 9.5°. The weighted mean is within about 5° of the combined STEREO A result, with a stated standard error of 3.4°, or still outside the mutual error bars. The result obtained with the combined ACE SWICS data set comes much closer to the STEREO PLASTIC result, now with a 0.3° fit error.

Journal of Physics: Conference Series **767** (2016) 012017 doi:10.1088/1742-6596/767/1/012017

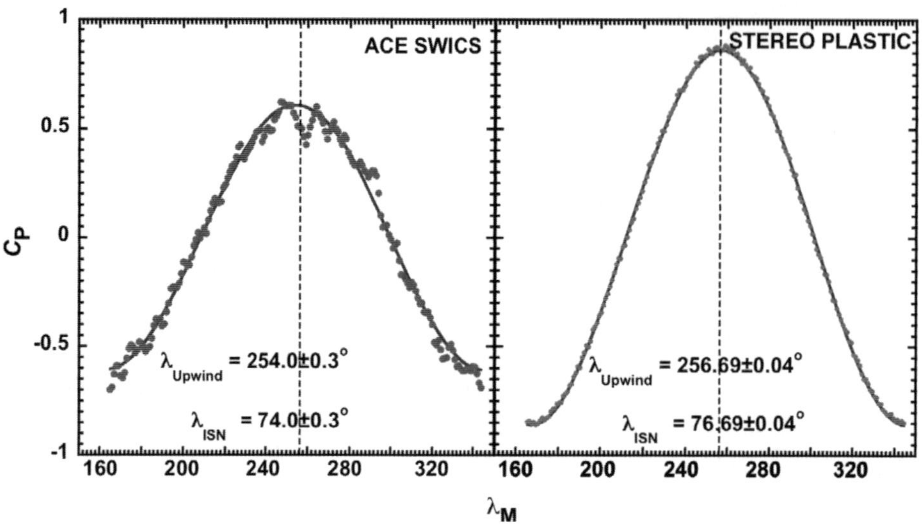

Figure 4: Pearson correlation coefficient as a function of λ_M for the combined ACE SWICS data for 2001 – 2010, excluding leap years, and STEREO A PLASTIC data for March 2007 – March 2014 in the same format as figure 3.

4.2. Discussion of the annual and combined results

Clearly, individual yearly data appear to be affected substantially by systematic effects, which pull the ISN flow direction obtained with the PUI cut-off method. These systematic effects are much stronger for the ACE SWICS data as can be expected from the fact that so far no provisions have been taken to adjust for the difference that ACE SWICS data are used in the spacecraft frame, while the STEREO PLASTIC data are transformed into the solar wind frame. Also the statistical fit errors and their standard counterparts for the weighted mean of the yearly results substantially understate the complete set of uncertainties involved. Systematic effects are visible also in the STEREO PLASTIC results and are definitely larger than the statistical uncertainties, solely from the fits or based on the Monte Carlo simulations in Section 3, but they are much smaller than for ACE SWICS.

Although the STEREO PLASTIC observations have been constrained to solar wind speeds <450 km/s to exclude CMEs and high speed solar wind streams for now, all PUI observations have been averaged over 1° in ecliptic longitude, or approximately one day, while the solar wind varies on time scales faster than that. Because the observations are binned in *w*' the observed cut-off is modulated over ecliptic longitude in $V_r/V_{sw,}$ thus retaining the memory of V_{sw} variations. In addition, fast solar wind compressions and rarefactions can lead to variations of the cut-off due to adiabatic changes of the PUI distribution [41]. Obviously, the ACE SWICS observations are affected more strongly. First, the SWICS data have not been constrained in solar wind speed to maintain high counting statistics and to see effects from solar changes more clearly. Second, ACE SWICS data are not transformed into the solar wind frame. CMEs and fast solar wind streams coincide with large-scale variations in the IMF orientation. Although the data have been constrained to IMF angles relative to the solar wind direction >75° the observed cut-off is an average over the SWICS field-of-view, with smaller IMF angles reducing the observed value for otherwise identical conditions. Therefore, we will identify the parameter dependencies of these variations in the future and then build them into the model of the PUI cut-off to compensate for these systematic effects. However, these steps are beyond the scope of the current study.

As can be seen from the right panel of figure 4, the resulting best-fit ISN flow longitude based on the 7-year STEREO A PLASTIC PUI cut-off observations from DOY 100 2007 through DOY 100

2013 is now λ_{ISN} = 76.69±0.04° compared with λ_{ISN} = 75.21±0.04° as reported in [25] based on a data set from DOY 100 2007 through the end of 2013. The difference by 1° must be broken into three separate components, out which the first two are due to errors in the original analysis. First, we identified and corrected an erroneous shift in the algorithm to obtain the mirrored function of the PUI cut-off by 1°, which had moved the maximum correlation coefficient by 0.5° toward smaller longitude. Second, there has been a misunderstanding in the interpretation of the longitudes of the 1° angular bins of the STEREO PLASTIC data, as shown in figure 2. The angle values were interpreted as for the bin centers, whereas they were defined for the starting edge of each bin, thus leading to another shift of all longitude values by 0.5° toward smaller longitude. The STEREO PLASTIC data in this analysis have been corrected consistently for both effects, leading to longitude values that are larger by 1°. The ACE SWICS data were binned using the center longitudes of the bins, and the analysis was performed solely with the corrected version of the algorithm.

The remaining difference of about 0.5° between the corrected value λ_{ISN} = 76.21±0.04° for the STEREO data set through the end of 2013 used in [25] and the result 76.69° reported here for the 7-year data set through DOY 100 2013 is due to an important systematic effect on the PUI cut-off analysis that is revealed by this difference. The average of the differences between the PUI cut-off values taken for the two different data sets shows a small, but noticeable, asymmetry about the ISN flow direction, which explains the direction difference by 0.5° and goes along with a variation in the average solar wind speed because in the current analysis the relevant radial ISN flow velocity component appears normalized to V_{SW} as $w_r = V_r/V_{SW}$. In other words, a slow change in V_{SW} over the year(s) used for the cut-off analysis may lead to a systematic effect on the resulting ISN flow direction. We find that the average solar wind speed mostly decreases over most of the 7 years of the STEREO PLASTIC data set. For decreasing solar wind speed, the resulting flow direction is shifted toward larger longitude and into the opposite direction for increasing average solar wind speed.

Note that the STEREO data set, with decreasing average solar wind speeds, is taken during the rise to the present solar maximum. Conversely, the ACE SWICS data set mostly covers the previous solar maximum and the declining phase, likely with the opposite behavior. Interestingly, the longitude obtained from the combined ACE SWICS data set appears noticeably smaller than the combined STEREO PLASTIC result. These findings reemphasize that the values shown in Table 2 contain significant systematic uncertainties and thus are not to be interpreted yet as new values for the ISN flow longitude. We have merely started to investigate how large the statistical and systematic uncertainties inherent in the data sets and this new analysis method are, and which effects are behind them. The influence of variations in V_{SW} on the PUI cut-off analysis can be mitigated by a modification in the normalization of the PUI distributions with emphasis on the radial component V_r of the ISN flow velocity at 1 AU rather than using w as in previous PUI investigations. In addition, the variation of the cut-off speed with V_{SW} is known and can be fully included in the model that is used for the analysis. This is beyond the scope of this paper and will be the subject of a follow-on analysis.

5. Conclusions and outlook

In this paper we have built on the finding that the cut-off speed of the interstellar PUI distribution has a maximum exactly in the upwind direction, with a symmetric decrease with longitude away from the upwind direction [36]. This behavior has led to a method, which uses a Pearson correlation analysis between the longitude variation of the cut-off and its mirror function to determine the ISN flow direction with a very small fit uncertainty [25].

We have investigated the influence of Poissonian fluctuations in the observed PUI count rates on the correlation result and found:

- A correlation method, in which a generic cut-off function is compared with the observed PUI spectra, produces results with substantially smaller uncertainties.
- A realistic purely statistical uncertainty for a comparable data set as used in [25] is about twice the reported fit uncertainty.
- A single-year statistical uncertainty with STEREO PLASTIC is likely ≈±0.2°.

Furthermore, we have investigated the overall influence of any systematic effects on the resulting ISN longitude by applying the mirror correlation method to ACE SWICS and STEREO PLASTIC data for individual years, obtaining weighted mean values of these individual results, and comparing them with results from the multi-year data sets. We have found:

- Individual year data sets include substantial systematic effects, likely due to the distribution of large-scale solar wind structures over the year, as one can see by inspection from the longitude variation of the PUI cut-off values.
- ACE observations are affected much more severely than STEREO observations, with standard deviations around the multi-year weighted mean results of 9.5° and 2.2°, respectively.
- The weighted mean of the individual year results for STEREO agrees with the multi-year correlation result within the standard error of the mean of 0.9°.
- The multi-year STEREO result is different by 1° over [25] because of errors in the analysis algorithm that have been corrected now.
- The weighted mean for ACE does not agree in a similar way with the multi-year result.
- The multi-year results of the two data sets and the weighted means come close to each other, but still fall outside the 1-σ uncertainties.
- Continuous slow variation of the average solar wind speed, as is expected over the course of the solar cycle, also leads to systematic variation of the resulting ISN flow longitude, which can be addressed within the model for the cut-off and a modification of the normalization of the PUI distributions.

These results re-emphasize the promise that, if the systematic effects have been identified and largely corrected for, the ISN flow longitude can be obtained with very small uncertainties of $\approx\pm 0.4°$ from STEREO for individual years and $\leq 0.15°$ with the entire growing data set. Systematic effects are much more prominent for the ACE SWICS data. Likely candidates for systematic effects are solar wind speed variations for both data sets, and, in addition, IMF direction changes for ACE SWICS. Although, the PUI distributions have been obtained for the highest time resolution as a function of v/V_{SW}, the implicit variation of V_r/V_{SW} is only taken into account for the entire data set thus far, and no adjustment for potential adiabatic changes of the cut-off due to rapid solar wind speed changes [41] have been made. The STEREO observations are largely immune to IMF directional changes as long as the initial PUI torus remains in the sensor field-of-view [39], which is one of the selection criteria for the data because there is no effect on the cut-off in the solar wind frame. The cut-off obtained with ACE SWICS data, on the other hand, is affected by such changes, because the initial torus distribution moves noticeably in v/V_{SW} in the spacecraft frame. This makes the ACE observations much more vulnerable to solar wind structures. However, with the variation of the initial torus distribution with IMF direction known from STEREO PLASTIC observations, this effect can be built into a forward model of the ACE SWICS PUI distributions and thus compensated for. For 2007 even a direction validation is possible because co-located observations with ACE and STEREO exist for this year.

Currently, the PUI cut-off analysis is still subject to systematic effects that apparently increase the uncertainties substantially, for the STEREO multi-year result likely of the order of the standard error of the weighted mean over the yearly results of 0.9°. It is evident from the improvements made with analyzing both multi-year data sets over individual years that most of the systematic effects are distributed over longitude in a stochastic manner from year to year, thus allowing a compensation. By taking into account solar wind speed variations on the highest time resolution and including IMF variations in the modeling of the SWICS PUI distributions the aforementioned effects will be largely compensated in the future. In addition, a full interplanetary parameter analysis, including solar wind, IMF, and wave power variations, will reveal further systematic effects, and a method used in [32], where they used dual-year data sets with all possible permutations in the combination of years, will allow a further reduction of the systematic effects.

Multi-year observations with ACE and STEREO provide for a precision determination of the ISN flow longitude, which is complementary to the precise determination of a 4-dimensional ISN parameter tube with IBEX [18, 19, 20, 21]. In the future, the combination of these observations will allow a direct comparison with astronomical observations of the physical state of the local interstellar cloud

[16] for any potential small-scale spatial variations and/or any temporal variations of the ISN parameters over the past decades. This analysis presents another stepping stone on the journey to explore the outer boundary of the heliosphere and the surrounding interstellar medium, for which Dr. Ed Stone has paved the way and continues to do so with his effective and tireless lead of key heliophysics missions, such as the Advanced Composition Explorer and the Voyager Interstellar Mission.

Acknowledgements: Support by NASA through ACE (NNX08AI11G), IBEX (NNG05EC85C), STEREO (NNX13AP52G) and GI grants NNX16AF79, NNX12AB32G, NNX11AO97G, as well as at the Universität Kiel by the DFG (Wi2139/7-1) and DLR (50 0C 1103) is acknowledged.

References

[1] Lallement, R. et al. 2005, *Science,* **307**, 1447.
[2] Izmodenov, V., D. Alexashov, & A. Myasnikov 2005, *A&A*, **437**, L35.
[3] Opher, M., E. C. Stone, & T. Gombosi 2007, *Science,* **316**, 875.
[4] Pogorelov, N.V., G.P. Zank & T. Ogino, 2006, *ApJ*, **644**, 1299.
[5] Pogorelov, N. V., et al. 2009, *ApJ. Lett.,* **695**, 31.
[6] Schwadron, N.A., et al. 2014, *Science* **343**, 988.
[7] Bertaux, J., & J. Blamont 1971, *A&A*, **11**, 200.
[8] Weller, C., & R. Meier 1974, *ApJ*, **193**, 471.
[9] Möbius, E., D. Hovestadt, & B. Klecker 1985, *Nature*, **318**, 426.
[10] Gloeckler, G., et al. 1992, *A&A*, **92**, 267.
[11] Witte, M., et al. 1996, *SSR*, **78**, 289.
[12] Möbius, E., et al. 2009, *Science* **326**, 969.
[13] Frisch, P.C., et al. 2013, *Science* **341**, 1080.
[14] Lallement, R. & J.-L. Bertaux 2014, *A&A*, **565**, A41.
[15] Frisch, P.C., et al. 2015, *ApJ*, **801**:61.
[16] Redfield, S, & J. Linsky 2008, *ApJ*, **673**:283.
[17] McComas, D.J., et al. 2012, *Science*, **336**, 1291.
[18] Bzowski, M., et al. 2015, *ApJS*, **220**:28.
[19] McComas, D. J., et al.,2015b, *ApJS*, **220**:22.
[20] Möbius, E., et al. 2015a, *ApJS*, **220**:24.
[21] Schwadron, N.A. et al. 2015, *ApJS*, **220**:25.
[22] Bzowski, M., et al. 2014, *A&A*, **569**, A8.
[23] Wood, B., H.-R. Mueller, & M. Witte 2015, *ApJ*, **801**, 62.
[24] Witte, M. 2004, *A&A* **426**, 835.
[25] Möbius, E., M.A. Lee & C. Drews 2015b, *ApJ*, **815**:20.
[26] Stone, E.C., et al. 2005, *Science* **309**, 2017.
[27] Stone, E.C., et al. 1998, *Space Sci. Rev.*, **86**, 1.
[28] Gloeckler, G., et al. 1998, *Space Sci. Rev.*, **86**, 497.
[29] Galvin, A.B., et al. 2008, *SSR*, **136**, 437.
[30] Möbius, E., D. Rucinski, D. Hovestadt, & B. Klecker 1995, *A&A,* **304**, 505.
[31] Gloeckler, G., et al. 2004, *A&A*, **426**, 845.
[32] Drews, C., et al. 2012, *JGR*. **117**, 9106.
[33] Möbius, E., et al. 2010, *AIP Conf. Proc.*, **1302**, 37.
[34] Möbius, E., et al. 1996, *Ann. Geophys.*, **14**, 492.
[35] Chalov, S., & H.-J. Fahr 2006, *Astron. Lett.* **32**, 487.
[36] Möbius, E., et al. 1999, *GRL*, **26**, 3181.
[37] Lee, M. A., et al. 2012, *ApJS,* **198**:13.
[38] Lee, M. A., et al. 2015, *ApJS*, **220**:23.
[39] Drews, C., et al. 2015, *A&A*, **575**, A97.
[40] Möbius, E., et al. 2016, *AIP Conf. Proc.*, **1720**, 09002.
[41] Saul, L., et al. 2003, *AIP Proc. of the 10th Solar Wind Conf.*, **679**, M. Velli, et al. eds., 778.

Journal of Physics: Conference Series **767** (2016) 012018 doi:10.1088/1742-6596/767/1/012018

A global view of the modulation of cosmic ray protons in the heliosphere for the solar minimum period up to 2009

M S Potgieter and E E Vos

Centre for Space Research, North-West University, 2520 Potchefstroom. South Africa

E-mail: Marius.Potgieter@nwu.ac.za

Abstract. The heliosphere was in a state of ideal solar minimum conditions for at least three years up to the end of 2009. During this period the highest ever recorded cosmic ray spectra were observed at the Earth. Fortunately, the PAMELA and Ulysses KET instruments simultaneously observed proton intensities for most of the period between July 2006 and June 2009, while Voyager 1 made observations in the outer heliosphere. This provides a good opportunity to compare the basic features of a comprehensive numerical model for the global modulation of cosmic rays in the heliosphere with these observations. Global gradients for protons are computed with the model for this prolonged solar minimum of cycle 23/24. This is done for both radial and latitudinal gradients, with the latter possible because Ulysses changed its position significantly in the heliocentric meridional plane during this period. The modulation model is set up for the conditions that prevailed during this unusual solar minimum period so that insight is gained on what role particle drifts played in establishing the observed gradients for this period. Good agreement is found between computed and observed gradients so that we conclude that the model gives a most reasonable representation of modulation conditions from the Earth to the heliopause for the mentioned period. These results can be used to refine the theory for diffusion, particle drifts and turbulence in the heliosphere.

1. Introduction

The solar minimum modulation period from 2006 to the end of 2009 had been unusual. It was expected that a new activity cycle would begin early in 2008, assuming a 10.5 year periodicity, but instead minimum modulation conditions had prevailed until the end of 2009. This was characterized by a much weaker heliospheric magnetic field (HMF) magnitude compared to previous cycles. The tilt angle of the wavy heliospheric current sheet (HCS) stayed below 15° for four consecutive years (except for a brief period in 2008). It was reported that cosmic rays (CRs) with a high rigidity reached record setting intensities during this time, e.g. [1, 2]. When the PAMELA proton spectra became available it turned out that record setting intensities were observed at the Earth at all reported energies (down to 80 MeV for PAMELA's range); see [3].

The mentioned observations were found quite puzzling from a drift theory point of view for CR modulation according to which CRs drift inwards or outwards along the HCS depending on the solar magnetic polarity cycle and the charge-sign of these particles. A higher flux of positively charged CRs should arrive at Earth during solar minimum periods with an $A > 0$ magnetic polarity because they then drift inwards primarily through the polar regions of the heliosphere. During $A < 0$ cycles, such as in 2006 to 2009, positively charged particles drift in towards the Earth mostly along the HCS, so that its waviness as represented by the tilt angle plays a more direct modulation role than during $A > 0$ cycles, e.g. [4,5]. It is therefore expected that proton spectra for these $A < 0$ cycles should be lower than spectra during A > 0 cycles at kinetic energy $E <\sim 1$ GeV. This is indeed the case, as was clearly illustrated by [6] who show that all previously observed proton spectra during solar minimum periods can be grouped together and categorized according to A > 0 and A < 0 polarity cycles without any exceptions. However, instead of grouping together with previous A < 0 cycles, the PAMELA proton spectrum for 2009 was even higher than the previously observed A > 0 spectra, as such quite a remarkable phenomenon. Of course, according to drift theory the upcoming A > 0 solar minimum period (perhaps in 2018-2019) should produce even higher CR spectra if solar activity would be again as low as in 2006 to 2009. This is expected because the Sun had been much quieter than before, even during the recent solar maximum activity period.

Particle drifts in the heliosphere are also recognized as the pre-eminent process behind the 22-year solar cycle observed in the CR latitudinal gradients; characteristically, drifts cause negative latitudinal gradients during A < 0 polarity epochs [7, 20, 27]. The model is used to compute such gradients for each year from 2006 to 2009 to be compared with values obtained by using corresponding PAMELA and Ulysses proton observations (Gieseler, private communication).

Changes in the HCS tilt angle, as for all other proxies of solar activity related to CR modulation, propagate outwards through the heliosphere at the solar wind speed so that when the Sun became quite already in 2006, these ideal solar minimum conditions settled throughout the heliosphere and lasted about three years. The heliosphere was therefore in a state of perfect solar minimum conditions for at least three years. This fact is used to simulate with a numerical model the CR proton intensities, and the corresponding radial and latitudinal gradients, throughout the heliosphere for the period 2006 to 2009. The modelling results are shown in comparison with CR observations, where available, made during this unusual and unprecedented solar minimum period.

2. The numerical model

A full three-dimensional (3D) model is used to compute the differential intensity of 10 MeV to 30 GeV protons in the equatorial plane of the heliosphere, at various radial distances along the trajectory of Voyager 1 and along the trajectory of the Ulysses spacecraft. It is based on the numerical solution of the well-known heliospheric transport equation (TPE; [8]):

$$\frac{\partial f}{\partial t} = -\left(\mathbf{V}+\left\langle \mathbf{v}_D\right\rangle\right).\nabla f + \nabla.\left(\mathbf{K}_S.\nabla f\right) + \frac{1}{3}\left(\nabla.\mathbf{V}\right)\frac{\partial f}{\partial \ln P}, \tag{2}$$

where $f(\mathbf{r},P,t)$ is the CR distribution function, P is rigidity, t is time, $\mathbf{r}$ is the position vector in three dimensions (3D), with the usual three coordinates r, θ, and φ specified in a heliocentric spherical coordinate system where the equatorial plane is at a polar angle of $\theta = 90°$ (heliolatitude of 0°). It is assumed that $\partial f / \partial t = 0$, which means that all short-term modulation effects (periods shorter than one solar rotation) are neglected, which is considered as a most reasonable assumption for the quiet solar minimum conditions during the mentioned period. Terms on the right hand side respectively represent convection, where $\mathbf{V}$ is the solar wind velocity; $\langle \mathbf{v}_D \rangle$ is the averaged particle drift velocity caused by gradients and curvatures in the global HMF, including the wavy HCS; with $\mathbf{K}_S$ the symmetry diffusion tensor; lastly, the term describing adiabatic energy changes which are treated the same everywhere in the heliosphere.

The numerical approach, the expressions for the solar wind speed and the HMF, the three diffusion coefficients (DCs) and the drift coefficient in terms of their spatial and rigidity dependence are described in great detail by [9] as well as the approach to simulate the modulation conditions throughout the heliosphere for the mentioned period. This approach produces the averaged modulation conditions in terms of the solar wind speed, the tilt angle and the HMF magnitude in the heliosphere up to about a year before the proton spectrum for a given time was observed at the Earth. The heliopause (HP) is positioned at 122 AU in the model to make it consistent with the reported Voyager 1 observations [10]. The position of the solar wind termination shock (TS) was shifted inwards from

~90 AU (2006) to ~80 AU (2009) from the Sun with deceasing solar activity; see also [11] and references therein from a cosmic ray modulation point of view, and [29] from an MHD point of view.

3. Results

3.1. The PAMELA proton spectra and modelled spectra up to the heliopause

Four PAMELA proton spectra are selected to be reproduced by the model, respectively averaged over 27 days for November 2006, December 2007, December 2008 and December 2009 and referred to as the 2006e, 2007e, 2008e and 2009e spectra, extending on the modelling reported by [9, 12]. By accounting for all of the appropriate modulation parameters in the model, and by carefully adjusting

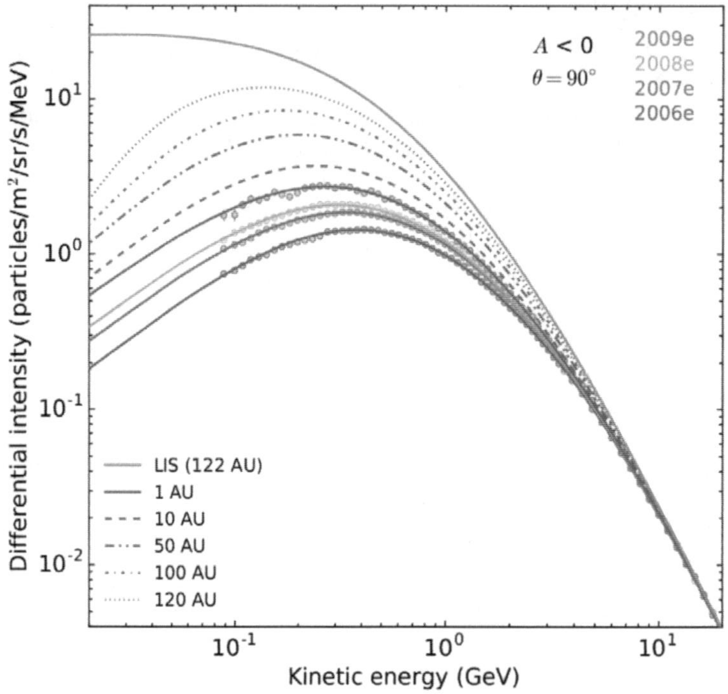

Figure 1. The observed PAMELA and computed proton spectra at the Earth (1 AU) for the periods 2006e, 2007e, 2008e and 2009e as described in the text. Corresponding computed spectra are also shown in the equatorial plane (θ = 90) at 10 AU, 50AU, 100 AU and 120 AU by the blue lines. The local interstellar spectrum (LIS; grey line) from [9] is specified at 122 AU.

the values of the DCs, each consecutive year-end PAMELA proton spectrum was reproduced satisfactory as shown in Figure 1. The values of the DCs and other modulation parameters are given by [9]; see their Tables 1 and 2 and Figure 12. Additional computed spectra in the equatorial plane are shown at 10 AU, 50 AU, 100 AU and 120 AU in relation to the newly constructed very local interstellar spectrum (LIS) for protons from [9]. Concerning the new LIS's based on PAMELA and Voyager 1 observations, see also [13] and [14].

Evidently, the proton spectra at the Earth became progressively softer from 2006 to 2009, reaching a maximum intensity at the end of 2009, with an accompanying shift of the spectral peak to lower energies. Predictions of intensity levels are made with the model for kinetic energy E < 80 MeV, where PAMELA measurements are unavailable. Noticeably, the modulated spectra bend into the characteristic E^{+1} slope as a result of adiabatic energy losses which is the dominant process at non-

Journal of Physics: Conference Series **767** (2016) 012018 doi:10.1088/1742-6596/767/1/012018

relativistic proton energies deep inside the heliosphere, e.g. [15]. The computed spectra at larger radial distances show how the maximum in the spectra progressively shifts to lower energies. Protons, with $E < 100$ MeV, evidently experience a significant amount of modulation overall but especially at large radial distances close the HP. This spatial dependent behavior is further discussed next.

3.2. The modelled radial and latitudinal profiles for CR protons

The departure point for computing the radial profiles is the PAMELA spectra as observed and reproduced by the model for the periods 2006e, 2007e, 2008e and 2009e. It is then computed how the corresponding radial profiles at a given E unfold outwards in the heliosphere. The focus is on obtaining a global view of the behaviour of CRs from 2006 up to 2009. A comparison is made with Voyager 1 observations at 182 MeV for this period as is shown in Figure 2. Concentrating on the computations first, the results illustrate how the intensity profiles change from 2006 to 2009 by increasing systematically as solar activity decreases. The radial trends of the profiles change at the TS,

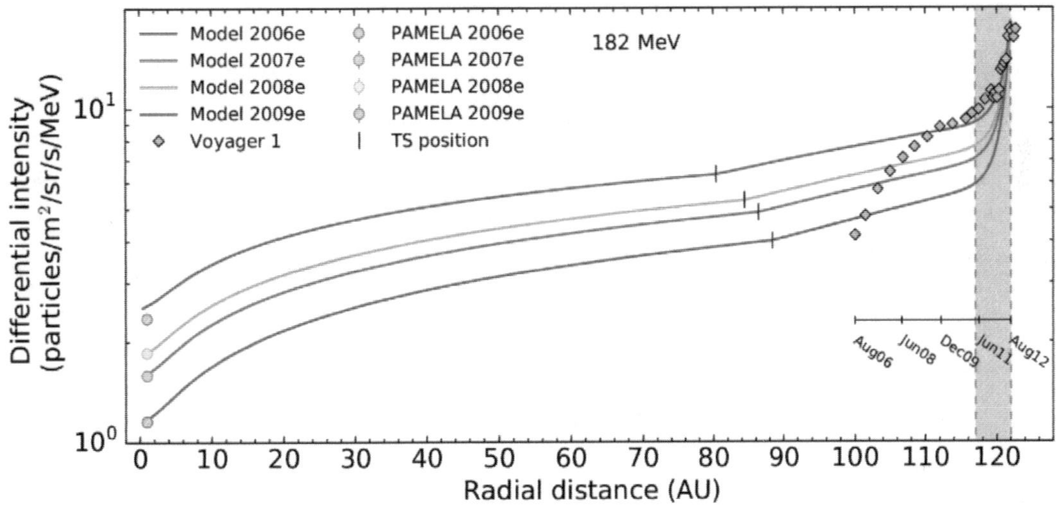

Figure 2. Computed radial intensities for 182 MeV protons at 56° colatitude based on the PAMELA measurements given by the coloured circles at 1 AU from 2006 (red line) and 2009 (blue line). The HP is fixed at 122 AU while the TS shifted its position as indicated by the short vertical black line. These profiles are compared to Voyager 1 measurements [16, 17] with time inset August 2006 to August 2012. The significance of the shaded region beyond ~116 AU is discussed in the text.

resuming inside the heliosheath with a different slope (larger radial gradients) towards the HP. Before discussing what happens beyond 117 AU inside the shaded region, first note that the way in which the Voyager 1 observed protons increased from 2006 to 2009 (see time inset), as it moved from ~100 AU to ~116 AU, corresponds very well with what the model predicts for this time period. In 2010, the observed proton intensity levels off at the position of Voyager 1 as solar minimum conditions as observed earlier at the Earth settled also in the outer heliosphere. But then, from ~ May 2011 onwards, the observed protons started to increase significantly and progressively more as the spacecraft approached the HP in August 2012 to eventually reached the observed value of the LIS at 182 MeV. From a modelling point of view, the straight forward approach of predicting the intensity globally up to 2009 based on the PAMELA spectra did not work any longer. To reproduce this increase, which is highly dependent on the energy of the CRs, the DCs in the model had to be reduced exponentially across this relatively narrow region. Although this has to be further investigated, it is evident that Voyager 1 experienced a region which may be called a HP modulation barrier for cosmic rays where the 182 MeV galactic protons increased by a factor of ~2, over just a few AU, to reach the corresponding LIS value; see also the discussion by [18, 32]. This is indicative of a very turbulent region associated with the HP, where instabilities as predicted with MHD models [29, 30], together with large magnetic field fluctuations [31] dominate.

As mentioned above, particle drifts in the heliosphere cause the 22-year solar cycle observed in the CR latitudinal gradients with negative latitudinal gradients during A < 0 polarity epochs. A negative latitudinal gradients means the CR intensity is higher in the equatorial plane than away from it; see [20] for illustrative examples.

De Simone et al. [19] used measurements from PAMELA and Ulysses to investigate the radial and latitudinal gradients of protons in the inner heliosphere during the A < 0 solar minimum leading up to 2009. Using an empirical approach to separate the radial and latitudinal gradients, they reported for the

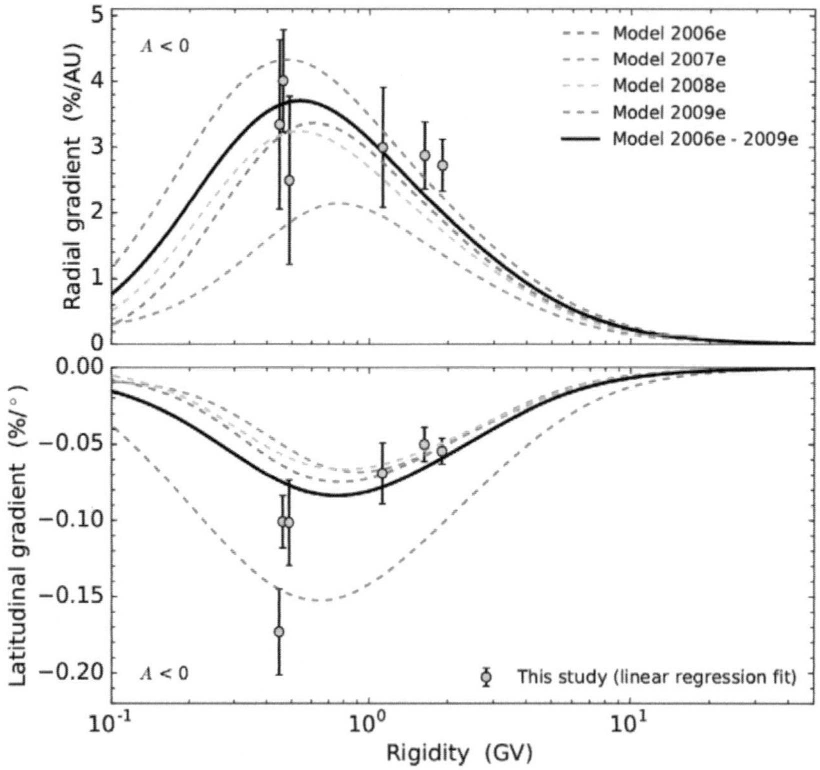

Figure 4. Top panel: Computed global radial gradients (G_r) between the positions of PAMELA (at the Earth) and along Ulysses' orbit, from July 2006 to June 2009, an A < 0 polarity cycle. Bottom panel: Similar but for the global latitudinal gradients (G_θ). In both panels the modelling gradients are given by the coloured dashed lines, with the solid black lines for the combined time period. These computed (modelled) gradients are compared to the corresponding gradients (grey symbols) calculated for this study from Ulysses KET and PAMELA proton observations.

rigidity interval 1.6 - 1.7 GV, a radial gradient G_r = (2.7 ± 0.2) %/AU and a latitudinal gradient of G_θ = (– 0.024 ± 0.005) %/degree, with the latter less negative than latitudinal gradients predicted by earlier drift models [5]. In hindsight, as discussed by [12], this was an indication that the modulation conditions during the minimum of cycle 23/24 were different and not indicative of drifts being overall less important in heliospheric modulation.

Numerical models can be used to compute local radial and latitudinal gradients for a given exact position anywhere in the heliosphere which elegantly illustrate the unique modulation characteristics of drifts, e.g. [20]. However, these local gradients are not useful if a comparison wants to be made with gradients calculated from observed CR intensities, mostly made far apart in space and at different times. For a meaningful comparison with PAMELA and Ulysses KET observations, a pragmatic, empirical method of analysis is used to calculate these global gradients, similar to the approach by [19], where details can be found; see also the PhD-thesis of Vos [21].

After applying the above mentioned procedure to every rigidity step of the model solutions, as well as to the available Ulysses KET and PAMELA observations, a comprehensive picture of the gradients G_r and G_θ emerges. This is shown in Figure 4, with G_r and G_θ given in the top and bottom panels,

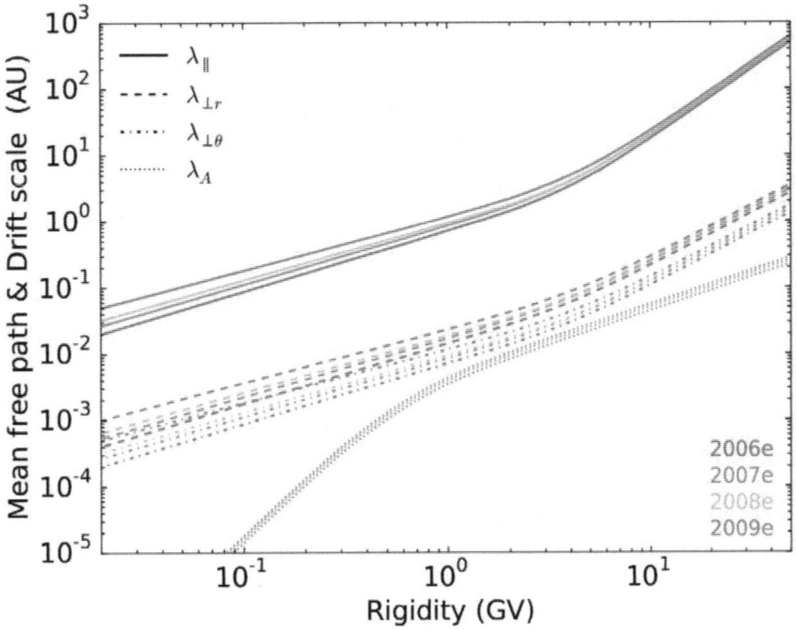

Figure 5. The time development of the rigidity dependence over the period 2006 to 2009 for the proton mean free paths (MFP) in AU; the parallel MFPs ($\lambda_{\parallel}$) are given by the four solid lines, the perpendicular MFPs in the radial ($\lambda_{\perp r}$) and polar ($\lambda_{\perp\theta}$) directions are given by the dashed and dashed-dotted lines, respectively. The drift scale (λ_A) is given by the four lowest dotted lines.

respectively. The model evidently produces positive G_r at all rigidities and negative G_θ for all four years, as expected for an $A < 0$ cycle but with G_θ less negative than previously predicted although still distinctively negative. According to this analysis, the model predicts that the largest G_r between the Earth and the position of Ulysses occurred during 2009, with a maximum of 4.25%/AU around 500 MV while the smallest is found for 2007, as a result of the fact that the latitudinal difference between the two spacecraft varied significantly during 2007 when Ulysses performed its fast latitude scan. Importantly, a consistency exists between the few observational values (symbols) and modelled values (lines). For G_θ, the most negative value is found for 2009, with – 0.15 %/degree around 600 MV, while the least negative G_θ is found for 2007. A characteristic of the model is that these gradients decrease significantly below ~400 MV because drifts decrease significantly with decreasing rigidity (see Figure 5). This serves to illustrate that when drifts are reduced, the negative latitudinal gradients will dissipate. The observational G_r of [19] is consistent with what we found, but G_θ is less negative than the – 0.065 %/degree combined modelling value. Otherwise, the observational and modelling values are in a satisfactory agreement.

3.3. The rigidity-time dependent mean free paths and drift scale for the period 2006 to 2009
It was shown above that the numerical model can reproduce the proton spectra observed between 2006 and 2009. In Figure 5 the development in time of the rigidity dependence of the proton mean free paths (MFP) is shown as required in the numerical approach to reproduce the mentioned spectra (as in Figure 1). The MFPs are presented in AU for diffusion parallel and perpendicular to the magnetic field lines at Earth and for the drift scale; the parallel MFPs ($\lambda_{\parallel}$) are given by the solid lines, perpendicular MFPs in the radial ($\lambda_{\perp r}$) and polar ($\lambda_{\perp\theta}$) directions by the dashed and dashed-dotted lines, respectively, and the drift scale (λ_A) by the dotted lines. The details of the approach are given by [9, 21]. The rigidity dependence of λ_A is similar to what [22] found when studying the global modulation of galactic Carbon with an independent numerical model.

The results shown in Figure 5 can be used to refine diffusion theory for the solar modulation of protons because it can be interpreted as empirical evidence of what represents very quiet solar modulation conditions in terms of: (1) How the three main DCs should scale with rigidity over a wide range. (2) Importantly, how this rigidity dependence changes over time for the perfect solar minimum period of 2006 to 2009.

4. Discussion and conclusions

The reported PAMELA proton spectra observed between mid-2006 and the end of 2009 enabled us to do a comprehensive numerical study of modulation conditions during this unusual solar minimum period as reported by [12] and extended by [9]. A newly constructed very LIS was used as an input spectrum that takes into account recent Voyager 1 observations at low energies [9]. See also the assessment of solar minimum proton spectra by [6] and the conclusion that the 2009 PAMELA proton spectrum was the highest recorded during the space era.

Based on these studies, the conclusion was made that the modulation minimum period of 2009 can be described as unusual; being relatively more diffusion dominated instead of being drift dominated as previous A < 0 polarity cycles seem to have been. However, our studies illustrate that drifts nevertheless played a notable role [27], especially because the HMF had decreased significantly until the end of 2009, in contrast to the moderate decreases observed during previous minimum periods.

For this report, we set out using the model of [9] with the parameters tuned to reproduce four year-end PAMELA proton spectra, and then applied the model to compute the radial dependence of the proton spectrum throughout the heliosphere from 2006 to 2009. Corresponding radial profiles were computed for each year along the Voyager 1 trajectory and compared to available relevant observations. It is found that the computed intensity levels are in agreement with solar minimum observations from Voyager 1 at multiple energies. In addition, the model, after some adjustments, could also reproduce the steep intensity increases observed when Voyager 1 crossed the HP region and comprehensively simulated with independent models by [23, 24]. In this context, we conclude that our model gives a most reasonable presentation of the cosmic ray radial profiles, from the Earth to the HP, for the years 2006 to 2009.

Fortunately, simultaneous observations from Ulysses KET and PAMELA are available between July 2006 and June 2009, so that we could study the global radial and latitudinal gradients in the inner heliosphere as well. To assure a meaningful comparison between our modelling results and observations, we applied an empirical method similar to that of [19] based on our simulations at the Earth and along Ulysses' orbit for this period. We found good agreement between the computed values and those calculated from observations for both the radial and latitudinal gradients. We conclude that the model also gives a most reasonable representation of the spatial gradients in the inner heliosphere for 2006 to 2009.

Our computations reflect that drifts indeed influence CR modulation during this unusual solar minimum. The notion that drifts were unimportant during the recent solar minimum [25] is not supported. Even though the observable effects of drifts were somewhat suppressed by the excess diffusion (larger DCs), drifts still maintain a strong presence as explained by [12]. We emphasize that the drift effects shown here, with a model tuned to the special conditions during the 2009 solar minimum, are indeed less than what drift models predicted previously for A < 0 cycles [26, 28]. Evidently, nobody had foreseen that the minimum modulation conditions for the A < 0 cycle of 2009 would be so different and unusual. The next solar minimum cycle may produce even more ideal minimum modulation conditions [6] since the Sun seems to be much quieter over the past decade than before.

Acknowledgments
The authors express their gratitude to the South African National Research Foundation (NRF) for providing partial research funding. EEV also thanks the South African National Space Agency (SANSA) for partial financial support during his PhD studies. They thank Jan Gieseler for discussions during his visit to South Africa and for providing them with updated Ulysses observations. They also thank Mirko Boezio, Valeria di Felice and Riccardo Munini as members of the PAMELA team for many insightful discussions of the PAMELA data.

References
[1] Heber B, Kopp A, Gieseler J, et al. 2009 *Astrophys. J.* **699** 1956
[2] Mewaldt R A, Davis A J, Lave K A, et al. 2010 *Astrophys. J. Lett.* **723** L1
[3] Adriani O, Barbarino G C, Bazilevskaya G A, et al. 2013 *Astrophys. J.* **765** 91
[4] Webber W R, Potgieter M S and Burger R A 1990, *Astrophys. J.* **349** 634.
[5] Potgieter M S 2014a *Adv. Space Res.* **53** 1415
[6] Strauss R D and Potgieter M S 2014 *Solar Phys.* **289** 3197
[7] Potgieter M S, le Roux J A and Burger R A 1989 *J. Geophys. Res.* **94** 2323
[8] Parker E N 1965 *Planet. Space Sci.* **13** 9
[9] Vos E E and Potgieter M S 2015 *Astrophys. J.* **815** 119
[10] Stone E C, Cummings A C, McDonald F B, et al. 2013 *Science* **341** 150
[11] Manuel R, Ferriera S E S and Potgieter M S 2015 *Astrophys. J.* **799** 223
[12] Potgieter M S, Vos E E, Boezio M, et al. 2014 *Solar Phys.* **289** 391
[13] Potgieter M S 2014b *Braz. J. Phys.* **44** 581
[14] Bisschoff D and Potgieter M S 2016 *Astrophys. Space Sci.* **361** 48
[15] Caballero-Lopez R A and Moraal H 2004 *J. Geophys. Res.* **109** A01101
[16] Webber W R, McDonald F B, Higbie P R and Heikkila B 2011 *arXiv* 1111.2377
[17] Webber W R, Higbie P R and McDonald F B 2013 *arXiv* 1308.1895
[18] Webber W R and Quenby J J 2015 *Astrophys. J.* **806** 138
[19] De Simone N, Di Felice V, Gieseler J, et al. 2011 *Astrophys. Space Sci. Trans.* **7** 425
[20] Ngobeni M D and Potgieter M S 2010 *Adv. Space Res.* **46** 391
[21] Vos E E, 2016 PhD thesis, North-West University, South Africa
[22] Ngobeni M D and Potgieter M S 2015 *Adv. Space Res.* **56** 1525
[23] Luo X, Zhang M, Potgieter M S, et al. 2015 *Astrophys. J.* **808** 82
[24] Strauss R D, Potgieter M S, Ferreira S E S, et al. 2013 *Astrophys. J. Lett.* **765** L18
[25] Cliver E W, Richardson I G and Ling A G 2013 *Space Sci. Rev.* **176** 3
[26] Potgieter M S 2013 *Living Rev. Solar Phys.* **10** 3
[27] Kóta, J 2013 *Space Sci. Rev.* **176** 391
[28] Heber B, Dröge W, Ferrando P, et al. 1996 *Astron. Astrophys.* **316** 538
[29] Washimi H, Zank G P, Hu Q, et al. 2011, *MNRAS* **416** 1475
[30] Avinash K, Zank G P, Dasgupta B and Bhadoria S 2014 *Astrophys. J.* **791** 102
[31] Burlaga L F, Ness N F, Florinski V and Heerikhuisen J, 2014 *Astrophys. J.* **792** 134
[32] Luo X, Potgieter M S, Zhang M, et al. 2016 *Astrophys. J.* in press

Modeling Secondary Neutral Helium in the Heliosphere

Hans-Reinhard Müller[1,2], **Eberhard Möbius**[3] **and Brian E. Wood**[4]

[1] Department of Physics and Astronomy, Dartmouth College, Hanover, NH 03755, USA
[2] Center for Space Plasma and Aeronomic Research, University of Alabama, Huntsville, AL 35805, USA
[3] Space Science Center and Department of Physics, University of New Hampshire, Durham, NH 03824, USA
[4] Naval Research Laboratory, Space Science Division, Washington, DC 20375, USA

E-mail: hans.mueller@dartmouth.edu

Abstract. An accurate, analytic heliospheric neutral test-particle code for helium atoms from the interstellar medium (ISM) is coupled to global heliospheric models dominated by hydrogen and protons from the solar wind and the ISM. This coupling enables the forward-calculation of secondary helium neutrals from first principles. Secondaries are produced predominantly in the outer heliosheath, upwind of the heliopause, by charge exchange of helium ions with neutral atoms. The forward model integrates the secondary production terms along neutral trajectories and calculates the combined neutral helium phase space density in the innermost heliosphere where it can be related to in-situ observations. The phase space density of the secondary component is lower than that of primary neutral helium, but its presence can change the analysis of primaries and the ISM, and can yield valuable insight into the characteristics of the plasma in the outer heliosheath.

1. Introduction

The detection of neutral interstellar helium atoms close to Earth has enabled a detailed characterization of key parameters of the local interstellar medium (LISM), including densities, velocity magnitude and direction, and LISM temperature. The LISM is the portion of the interstellar medium with which the Sun and its solar wind interacts, and it is thought to be a representative sample of the so-called Local Interstellar Cloud, the interstellar cloud that the heliosphere currently traverses on its galactic orbit.

The preference on studying helium as a key heliospheric neutral particle species is owed to helium being the species that suffers the least changes (losses, processing) on its way from the LISM to the detectors in the inner solar system, while having a high abundance in the ISM as compared to heavier elements. There are several ways to detect neutral helium atoms of interstellar origin in the innermost heliosphere, and this paper focuses on the direct detection through instruments on board of the *Ulysses* and *IBEX* missions. Both missions have a high sensitivity to helium, very good directional accuracy, and good energy resolution. The helium fluxes measured at the spacecraft hence can be related well to the neutral helium velocity distribution function (VDF) in the inner heliosphere.

With the help of archival *Ulysses* data and data collected by *IBEX* since 2009, it was possible to constrain the LISM parameters through the analysis of primary interstellar helium whose source is the LISM itself (Bzowski et al. 2015, Möbius et al. 2015, Wood, Müller & Witte 2015, McComas et al. 2015). A secondary component of neutral helium was noticed as well, albeit at a lower flux level. Efforts have been undertaken to separate the primary and the secondary signal from each other (Kubiak et al. 2014, 2016). One of the aims of such studies is to shore up the accuracy of the determination of primary neutral helium (and hence the pristine LISM itself) even further than with the prior analyses where the perturbation of the primaries due to the presence of secondaries was neglected.

Another, even stronger motivation for investigating the secondary helium component is rooted in the physics of their creation: Secondary helium is born predominantly through charge exchange in the outer heliosheath between the interstellar bow shock and the heliopause. The charge exchange collision converts a shock-decelerated, singly ionized He^+ ion into a secondary neutral helium atom, while maintaining the original momentum of the parent ion; the neutral can be detected by *Ulysses* and *IBEX* just like any other neutral helium atom. Secondaries therefore open a very intriguing window of remote sensing of the plasma conditions in the outer heliosheath, including the ion velocity distribution, using neutral detectors as telescopes taking advantage of this window. Thus far only *Voyager 1* has pushed into the outer heliosheath and is taking measurements very close to the heliopause. Both *Voyager* 1 and 2, when it crosses the heliopause, have limited instrumental capabilities to diagnose the outer heliosheath ion population and can provide only local samples, while, through secondary neutrals, IBEX provides an image over a large portion of the sky.

2. Framework of Determining Secondary Helium

2.1. Physics Context

Through an assessment of cross sections and the characteristic parameters of collision partners, three channels of charge exchange (c.x.) have been identified as the most important ones for the creation of secondary helium in the heliosphere (Bzowski et al. 2012):

He+He$^+$ → He$^+$+He — simple helium c.x. between primary neutral helium and helium ions; dominant in the outer heliosheath where both particle species are abundant; particularly He ions in the LISM have densities comparable to those of neutral He (Slavin & Frisch 2008).

He+He^{++} → He^{++}+He — double c.x. which has a large cross section; dominant in the solar wind where there are plenty of α particles, with a $1/r^2$ falloff; very little α in the LISM.

He$^+$+H → He+p — c.x. with neutral hydrogen; the cross section is low, but there is a higher density of neutral H partners than He, especially in the hydrogen wall that is co-located with the outer heliosheath.

The third reaction turns out to be much less important than the first one (Bzowski et al. 2012), and the second one is not relevant for spacecraft in the inner solar system, such as *Ulysses* and *IBEX*, which avoid pointing their detectors sunward, i.e. the expected arrival direction of these particular secondaries (neutralized solar wind).

Focusing on the dominant c.x. interaction $He+He^+ \rightarrow He^+ + He$, it is a generally valid assumption that the newly born neutral bears the velocity characteristics of the former He^+ ion. In the outer heliosheath this means that the secondaries are slower and correspond to a higher temperature, than the primaries because they originate from heated and decelerated ions. The lower velocity translates to stronger gravitational deflection for secondaries that reach the innermost heliosphere, and to a lower energy than the primaries, making them potentially harder to detect.

For simplicity, different ion species in interstellar plasma and solar wind plasma are assumed to be in equilibrium so that the velocity state of a He^+ ion is estimated with the help of the general plasma state, the latter being dominated by protons. The velocity state of the primary neutral helium can be estimated from prior test-particle simulations of this species (Müller 2012; Müller & Cohen 2012; Müller et al. 2013), and for the remote distances of the outer heliosheath, the primary helium can at first be well approximated by a shifted Maxwellian.

Once a secondary helium atom has been created through c.x., it is transported forward in a test-particle code (section 2.2 below) that has been developed for the accurate calculation of primary He in the heliosphere. Before reaching the detector, secondaries and primaries alike are subjected to the same types of losses on their trajectory from their source to the detector. In contrast to primary He with their "point source at infinity", the secondaries have distributed, non-zero production terms along the entire Keplerian trajectory that need to be integrated to get the right flux at a point of interest like the spacecraft detector. Production and losses together give a net phase space density at the velocity that this trajectory corresponds to at the detector.

Repeating the procedure for all different arrival directions, and different possible velocity magnitudes, is equivalent to testing a whole variety of trajectories that all intersect at the detector, each single one with integrated production and loss terms. This characterizes an entire velocity distribution function (VDF) of secondaries at the location of the spacecraft which can be converted into predicted fluxes. This method of calculating the secondary VDF operates from first principles and makes no assumptions about the shape of the VDF at any reference distance. As creation and loss calculations depend on the global heliosphere background model, the results for secondaries will depend on the background heliosphere model sensitively, more so than the primary He calculations. The secondary results, just like the primary ones, also depend on the accuracy of our knowledge of the c.x. cross sections, and on the underlying assumptions listed above, including the assumptions about the VDF (both neutral and ionized) in the pristine LISM being Maxwellian, and the plasma components being in equilibrium.

2.2. Calculation Method

2.2.1. Dartmouth test-particle forward modeling. The Dartmouth test-particle code and its results when applied to primary helium have been described elsewhere (Müller 2012; Müller & Cohen 2012; Müller et al. 2013; Wood & Müller 2015; Wood et al. 2015). In summary, the calculation method is an ab-initio modeling that assumes a fixed neutral helium VDF in the undisturbed LISM and transports neutral particles from a distant reference distance to the location of spacecraft such as *Ulysses* and *IBEX*. For each particle, quantities that are conserved along the entire trajectory (total energy, angular momentum, and eccentricity vector; Müller & Cohen (2012)), are used to determine in one algebraic step the neutral He velocity at a desired point in space, when the velocity in the pristine LISM ("at infinity") is given. During transport to the position of the spacecraft, the distribution will be primarily affected by Sun's gravity, and through photoionization by solar EUV photons. Moreover, c.x. losses and those by electron impact ionization contribute changes. The latter two loss channels require the knowledge of the global plasma heliosphere background model which is obtained from an MHD model independent from the neutral He atoms (e.g., Müller et al. 2008); in this way the test-particle code is coupled to the global heliosphere model. The repetition of the trajectory method for all trajectories that intersect at the desired point of interest yields a complete primary neutral He VDF at that point. Sample figures of these primary VDFs are shown in the references cited above.

For exploration purposes, an axisymmetric global heliosphere model is chosen as background model here, to begin with the simplest possibility. A heliocentric Cartesian coordinate system is defined, with the x axis antiparallel to the LISM flow, and y and z perpendicular coordinates. The parameters characterizing interstellar matter and the solar wind are given in Table 1; they are just an example setup (proof of concept), without the claim of being the most appropriate or realistic parameters. The parameters correspond to a two-shock heliosphere with a weak interstellar bow shock present, hence in what follows, we make reference to a bow shock even though the presence of such a shock is not universally accepted.

Journal of Physics: Conference Series **767** (2016) 012019 doi:10.1088/1742-6596/767/1/012019

Table 1. Parameters, both for LISM and the solar wind, of the global heliosphere background model and the neutral test-particle code. Units for speed, temperature T, and number densities n are km/s, K, and cm^{-3}, respectively. Symbols p, H, and He signify protons, hydrogen, and helium, the latter in various charge states. The He^{++}/p and He^{+}/p columns represent plasma abundance ratios.

	speed	T	n_p	He^{++}/p	He^{+}/p	n_H	n_{He}
LISM	26.3	7000	0.047	0.0079	0.36	0.216	0.018
SW at 1 AU	400	10^5	5.0	0.04	0.001	-	-

2.2.2. Secondary helium modeling. The calculation of secondary VDF follows the exact same strategy as the one for primaries: The VDF at a location of interest is assembled as a comprehensive scan in velocity space. Each individual velocity that is picked, determines a unique Keplerian trajectory along which the integration of secondary production terms is carried out. The secondary production calculation at an arbitrary point of the trajectory requires the knowledge of two collision partners, the helium ion and the neutral. The helium ion will become the secondary neutral after c.x., so it needs to be taken from a local plasma Maxwellian at exactly the velocity state that corresponds to what the neutral Kepler trajectory requires at that location. This gives an overall phase space density factor; often it is quite small because it represents a very improbable ion from the tail of the Maxwellian distribution, and overall strong secondary fluxes are only obtained when trajectory velocities coincide with the peak of local plasma Maxwellians. For simplicity, the width of the local He^{+} Maxwellian (thermal velocity) is derived here from the local plasma temperature. As has been emphasized by, e.g., Zank et al. (1996, 2013), heliosphere plasmas are heated by charge exchange, yet the equilibration time scales are too long to allow an actual equilibration in the heliosphere, including in the outer heliosheath (Zank et al. 2014). The majority of global heliosphere models which track protons in the heliosphere as a single MHD fluid component, summarizes the contributions of thermal protons and non-thermal (such as pickup) ions in the form of a single, effective plasma temperature. The cooler thermal plasma component dominates by number and hence should be used for calculating charge exchange; using the effective plasma temperature currently systematically overestimates the width of the local He^{+} Maxwellian. This issue will be addressed in future work.

For the other collision partner, i.e. the neutral atom, the neutral distribution function needs to be integrated in the form of a Boltzmann-style production integral, where the integrand depends on the relative neutral-ion velocity during the integration, as well as the cross section which itself depends on that relative velocity (e.g., Müller et al. 2000). Together with the standard loss calculation, the integration of the production terms yields the secondary phase space density at this investigated VDF velocity, at the point of interest under investigation. Scanning through all possible VDF velocities in this manner finally results in a complete secondary neutral VDF (example in section 3).

For the primary VDF calculations it was advantageous to determine first the locations of the peaks (both direct and indirect) in phase space density (PSD), and then obtain the VDFs in the velocity-vicinity of these peaks (Müller 2012; Müller & Cohen 2012). For the secondary VDF, this is not possible as it represents a distributed source rather than a "point" source at infinity. For this reason, the entire possible velocity space should in principle be scanned. In practice, one can however identify some promising velocity regions and hence restrict the velocity scanning.

The main production site for secondaries is the outer heliosheath. There is no production of secondaries upwind of the bow shock, and negligible production inside the heliopause. This results in secondary VDFs whose maxima are quite close to the primary direct and indirect VDFs. Furthermore, there is a sharp cutoff toward lower velocities: For a given point of interest, there is a minimum velocity below which the total energy becomes negative, corresponding to closed, bound trajectories as opposed to the open ones that reach infinity and have non-zero kinetic energy even at infinity. Closed trajectories (Kepler ellipses) can still have aphelia that are distant enough to be in the outer heliosheath and beyond and therefore have secondary production along parts of their trajectory. When the aphelia are interior to the heliopause, then there are practically no secondaries produced along the entire orbit, leading to a precipitous drop in phase space density (PSD) for these velocities. This criterion forms a sphere in velocity space around the origin; the spherical volume within has practically zero secondaries, and there is no need to scan this velocity volume (which of course also lacks primaries). For this reason, the scan strategy for the secondary VDF is carried out in spherical polar coordinates in velocity, with the inner radius (minimum velocity) as outlined. Far away from direct and indirect primary peaks, the secondary VDF also drops to very low levels, hence the scanning strategy stops at an outer radius (maximum velocity). For more comprehensive studies, VDF scanning should be carried further to velocities that correspond to solar wind velocities, to identify side peaks in the neutral secondary VDF generated by double charge exchange of solar wind alpha particles (second c.x. reaction listed in section 2.1).

3. Results and Discussion

As a sample of the viability and the success of the method outlined above, Figure 1 contains results representing the secondary neutral helium VDF at a location of interest (0,1) that is 1 AU from the Sun. The location is perpendicular to the ISM flow. While the actual VDF is an object in three-dimensional velocity space, Figure 1 presents a 2D slice through it, at a central slice plane that contains the peaks of the VDF.

The figure shows the concept of the VDF not having been calculated inside of a minimum velocity (here, about 40 km/s); this entire region represents closed ellipses that do not reach outside of the heliopause and therefore do not contain any secondary production terms. As can be seen from the phase space densities both in color and as contour lines, the transition to this regime is quite steep, which is an indication that the aphelion distance is a steep function of velocity magnitude here.

In the velocity region indicated by the inner and outer circular bounds (a thick velocity shell in 3D), the black contour lines identify the primary direct VDF, peaking at (-45, -20) velocity which is the ISM velocity vector (-26.3, 0) transported into the solar gravity well to the point (0, 1), and the primary indirect VDF peaking at (20, 45) which corresponds to particles that have run from the ISM through perihelion and were on their way out when they intersected with the location (0, 1). Losses for the indirect peak are stronger because of the trip around the Sun, resulting in depletion so that the normalized phase space density $f = 0.1$ contour is already missing (the respective outermost contours correspond to $f = 10^{-4}$; all these numbers are normalized so that the PSD in the pristine LISM is 1).

The color scheme represents the secondary neutral He phase space density with the same normalization as the primary one, i.e. directly comparable to the primaries. The color map spans 5 orders of magnitude, and starts where the primary contours end. The secondary PSDs are quite low, yet they organize themselves in the vicinity of the direct and indirect primary peaks. Because of the general nature of secondaries being slower, it makes sense that the secondary peaks occur at lower velocities, (-34, -25) for the direct secondary peak and (29, 32) for the indirect secondary peak. The appearance also confirms that secondaries are more deflected when compared to primaries. The actual peak value for the direct secondary peak is $f_2 = 2.8 \times 10^{-5}$ (for the indirect, $f_2 = 3.4 \times 10^{-6}$). These numbers, which are normalized to the neutral helium PSD in the LISM, are quite low; a recent *IBEX* analysis determined secondary flux levels to be

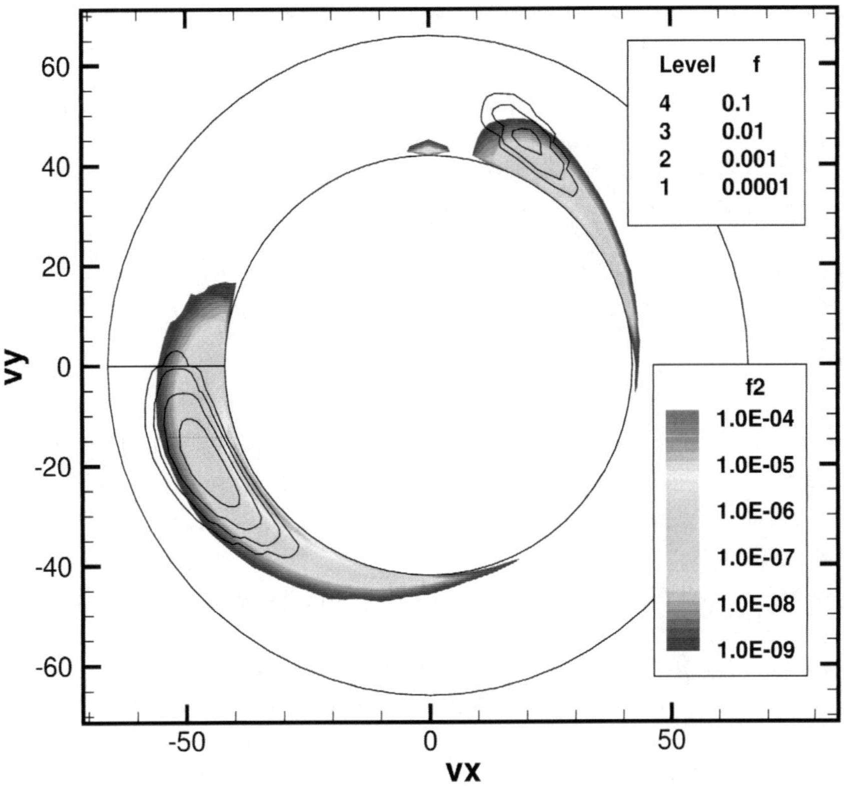

Figure 1. Example calculation of primary (black contours, f) and secondary (color contours, f_2) particle distributions at the point (0,1) sidewind, for one particular global heliospheric model. The values are normalized to the primary LISM value. While the secondary neutrals are at lower densities, their addition to the overall VDF still yields a shift in bulk velocity and temperature. The inner cutoff radius expresses a speed for which trajectories no longer connect with the outer heliosheath.

5.7% of the interstellar primary value (Kubiak et al. 2016). The discrepancy between the *ab-initio* model of secondaries presented here and the *IBEX* observations deserves future attention, including the exploration of the parameters of Table 1 and of the other assumptions made for this initial version of the calculations.

A modeled trajectory in the actual code that represents the secondary direct peak particles of Figure 1, enters the global background simulation domain at location (1000 AU, -28 AU) with a positive v_y component, traverses the outer heliosheath on a trajectory close to the symmetry axis and essentially parallel to the shocked plasma, runs through the point of interest (0, 1), experiences a perihelion at $r = 0.66$ AU, and exits the simulation domain on the outgoing hyperbola leg at location (780 AU, -620 AU).

Around the direct peak, the primary and secondary levels become comparable, at least away from the primary peak, which supports the statement that secondaries have the potential to disguise as tails of the primaries and hence throw off the analysis of the primaries when secondaries are neglected. The addition of the secondary component to the primary VDF can shift and broaden the overall VDF. These parameter changes will depend on the location of interest.

Once the combined He velocity distribution is known at the location of the spacecraft, a simulated He flux map, comprised of both primaries and secondaries, can be computed for comparison with the observations. This simulated flux map must take into account the energy efficiency and detector geometry function for the spacecraft detector involved. It will be the comparison of synthesized with measured fluxes from *Ulysses* and *IBEX* that will allow a determination of the plasma characteristics of the outer heliosheath through model parameter variations and identifying the best match between model and experiment.

Acknowledgments

The authors acknowledge funding for parts of this work by NASA grants NNX10AC44G, NNH13AV19I, NNX14AC10G, and the *IBEX* contract NNG05EC85C.

References

Bzowski, M., Kubiak, M. A., et al. 2012, ApJS 198, 12, doi:10.1088/0067-0049/198/2/12

Bzowski, M., Swaczyna, P., et al. 2015, ApJS 220, 28, doi: 10.1088/0067-0049/220/2/28

Kubiak, M. A.; Bzowski, M.; Sokol, J. M.; Swaczyna, P.; et al. 2014, ApJS 213, 29, doi 10.1088/0067-0049/213/2/29

Kubiak, M. A.; Swaczyna, P.; Bzowski, M.; Sokol, J. M.; et al. 2016, ApJS 223, 25, doi: 10.3847/0067-0049/223/2/25

McComas, D. J., et al. 2015, ApJS 220, 22, doi 10.1088/0067-0049/220/2/22

Möbius, E., Bzowski, M., et al. 2015, ApJS 220, 24, doi 10.1088/0067-0049/220/2/24

Müller, H.-R. 2012, in Numerical Modeling of Space Plasma Flows (Astronum 2011), N.V. Pogorelov, J.A. Font, E. Audit & G.P. Zank (Eds.), ASP Conference Series 459, 228-233 (http://arxiv.org/abs/1205.1555)

Müller, H.-R., Bzowski, M., Möbius, E., & Zank, G. P. 2013, in: Solar Wind 13, G.P. Zank et al. (Eds.), AIP Conference Series 1539, 348-351, doi: 10.1063/1.4811058

Müller, H.-R., & Cohen, J. H. 2012, in Physics of the Heliosphere: A 10-year Retrospective, ed. J. Heerikhuisen & G. Li, AIP Conference Series 1436, 233-238, doi: 10.1063/1.4723613

Müller, H.-R., Florinski, V., Heerikhuisen, J., Izmodenov, V. V., Scherer, K., Alexashov, D., & Fahr, H.-J. 2008, A&A, 491, 43, doi 10.1051/0004-6361:20078708

Müller, H.-R., Zank, G. P., & Lipatov, A. S. 2000, J. Geophys. Res., 105, 27419, doi 10.1029/1999JA000361

Slavin, J. D., & Frisch, P. C. 2008, A&A, 491, 53, doi 10.1051/0004-6361:20078101

Wood, B. E., & Müller, H.-R. 2015, in Linear and Nonlinear Particle Energization Throughout the Heliosphere and Beyond, Journal of Physics: Conference Series 642, article id. 012029, doi 10.1088/1742-6596/642/1/012029

Wood, B. E., Müller, H.-R., & Witte, M. 2015, ApJ, 801, 62, doi 10.1088/0004-637X/801/1/62

Zank, G. P.; Heerikhuisen, J.; Wood, B. E.; et al. 2013, ApJ, 763, 20, doi 10.1088/0004-637X/763/1/20

Zank, G. P., Hunana, P., Mostafavi, P., & Goldstein, M. L. 2014, ApJ, 797, 87, doi 10.1088/0004-637X/797/2/87

Zank, G. P., Pauls, H. L., Williams, L. L., & Hall, D. T. 1996, J. Geophys. Res., 101, 21639, doi 10.1029/96JA02127

Pickup Ion Effect of the Solar Wind Interaction with the Local Interstellar Medium

N. V. Pogorelov,[1] M. C. Bedford,[1] I. A. Kryukov,[2,1] and G. P. Zank[1]

[1]Department of Space Science, University of Alabama in Huntsville, Huntsville, AL 35805, U.S.A.

[2]Institute for Problems in Mechanics, Russian Academy of Sciences, Moscow, Russia

E-mail: `nikolai.pogorelov@uah.edu`

Abstract. Pickup ions are created when interstellar neutral atoms resonantly exchange charge with the solar wind (SW) ions, especially in the supersonic part of the wind, where they carry most of the plasma pressure. Here we present numerical simulation results of the 3D heliospheric interface treating pickup ions as a separate proton fluid. To satisfy the fundamental conservation laws, we solve the system of equations describing the flow of the mixture of electrons, thermal protons, and pickup ions. To find the density and pressure of pickup ions behind the termination shock, we employ simple boundary conditions that take into account the *Voyager* observations that showed that the decrease in the kinetic energy of the mixture at the termination shock predominantly contributed to the increase in the pressure of pickup ions. We show that this model adequately describes the flow of the plasma mixture and results in a noticeable decrease in the heliosheath width.

1. Introduction

The solar wind (SW) interaction with the local interstellar medium (LISM) is strongly affected by charge exchange between ions, predominantly protons, and neutral atoms, predominantly neutral hydrogen (H). The importance of charge exchange had been acknowledged before the first quantitative model of the SW–LISM interaction was put forward [1, 2, 3, 4]. The reason for this is based on the partial ionization of the LISM. In particular, the neutral H density in the LISM is about three times greater than the proton density. In principle, there is no direct measurement of latter density because the plasma instrument onboard *Voyager* 1 ($V1$), which is believed to have been traversing the LISM since July 2012 [5, 6, 7, 8, 9]. The electron density has been inferred to be about 0.08 cm^{-3} from the plasma wave instrument observations [10]. The neutral H density in the unperturbed LISM is usually derived from numerical simulations, in particular, to satisfy the condition that the H density be 0.08–0.09 cm^{-3} in the inner heliosphere. These values are related to pickup ion (PUI) measurements by *Ulysses* [11, 12] and SW ion deceleration at *Voyager* 2 ($V2$) due to charge exchange [13]. PUIs are secondary ions that are born when primary ions experience charge exchange with neutral atoms. As PUIs born from different H atoms in different regions of the SW–LISM interaction have substantially different properties, they should ideally be modeled as multiple populations, e.g., as in [14], and kinetically. The SW–LISM interface is usually, for convenience of interpretation, subdivided into 4 regions: (1) region 0 is in the LISM unperturbed by the presence of the heliosphere; (2) region 1 is the LISM plasma region modified due to the flow deceleration at the heliopause (HP) – a tangential discontinuity which separates the SW from the LISM, if ideal MHD terminology is used; (3) region 2 is between the HP and the heliospheric termination shock (TS); and (4) region 3 is the supersonic SW. A PUI created in any of these regions experiences the action of an electric field that acts on it in the plasma frame until the PUI bulk velocity becomes equal to the plasma bulk velocity [15]. The newly created PUIs form essentially a ring-beam distribution with the speed ranging between 0 and $2V_{SW}$ in the Sun's inertial frame [16]. Such distribution function is unstable to a number of instabilities (see [16] and references therein), which results in a spherical shell distribution that becomes filled as the SW propagates outwards [17, 18]. The heliosphere beyond the ionization cavity is dominated thermally by PUIs [19, 20, 21, 22, 23]. According to [24, 25], the inner heliosheath (IHS) pressure contributed by energetic PUIs and anomalous cosmic rays far exceeds that of the thermal background plasma and magnetic field. The IHS here coincides with region 2 introduced above.

A number of one-plasma-fluid models [26, 27, 28, 29, 30, 31, 32, 33] take into account the effect PUIs by assuming that they are in thermal equilibrium with the background plasma. Although this is not true, the conservation laws of mass, momentum, and energy for the mixture of electrons, ions, and PUIs are still satisfied approximately. According to [22, 23], such approaches may be correct only if the heat conduction tensor and the dissipation coefficient due to ion-PUI interaction are both zero. This happens when PUIs are completely coupled with the thermal ions and the scattering time is infinitely small. It was shown in [34] that the effect of PUIs can be quantified if they are treated as a separate fluid. In [35, 36, 37, 38], time-dependent simulations of the supersonic SW are presented with PUIs treated as a separate fluid and turbulence effects taken into account. These models still make the approximation of instantaneous isotropization, i.e., that the PUI description can be described as a filled shell, thus neglecting the nearly-isotropic character of the PUI distribution function when a finite scattering time is included [22].

In principle, any PUI fluid model is an approximation of their kinetic behavior. Such a model was developed by [14] on the assumptions that PUIs are isotropic away from discontinuities and the TS is a perpendicular shock. However, proper boundary conditions are necessary to describe the PUI crossing the TS. Paper [39] investigates the evolution of the PUI distribution function (in a pitch-angle-averaged approximation) together with the PUI-generated waves PUIs that heat the thermal SW ions.

The extension of a 2-fluid SW model developed for the supersonic flows is not straightforward. This is because the pressure equation for PUIs used in such models is not valid across the TS. Instead, some boundary conditions should be developed for the PUI density and pressure, ρ_{PUI} and p_{PUI}. Such boundary conditions are discussed in [40, 41, 42, 43, 44, 45]. It is known, e.g., the most of the SW kinetic energy was absorbed by PUI when the TS was crossed by $V1$ [46] (see also predictions in [41]). One would not expect a simple finite approximation of p_{PUI} across the TS to satisfy such boundary conditions. In principle, as explained, e.g., in [47], one should either write out and solve a system of conservation laws across of a discontinuity or specify the boundary conditions at it. These approaches are equivalent, because the number of possible conservation laws is infinite, which results in an infinite number of shock boundary conditions. It is understood that only the fundamental physical conservation laws are appropriate for the description of magnetized fluid flows. In contrast, [48] solve non-conservative pressure equations everywhere in the SW–LISM interaction region, which violates the basic principles required to solve systems of hyperbolic equations. Moreover, the turbulence model of [49] used in [48] is based on an Alfvén-mode approximation and is invalid in the IHS. Moreover, the source terms used in [48] are valid only for "cold" plasma and are not applicable to the hot IHS plasma.

In this paper, we use simple boundary conditions at the TS which allows us to demonstrate that the IHS width decreases when the PUI fluid is treated separately.

2. Governing Equations

We present the system of equations that governs the flow of the plasma mixture (thermal protons, PUIs, and electrons), neutral atom populations, and PUIs. Subscripts for charge exchange source terms are identified by the subscripts a, which is a fixed integer for the corresponding population of neutral atoms, p for thermal protons, and I for pickup protons:

Plasma mixture

$$\frac{\partial \rho}{\partial t} + \nabla \cdot (\rho \mathbf{u}) = 0 = D_\Sigma, \tag{1}$$

$$\frac{\partial \rho \mathbf{u}}{\partial t} + \nabla \cdot \left(\rho \mathbf{u}\mathbf{u} + p^* \mathbf{I} - \frac{1}{4\pi}\mathbf{B}\mathbf{B} \right) = \mathbf{M}_\Sigma, \tag{2}$$

$$\frac{\partial E}{\partial t} + \nabla \cdot \left((E + p^*)\mathbf{u} - \frac{1}{4\pi}(\mathbf{B}\cdot\mathbf{u})\mathbf{B} \right) = E_\Sigma, \tag{3}$$

$$\frac{\partial \mathbf{B}}{\partial t} + \nabla \cdot (\mathbf{u}\mathbf{B} - \mathbf{B}\mathbf{u}) = 0. \tag{4}$$

Neutrals

$$\frac{\partial \rho_a}{\partial t} + \nabla \cdot (\rho_a \mathbf{u}_a) = D_a, \tag{5}$$

$$\frac{\partial \rho_a \mathbf{u}_a}{\partial t} + \nabla \cdot (\rho_a \mathbf{u}_a \mathbf{u}_a + p_a \mathbf{I}) = \mathbf{M}_a, \tag{6}$$

$$\frac{\partial e_a}{\partial t} + \nabla \cdot ((e_a + p_a)\mathbf{u}_a) = E_a. \tag{7}$$

Pick-up ions

$$\frac{\partial \rho_\mathrm{I}}{\partial t} + \nabla \cdot (\rho_\mathrm{I} \mathbf{u}) = D_\mathrm{I}, \tag{8}$$

$$\frac{\partial p_\mathrm{I}}{\partial t} + \nabla \cdot (p_\mathrm{I} \mathbf{u}) + (\gamma - 1) p_\mathrm{I} \nabla \cdot \mathbf{u} = P_\mathrm{I} \tag{9}$$

Here p is the pressure, ρ the mass density, $\mathbf{u}$ the bulk velocity vector, $\mathbf{B}$ the magnetic filed vector, e the total energy density (includes the magnetic energy), p^* the total pressure of the mixture (including magnetic energy), and $\mathbf{I}$ is the identity tensor. The source terms D, $\mathbf{M}$, E,a nd P are defined later. The subscript Σ refers to the mixture properties.

The following quantities are used in the definition of the source terms [50]:

$$U^\rho_{ij} = \left[\frac{4}{\pi}(v^2_{T_i} + v^2_{T_j}) + (\mathbf{u}_i - \mathbf{u}_j)^2 \right]^{1/2}, \tag{10}$$

$$U^m_{ij} = \left[\frac{16}{\pi} v^2_{T_i} + \frac{9\pi}{4} v^2_{T_j} + 4(\mathbf{u}_i - \mathbf{u}_\mathrm{j})^2 \right]^{1/2}, \tag{11}$$

$$U^e_{ij} = \left[\frac{4}{\pi} v^2_{T_i} + \frac{64}{9\pi} v^2_{T_j} + (\mathbf{u}_i - \mathbf{u}_j)^2 \right]^{1/2}, \tag{12}$$

$$\lambda_{ij}(U^\rho_{ij}) = \sigma_\mathrm{cx}(U^\rho_{ij}) = \left(13.493 - 0.531 \log U^\rho_{ij} \right)^2 \left(1 - \exp \frac{-2.94 \times 10^9}{U^\rho_{ij}} \right)^{4.5} \tag{13}$$

Here the subscripts i and j can have values of a and either p or I. The quantity v_T is the thermal velocity and λ_{ij} is the charge exchange cross section of particles of type i with particles of type j.

The source terms for the density, momentum, energy, and PUI pressure equations have the form

$$H^{\rho}_{ak} = \lambda_k \rho_k \rho_a U^{\rho}_{ak}, \qquad k = \mathrm{p}, \mathrm{I} \tag{14}$$

$$\mathbf{H}^m_{ak} = \lambda_k \rho_k \rho_a \left(U^{\rho}_{ak}\mathbf{u} + \frac{v^2_{T_k}}{U^m_{ak}}(\mathbf{u} - \mathbf{u}_a) \right), \tag{15}$$

$$\mathbf{H}^m_{ka} = \lambda_k \rho_k \rho_j \left(U^{\rho}_{ka}\mathbf{u}_a - \frac{v^2_{T_a}}{U^m_{ka}}(\mathbf{u} - \mathbf{u}_a) \right), \tag{16}$$

$$H^e_{ak} = \lambda_k \rho_k \rho_a \left(\frac{1}{2}u^2 U^{\rho}_{ak} + \frac{3}{4}v^2_{T_k} U^e_{ak} + \frac{v^2_{T_k}}{U^m_{ak}}\mathbf{u}\cdot(\mathbf{u} - \mathbf{u}_a) \right), \tag{17}$$

$$H^e_{ka} = \lambda_k \rho_k \rho_a \left(\frac{1}{2}u_a^2 U^{\rho}_{ka} + \frac{3}{4}v^2_{T_a} U^e_{ka} - \frac{v^2_{T_a}}{U^m_{ka}}\mathbf{u}_a\cdot(\mathbf{u} - \mathbf{u}_a) \right), \tag{18}$$

$$H^p_{ak} = \lambda_k \rho_k \rho_a (\gamma - 1)\frac{3}{4}v^2_{T_k} U^e_{ak}, \tag{19}$$

$$H^p_{ka} = \lambda_k \rho_k \rho_a (\gamma - 1) \left(\frac{1}{2}U^{\rho}_{ka}(\mathbf{u} - \mathbf{u}_a)^2 + \frac{v^2_{T_a}}{U^m_{ka}}(\mathbf{u} - \mathbf{u}_a)^2 + \frac{3}{4}v^2_{T_a} U^e_{ka} \right) \tag{20}$$

We will assume here (this assumption can be lifted later) that there are no superthermal electrons. Thus, introducing the number density n, we obtain

$$p = n_{\mathrm{e}}kT_{\mathrm{e}} + n_{\mathrm{p}}kT_{\mathrm{p}} + n_{\mathrm{I}}kT_{\mathrm{I}}, \qquad n_{\mathrm{e}} = n_{\mathrm{p}} + n_{\mathrm{I}} = n,$$

which reduces to

$$p = nkT_{\mathrm{e}} + (n - n_{\mathrm{I}})kT_{\mathrm{p}} + n_{\mathrm{I}}kT_{\mathrm{I}}. \tag{21}$$

Assuming $T_{\mathrm{e}} = T_{\mathrm{p}}$, we obtain

$$\frac{p}{n} = (2 - \frac{n_{\mathrm{I}}}{n})kT_{\mathrm{p}} + \frac{n_{\mathrm{I}}}{n}kT_{\mathrm{I}}. \tag{22}$$

We now introduce the proton mass m and thermal ion velocity $v^2_T = \frac{2kT}{m}$, which can be expressed as

$$v^2_{T_p} = \frac{\frac{2p}{\rho} - \frac{n_{\mathrm{I}}}{n}v^2_{T_{\mathrm{I}}}}{2 - \frac{n_{\mathrm{I}}}{n}}. \tag{23}$$

It is easy to notice that if $n_{\mathrm{I}} = 0$, then $v^2_{T_p} = \frac{p}{\rho}$, recovering the formula for the one-ion fluid that takes into account the electron pressure. Obviously, $v^2_{T_a} = \frac{2p}{\rho}$ for neutral atoms.

Using the results of [14], each ion and neutral atom is identified as a thermal proton, pickup proton, or a neutral population atom. This is a simplification, but the best that can be done without adding more PUI fluids or using kinetic models directly.

Let us consider reactions between different species. In our extended notation, p_0, p_1, and H_i stand for thermal protons, pickup protons, and neutral atoms of population i. In region 3, a charge exchange results in a population 3 neutral atom and a pickup ion, except for thermal ion charge exchange with a population 3 neutral. The latter produces a thermal proton. So in region 3, PUIs are mostly created and thermal ions are mostly destroyed. In region 2, we assume that

Journal of Physics: Conference Series **767** (2016) 012020 doi:10.1088/1742-6596/767/1/012020

PUIs are only destroyed. This is because the population of PUIs born in the IHS is substantially different from those that have crossed the TS. Since we did not introduce additional populations of PUIs, all charge exchanges in region 2 result in thermal protons and population 2 neutrals. For the purpose of this study, we ignore PUIs created in the outer heliosheath beyond the TS. Thus, in region 1, PUIs and thermal LISM ions are indistinguishable. All charge exchanges produce thermal protons and population 1 neutrals.

In the following, the subscript order and sign depend on the reaction: in region 2, for example, $\mathbf{H}^m_{\mathrm{I}1}$ means the source term contribution to the momentum equation due to charge exchange of a population 1 neutral atom with a PUI. The formula below says that this results in a thermal ion, so it is subtracted from $\mathbf{M}_1$ and added to $\mathbf{M}_\Sigma$. The corresponding $H^m_{1\mathrm{I}}$ stands for the momentum lost by a PUI when it exchanges charge with a population 1 neutral atom, so it is subtracted from $\mathbf{M}_\Sigma$ and added to $\mathbf{M}_2$.

The sum of all mixture and neutral source terms should be zero, but not for the PUI pressure and density. So for region 3, the PUI source term includes positive contributions from thermal protons exchanging charge with populations 1 and 2 (which create PUIs), nothing from thermal protons exchanging charge with population 3 (no PUIs created or destroyed), both positive and negative contributions from PUIs exchanging charge with populations 1 and 2 (one PUI destroyed and another created), and only negative contributions from PUIs exchanging charge with population 3 (PUI destroyed).

In summary,

Region 3

$$
\begin{aligned}
p_0 + H_1 &\to p_1 + H_3, \quad & p_1 + H_1 &\to p_1 + H_3,\\
p_0 + H_2 &\to p_1 + H_3, \quad & p_1 + H_2 &\to p_1 + H_3,\\
p_0 + H_3 &\to p_0 + H_3, \quad & p_1 + H_3 &\to p_0 + H_3.
\end{aligned}
$$

This implies that

$$D_\Sigma = 0, \quad D_1 = -H^\rho_{\mathrm{p}1} - H^\rho_{\mathrm{I}1}, \quad D_2 = -H^\rho_{\mathrm{p}2} - H^\rho_{\mathrm{I}2},$$

$$
\begin{aligned}
D_3 = H^\rho_{\mathrm{p}1} + H^\rho_{\mathrm{p}2} + H^\rho_{\mathrm{I}1} + H^\rho_{\mathrm{I}2}, \quad D_{\mathrm{I}} = H^\rho_{\mathrm{p}1} + H^\rho_{\mathrm{p}2} - H^\rho_{\mathrm{I}3},\\
\mathbf{M}_\Sigma = -H^m_{1\mathrm{p}} + H^m_{\mathrm{p}1} - H^m_{2\mathrm{p}} + H^m_{\mathrm{p}2} - H^m_{3\mathrm{p}} + H^m_{\mathrm{p}3} - H^m_{1\mathrm{I}} + H^m_{\mathrm{I}1} - H^m_{2I} + H^m_{I2} - H^m_{3\mathrm{I}} + H^m_{\mathrm{I}3},\\
\mathbf{M}_1 = -H^m_{\mathrm{p}1} - H^m_{\mathrm{I}1}, \quad \mathbf{M}_2 = -H^m_{\mathrm{p}2} - H^m_{\mathrm{I}2},\\
\mathbf{M}_3 = -H^m_{\mathrm{p}3} + H^m_{3\mathrm{p}} + H^m_{1\mathrm{p}} + H^m_{2\mathrm{p}} - H^m_{\mathrm{I}3} + H^m_{3\mathrm{I}} + H^m_{1\mathrm{I}} + H^m_{2\mathrm{I}},\\
E_\Sigma = -H^e_{1\mathrm{p}} + H^e_{\mathrm{p}1} - H^e_{2\mathrm{p}} + H^e_{\mathrm{p}2} - H^e_{3\mathrm{p}} + H^e_{\mathrm{p}3} - H^e_{1\mathrm{I}} + H^e_{\mathrm{I}1} - H^e_{2\mathrm{I}} + H^e_{\mathrm{I}2} - H^e_{3\mathrm{I}} + H^e_{\mathrm{I}3},\\
E_1 = -H^e_{\mathrm{p}1} - H^e_{\mathrm{I}1}, \quad E_2 = -H^e_{\mathrm{p}2} - H^e_{\mathrm{I}2},\\
E_3 = -H^e_{\mathrm{p}3} + H^e_{3\mathrm{p}} + H^e_{1\mathrm{p}} + H^e_{2\mathrm{p}} - H^e_{\mathrm{I}3} + H^e_{3\mathrm{I}} + H^e_{1\mathrm{I}} + H^e_{2\mathrm{I}},\\
P_{\mathrm{I}} = H^{\mathrm{p}}_{\mathrm{p}1} + H^{\mathrm{p}}_{\mathrm{p}2} + H^{\mathrm{p}}_{\mathrm{I}1} - H^{\mathrm{p}}_{1\mathrm{I}} + H^{\mathrm{p}}_{\mathrm{I}2} - H^{\mathrm{p}}_{2\mathrm{I}} - H^{\mathrm{p}}_{3\mathrm{I}}.
\end{aligned}
$$

Journal of Physics: Conference Series **767** (2016) 012020 doi:10.1088/1742-6596/767/1/012020

Region 2

$$p_0 + H_1 \to p_0 + H_2, \quad p_1 + H_1 \to p_0 + H_2,$$
$$p_0 + H_2 \to p_0 + H_2, \quad p_1 + H_2 \to p_0 + H_2,$$
$$p_0 + H_3 \to p_0 + H_2, \quad p_1 + H_3 \to p_0 + H_2.$$

This implies that

$$D_\Sigma = 0, D_1 = -H^\rho_{\mathrm{p1}} - H^\rho_{\mathrm{I1}}, \quad D_2 = H^\rho_{\mathrm{p1}} + H^\rho_{\mathrm{p3}} + H^\rho_{\mathrm{I1}} + H^\rho_{\mathrm{I3}},$$

$$D_3 = -H^\rho_{\mathrm{p3}} - H^\rho_{\mathrm{I3}}, \quad D_{\mathrm{I}} = -H^\rho_{\mathrm{I1}} - H^\rho_{\mathrm{I2}} - H^\rho_{\mathrm{I3}},$$

$$\mathbf{M}_\Sigma = -H^m_{\mathrm{1p}} + H^m_{\mathrm{p1}} - H^m_{\mathrm{2p}} + H^m_{\mathrm{p2}} - H^m_{\mathrm{3p}} + H^m_{\mathrm{p3}} - H^m_{\mathrm{1I}} + H^m_{\mathrm{I1}} - H^m_{\mathrm{2I}} + H^m_{\mathrm{I2}} - H^m_{\mathrm{3I}} + H^m_{\mathrm{I3}},$$

$$\mathbf{M}_1 = -H^m_{\mathrm{p1}} - H^m_{\mathrm{I1}}, \quad \mathbf{M}_2 = -H^m_{\mathrm{p2}} + H^m_{\mathrm{2p}} + H^m_{\mathrm{1p}} + H^m_{\mathrm{3p}} - H^m_{\mathrm{I2}} + H^m_{\mathrm{2I}} + H^m_{\mathrm{1I}} + H^m_{\mathrm{3I}},$$

$$\mathbf{M}_3 = -H^m_{\mathrm{p3}} - H^m_{\mathrm{I3}},$$

$$E_\Sigma = -H^e_{\mathrm{1p}} + H^e_{\mathrm{p1}} - H^e_{\mathrm{2p}} + H^e_{\mathrm{p2}} - H^e_{\mathrm{3p}} + H^e_{\mathrm{p3}} - H^e_{\mathrm{1I}} + H^e_{\mathrm{I1}} - H^e_{\mathrm{2I}} + H^e_{\mathrm{I2}} - H^e_{\mathrm{3I}} + H^e_{\mathrm{I3}},$$

$$E_1 = -H^e_{\mathrm{p1}} - H^e_{\mathrm{I1}},$$

$$E_2 = -H^e_{\mathrm{p2}} + H^e_{\mathrm{2p}} + H^e_{\mathrm{1p}} + H^e_{\mathrm{3p}} - H^e_{\mathrm{I2}} + H^e_{\mathrm{2I}} + H^e_{\mathrm{1I}} + H^e_{\mathrm{3I}},$$

$$E_3 = -H^e_{\mathrm{p3}} - H^e_{\mathrm{I3}}, \quad P_{\mathrm{I}} = -H^{\mathrm{p}}_{\mathrm{1I}} - H^{\mathrm{p}}_{\mathrm{2I}} - H^{\mathrm{p}}_{\mathrm{3I}}.$$

Region 1

$$p_0 + H_1 \to p_0 + H_1, \quad p_0 + H_2 \to p_0 + H_1, \quad p_0 + H_3 \to p_0 + H_1.$$

This implies that

$$D_\Sigma = 0, \quad D_1 = H^\rho_{\mathrm{p2}} + H^\rho_{\mathrm{p3}},$$

$$D_2 = -H^\rho_{\mathrm{p2}}, \quad D_3 = -H^\rho_{\mathrm{p3}}, \quad D_{\mathrm{I}} = 0,$$

$$\mathbf{M}_\Sigma = -H^m_{\mathrm{1p}} + H^m_{\mathrm{p1}} - H^m_{\mathrm{2p}} + H^m_{\mathrm{p2}} - H^m_{\mathrm{3p}} + H^m_{\mathrm{p3}},$$

$$\mathbf{M}_1 = -H^m_{\mathrm{p1}} + H^m_{\mathrm{1p}} + H^m_{\mathrm{2p}} + H^m_{\mathrm{3p}}, \quad \mathbf{M}_2 = -H^m_{\mathrm{p2}}, \quad \mathbf{M}_3 = -H^m_{\mathrm{p3}},$$

$$E_\Sigma = -H^e_{\mathrm{1p}} + H^e_{\mathrm{p1}} - H^e_{\mathrm{2p}} + H^e_{\mathrm{p2}} - H^e_{\mathrm{3p}} + H^e_{\mathrm{p3}},$$

$$E_1 = -H^e_{\mathrm{p1}} + H^e_{\mathrm{1p}} + H^e_{\mathrm{2p}} + H^e_{\mathrm{3p}}, \quad E_2 = -H^e_{\mathrm{p2}}, \quad E_3 = -H^e_{\mathrm{p3}}, \quad P_{\mathrm{I}} = 0.$$

The following source terms are used in the pressure equation:

$$H^{\mathrm{p}}_{jk} = (\gamma - 1)\left(H^e_{jk} - \mathbf{u} \cdot \mathbf{H}^m_{jk} + \frac{1}{2}u^2 H^\rho_{jk}\right),$$

$$H^{\mathrm{p}}_{ak} = (\gamma - 1)\lambda_{ak}\frac{3}{4}v^2_{T_k}U^e_{ak},$$

3. Numerical results

Here we present the results of our steady-state calculation only of the SW–LISM interaction. The simulation is performed on a spherical grid with adaptive mesh refinement near the TS. The following boundary conditions have been used. The SW is assumed to be spherically symmetric at 1 AU: the plasma density is 7.4 cm^{-3}, radial velocity component 450 km/s, temperature 70,000 K, and the radial component of the magnetic field vector 49.8 μG. The heliospheric magnetic field is assumed to be the Parker field at the inner boundary.

In the unperturbed LISM: the neutral H density is 0.172 cm^{-3}, plasma number density 0.082 cm^{-3}, velocity $V_\infty = 26.4$ km/s, temperature 8,000 K, magnetic field strength $B_\infty = 3\ \mu$G. Simulations are performed in the heliospheric coordinate system x, y, and z, where the z-axis is aligned with the Sun's rotation axis, the x-axis belongs to the plane formed by the z-axis and the LISM velocity vector, $\mathbf{V}_\infty$, and directed upstream into the LISM, and the y-axis completes the right coordinate system. The directions of $\mathbf{V}_\infty$ and $\mathbf{B}_\infty$ are given by the formulae:

$$\mathbf{V}_\infty = V_\infty(-0.996, 0, -0.089), \quad \mathbf{B}_\infty = B_\infty(0.692, -0.477, 0.541).$$

We assumed a multi-fluid model for H atoms, as described in the Introduction. However, the extension of this approach to kinetic neutrals is straightforward. Although the governing system of equations is written in Cartesian coordinates (and we solve it this way), the grid is spherical. This approach allows us to avoid any geometrical source terms. This is important to ensure conservation laws hold in a flow that is characterized by the presence of shocks and the heliopause, which our ideal-MHD approach treats as a tangential discontinuity. In principle, the presence of source terms means that the conservation laws hold only approximately, which affects the strengths of numerically-obtained discontinuities and therefore their speed. The system of equations is solved using a Godunov-type method, where the fluxes of mass, momentum, energy, and magnetic field are obtained as solutions of an MHD Riemann problem in the linear approximation [47]. The scheme has second order of accuracy in space and time. Both MHD and Euler equations are solved in the framework of the Multi-Scale Fluid-Kinetic Simulation Suite (MS-FLUKSS) [51, 52]. MS-FLUKSS is built on the Chombo adaptive mesh refinement (AMR) framework [53]. The spherical volume extension of this framework was implemented for MS-FLUKSS in [54].

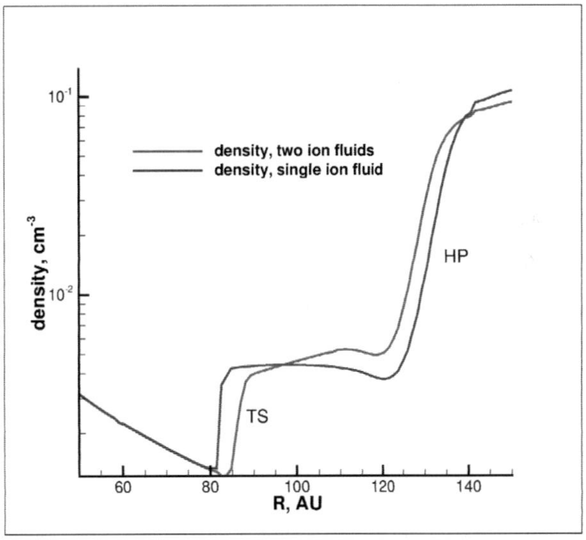

Figure 1. Ion density distributions along the *V1* trajectory shows that the TS moves farther from the Sun while the HP becomes closer to it, if PUIs are treated as a separate plasma component.

While a number of different approaches have been developed to describe the boundary conditions for PUIs across the TS which describe the energy transfer in this region [42, 43, 44, 45], here we use a simplified approach based on the assumption that thermal protons experience only adiabatic compression, i.e., the entropy of the ion fluid is preserved. We also assume that the density ratio behind and in front of the TS is the same for PUIs and thermal ions, so it can be determined from the corresponding ratio in the mixture. These two assumptions ensure a preferential energy influx into the PUI fluid. More accurate approaches will be implemented in the future. Any analytic boundary conditions

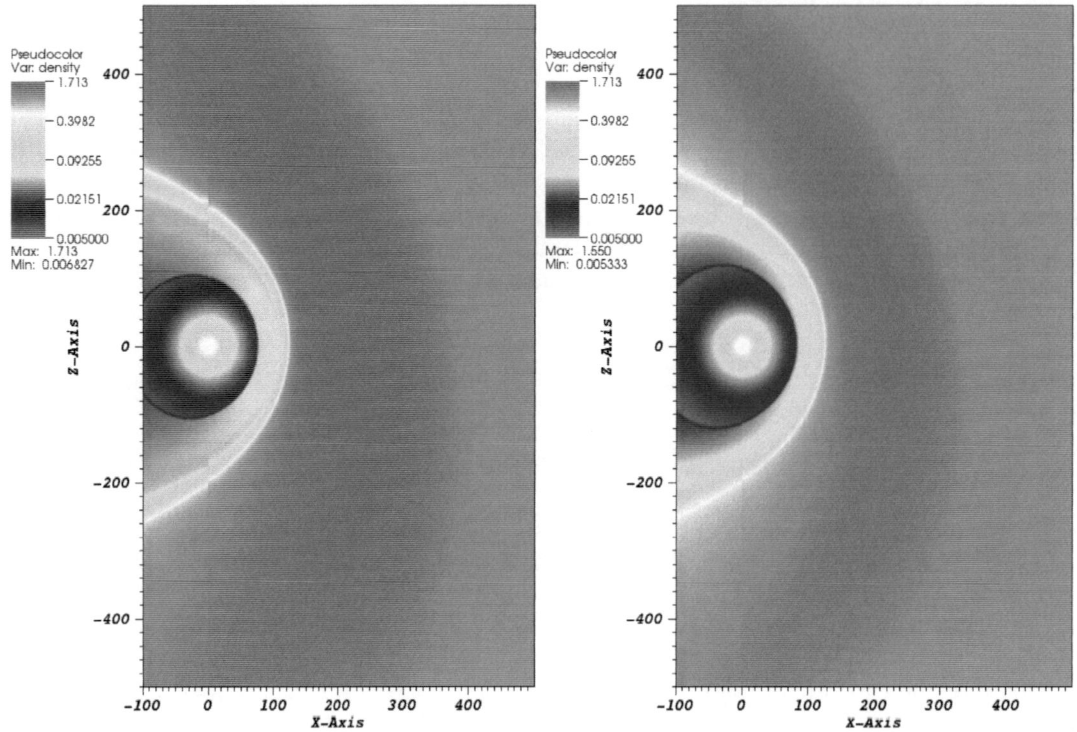

Figure 2. Plasma density distributions in the meridional plane for the single-ion-fluid model (left panel) and for the case when the PUI and thermal ion fluids behave as co-moving, but distinguishable components (right panel)

can be implemented in a manner similar to that presented here. To be sure that we choose proper shock conditions at the TS, or in other words, to take into account TS smearing over 2-3 computational cells, we impose these boundary conditions on the cell centers located two cells upstream and downstream.

Figure 1 shows the plasma density distribution in the direction of the Voyager 1 trajectory for the single-ion-fluid model (*Case* 1) and for the case when the PUI and thermal ion fluids behave as co-moving, but distinguishable components (*Case* 2). Figure 2 shows the plasma density distributions (dimensionless units with the scale of LISM plasma density) in the meridional plane (the xz-plane) for Case 1 (left panel) and Case 2 (right panel). It is clear from Fig. 1 that the HP moves inward towards the Sun and the TS moves outward away from the Sun with the inclusion of PUIs as a distinct component, i.e., the heliosheath width becomes quite significantly narrower when PUIs are treated distinctly from the thermal SW plasma. This result is consistent with that in [14] and addresses the observation made by $V1$ that the HP location was closer than expected [5, 6, 7, 8, 9], one interpretation of which is that the IHS is narrower than predicted by conventional models.

4. Conclusions

We have performed a SW–LISM simulations based on treating of PUIs as an individual plasma component and compared our results with our "standard" model where PUIs are immediately assimilated with background thermal protons. The neutral H flow was modeled using a multi-fluid approach. To make our simulations consistent with $V2$ observations at the TS, we employed simple boundary conditions for the PUI pressure and density that ensured the preferential heating of PUIs across the TS. Our results show that treating PUIs as a separate plasma component indeed makes the inner heliosheath about 10 AU thinner that in a single-ion-fluid model. It should be understood, however, that while the HP was at 120 AU when V1 crossed it in August 2012, we have no information about the TS location at the same time. It is therefore possible that the heliosheath width is not equal to 28 AU, which one would predict on assuming that neither the HP nor the TS changed their position between 2004, when V1 crossed the TS, and 2012, when it left the heliosphere and entered the LISM. Most time-dependent SW–LISM models predict substantial variations in the TS heliocentric distance as a function of time [55, 56, 57, 58, 59, 60].

Future work will focus on using better boundary conditions for PUIs at the TS, the implementation of a model with a kinetic treatment of H atoms, and including an extended treatment of the PUIs that goes beyond the assumption of complete isotropization.

Acknowledgments

This work was supported, in part, by NASA grants NNX14AJ53G, NNX14AF41G, NNX14AF43G,NNX15AN72G, and NNX16AG83G, and DOE Grant DE-SC0008334. This work was also partially supported by the IBEX mission as a part of NASA's Explorer program. We acknowledge NSF PRAC award OCI-1144120 and related computer resources from the Blue Waters sustained-petascale computing project. Supercomputer time allocations were also provided on SGI Pleiades by NASA High-End Computing Program award SMD-15-5860 and on Stampede by NSF XSEDE project MCA07S033.

Bibliography

[1] Blum P W and Fahr H J 1969 *Nature* **223** 936–937
[2] Wallis M 1971 *Nature Physical Science* **233** 23–25
[3] Wallis M K 1975 *Nature* **254** 202
[4] Holzer T E 1977 *Reviews of Geophysics and Space Physics* **15** 467–490
[5] Stone E C, Cummings A C, McDonald F B, Heikkila B C, Lal N and Webber W R 2013 *Science* **341** 150–153
[6] Burlaga L F, Ness N F and Stone E C 2013 *Science* **341** 147–150
[7] Krimigis S M, Decker R B, Roelof E C, Hill M E, Armstrong T P, Gloeckler G, Hamilton D C and Lanzerotti L J 2013 *Science* **341** 144–147
[8] Webber W R and McDonald F B 2013 *Geophys. Res. Lett.* **40** 1665–1668
[9] Burlaga L F and Ness N F 2014 *Astrophys. J.* **784** 146
[10] Gurnett D A, Kurth W S, Burlaga L F and Ness N F 2013 *Science* **341** 1489
[11] Gloeckler G, Fisk L A, Geiss J, Hill M E, Hamilton D C, Decker R B and Krimigis S M 2009 *Space Sci. Rev.* **143** 163–175
[12] Bzowski M, Möbius E, Tarnopolski S, Izmodenov V and Gloeckler G 2009 *Space Sci. Rev.* **143** 177–190
[13] Richardson J D, Liu Y, Wang C and McComas D J 2008 *A&A* **491** 1–5
[14] Malama Y G, Izmodenov V V and Chalov S V 2006 *Astron. Astrophys.* **445** 693–701
[15] Parker E N 1965 *Planet. Space Sci.* **13** 9–49
[16] Vasyliunas V M and Siscoe G L 1976 *J. Geophys. Res. Space Phys.* **81** 1247–1252
[17] Isenberg P A 1987 *J. Geophys. Res. Space Phys.* **92** 1067–1073
[18] Williams L L and Zank G P 1994 *J. Geophys. Res. Space Phys.* **99** 19
[19] Burlaga L F, Ness N F, Belcher J W, Szabo A, Isenberg P A and Lee M A 1994 *J. Geophys. Res. Space Phys.* **99** 21
[20] Richardson J D, Paularena K I, Lazarus A J and Belcher J W 1995 *Geophys. Res. Lett.* **22** 1469–1472
[21] Zank G P 1999 *Space Sci. Rev.* **89** 413–688
[22] Zank G P, Hunana P, Mostafavi P and Goldstein M L 2014 *Astrophys. J.* **797** 87
[23] Zank G P 2015 *Ann. Rev. Astron. Astrophys.* **53** 449–500
[24] Decker R B, Krimigis S M, Roelof E C, Hill M E, Armstrong T P, Gloeckler G, Hamilton D C and Lanzerotti L J 2008 *Nature* **454** 67–70
[25] Decker R B, Krimigis S M, Roelof E C and Hill M E 2015 *Journal of Physics Conference Series* **577** 012006
[26] Baranov V B and Malama Y G 1993 *J. Geophys. Res. Space Phys.* **98** 15
[27] Pauls H L, Zank G P and Williams L L 1995 *J. Geophys. Res. Space Phys.* **100** 21595–21604

[28] Pogorelov N V, Zank G P and Ogino T 2006 *Astrophys. J.* **644** 1299–1316
[29] Heerikhuisen J, Florinski V and Zank G P 2006 *Journal of Geophysical Research (Space Physics)* **111** A06110
[30] Heerikhuisen J, Pogorelov N V, Zank G P and Florinski V 2007 *Astrophys. J. Lett.* **655** L53–L56
[31] Izmodenov V, Alexashov D and Myasnikov A 2005 *A&A* **437** L35–L38
[32] Izmodenov V V, Malama Y G, Ruderman M S, Chalov S V, Alexashov D B, Katushkina O A and Provornikova E A 2009 *Space Sci. Rev.* **146** 329–351
[33] Opher M, Stone E C and Liewer P C 2006 *Astrophys. J. Lett.* **640** L71–L74
[34] Isenberg P A 1986 *J. Geophys. Res. Space Phys.* **91** 9965–9972
[35] Usmanov A V and Goldstein M L 2006 *Journal of Geophysical Research (Space Physics)* **111** A07101
[36] Usmanov A V, Goldstein M L and Matthaeus W H 2012 *Astrophys. J.* **754** 40
[37] Kryukov I A, Pogorelov N V, Zank G P and Borovikov S N 2012 Numerical modeling of the solar wind turbulence *American Institute of Physics Conference Series* (*American Institute of Physics Conference Series* vol 1436) ed Heerikhuisen J, Li G, Pogorelov N and Zank G pp 48–54
[38] Detman T R, Intriligator D S, Dryer M, Sun W, Deehr C S and Intriligator J 2011 *Journal of Geophysical Research (Space Physics)* **116** A03105
[39] Gamayunov K V, Zhang M, Pogorelov N V, Heerikhuisen J and Rassoul H K 2012 *Astrophys. J.* **757** 74
[40] Chalov S V and Fahr H J 1996 *Solar Phys.* **168** 389–411
[41] Zank G P, Pauls H L, Cairns I H and Webb G M 1996 *J. Geophys. Res. Space Phys.* **101** 457–478
[42] Fahr H J and Chalov S V 2008 *A&A* **490** L35–L38
[43] Zank G P, Heerikhuisen J, Pogorelov N V, Burrows R and McComas D 2010 *Astrophys. J.* **708** 1092–1106
[44] Fahr H J, Siewert M and Chashei I 2012 *Astrophys. Space Sci.* **341** 265–276
[45] Fahr H J and Siewert M 2013 *A&A* **558** A41
[46] Richardson J D, Kasper J C, Wang C, Belcher J W and Lazarus A J 2008 *Nature* **454** 63–66
[47] Kulikovskii A G, Pogorelov N V and Semenov A Y 2001, *Mathematical Aspects of Numerical Solution of Hyperbolic Systems* Chapman & Hall/CRC Press (London, UK / Boca Raton, Florida, USA)
[48] Usmanov A V, Goldstein M L and Matthaeus W H 2016 *Astrophys. J.* **820** 17
[49] Breech B, Matthaeus W H, Minnie J, Bieber J W, Oughton S, Smith C W and Isenberg P A 2008 *Journal of Geophysical Research (Space Physics)* **113** A08105
[50] Zank G P, Pauls H L, Williams L L and Hall D T 1996 *J. Geophys. Res. Space Phys.* **101** 21639–21656
[51] Pogorelov N V, Borovikov S N, Bedford M C, Heerikhuisen J, Kim T K, Kryukov I A and Zank G P 2013 Modeling Solar Wind Flow with the Multi-Scale Fluid-Kinetic Simulation Suite *Numerical Modeling of Space Plasma Flows (ASTRONUM-2012)* (*Astronomical Society of the Pacific Conference Series* vol 474) ed Pogorelov N V, Audit E and Zank G P p 165
[52] Pogorelov N V, Borovikov S N, Heerikhuisen J, Kim T K, Kryukov I A and Zank G P 2014 MS-FLUKSS and Its Application to Modeling Flows of Partially Ionized Plasma in the Heliosphere *Proceedings of the 2014 Annual Conference of the Extreme Science and Engineering Discovery Environment (XSEDE'14)*, paper a-22, Association for Computing Machinery (ACM), New York, NY
[53] Colella P, Bell J, Keen N, Ligocki T, Lijewski M and van Straalen B 2007 *Journal of Physics Conference Series* **78** 012013
[54] Borovikov S N, Heerikhuisen J, Pogorelov N V, Kryukov I A and Zank G P 2008 Modeling the Heliosphere with an Adaptive MHD-kinetic Code *Numerical Modeling of Space Plasma Flows* (*Astronomical Society of the Pacific Conference Series* vol 385) ed Pogorelov N V, Audit E and Zank G P p 197
[55] Pogorelov N V 2000 *Astrophys. Space Sci.* **274** 115–122
[56] Zank G P and Müller H R 2003 *Journal of Geophysical Research (Space Physics)* **108** 1240
[57] Izmodenov V, Malama Y and Ruderman M S 2005 *A&A* **429** 1069–1080
[58] Pogorelov N V, Borovikov S N, Zank G P and Ogino T 2009 *Astrophys. J.* **696** 1478–1490
[59] Washimi H, Zank G P, Hu Q, Tanaka T, Munakata K and Shinagawa H 2011 *MNRAS* **416** 1475–1485
[60] Pogorelov N V, Suess S T, Borovikov S N, Ebert R W, McComas D J and Zank G P 2013 *Astrophys. J.* **772** 2

Sub-Auroral Polarization Streams: A complex interaction between the magnetosphere, ionosphere, and thermosphere

Joachim Raeder[1], William D. Cramer[1], Joseph Jensen[1], Timothy Fuller-Rowell[2], Naomi Maruyama[2], Frank Toffoletto[3], Hien Vo[3]

[1]Space Science Center, University of New Hampshire, Durham, NH 03824, USA
[2]CIRES, University of Colorado, Boulder, CO 80303, USA
[3]Rice University, Houston, TX 77005, USA
[4]Universidad del Turabo, PR

E-mail: J.Raeder@unh.edu

Abstract. We use a novel coupled model of the magnetosphere-ionosphere-thermosphere system consisting of the OpenGGCM magnetosphere model, the CTIM ionosphere-thermosphere model, and the RCM model for the drift physics of the inner magnetosphere to study Sub-Auroral Polarization Streams (SAPS). We use the model to simulate the March 17, 2013 "St. Patrick's day" geomagnetic storm. It had been previously established that SAPS occur during this storm. We demonstrate that the model can reproduce the salient features of SAPS. In particular, the model reproduces the strong northward electric field associated with westward flows, well equatorward of the convection cells. The flows are located just equatorward of the lower boundary of electron precipitation and thus in a region of low ionosphere conductance. Associated with the flows is a deepening through in electron density. The SAPS extend from post-noon to post-midnight, and with increasing local time they extend to lower and lower latitudes. This had been previously fond in a statistical sense, but the model finds that this behavior is also true at given times. We thus conclude that self-consistent modeling of SAPS is now possible, which enables us to study their properties in more detail.

1. Introduction

Sub-auroral Polarization Streams [SAPS] are a feature of the ionosphere that develops mostly during geomagnetic active times [*Foster et al.*, 2007].

Foster and Vo [2002] have shown that SAPS statistically occur primarily in the afternoon sector and that they are more intense as geomagnetic activity increases. The primary characteristics of SAPS is an intense northward electric field that is well equatorward of the convection pattern. This electric field channel is a few degrees wide and can have peak values around 100 mV/m. This corresponds to plasma drift velocities of ~2000 m/s. Along with the high speed flow channel, the mid-latitude trough in the electron density deepens. Coincident measurements of electric field, field-aligned currents (FACs), and particle precipitation show that the SAPS electric field is located equatorward of plasma sheet electron precipitation overlapping with region-2 (R2) FAC. Because of the lack of sufficient precipitation, the ionosphere conductance is low. This leads to the simple conjecture that the strong SAPS electric field occurs because the magnetosphere imposes the current, which in a simple electrical circuit picture requires an increased voltage, i.e., stronger electric field, to support that current. At the same time, increased recombination decreases the ionosphere electron content, which further decreases ionosphere conductance [*Rodger*, 2008].

An increase of the relative flow velocity between the ions and the neutrals from 1000 km/s to 2000 km/s increases the recombination rate by about an order of magnitude [*Schunk et al.*, 1976]. Thus, there is possibly a positive feedback loop that may enhance the SAPS effect. However, this picture depends on assumptions, for example, that the magnetosphere is a current generator, and that the electron depletion occurs at altitudes where is has a significant effect on conductance.

Even though a rather straight-forward theory may suffice to explain SAPS, at least in a climatological sense, it is quite difficult to model them self-consistently, in particular during geomagnetically active times. Although SAPS are primarily a ionosphere phenomenon, they are driven by the magnetosphere that produces the FAC and electron precipitation, and they require fully three-dimensional modeling of the ionosphere-thermosphere (IT) system. Furthermore, R2 currents are believed to be driven by pressure gradients in the inner magnetosphere, which require the inclusion of the proper drift physics that produces the ring current. In the past each of these systems has been modeled separately, for example CTIM exists as a stand-alone model [*Fuller-Rowell et al.*, 1996], MHD based magnetosphere models exist for a few decades now [*Raeder*, 2003; *Lyon et al.*, 2004; *Toth et al.*, 2005]. Likewise, models of the inner magnetosphere alone have been used for a long time [*Toffoletto et al.*, 2003; *Fok et al.*, 2001; *Jordanova*, 2003; *Kozyra et al.*, 1998]. Pairwise coupled models are now common, such as OpenGGCM-CTIM [*Raeder et al.*, 2001a], or coupled models of the inner and outer magnetosphere [*Toffoletto et al.*, 2004; *Hu et al.*, 2010], as well as models of the coupled inner magnetosphere-IT system [*Maruyama et al.*, 2007]. However, coupled simulations with full feedbacks between all of the outer magnetosphere, inner magnetosphere, and the IT systems are relatively new and have not yet been extensively tested.

Here, we present first results from one such model, i.e., the coupled OpenGGCM (magnetosphere), CTIM (IT system), and Rice Convection Model (RCM, inner magnetosphere). As one of the first significant results of that model we show that it can produce the salient features of SAPS.

2. The OpenGGCM-CTIM-RCM Model

The OpenGGCM is a global coupled model of Earth's magnetosphere, ionosphere, and thermosphere. The magnetosphere part solves the MHD equations as an initial-boundary-value problem. The MHD equations are only solved to within $\sim 3\ R_E$ of Earth. The region within $3\ R_E$ is treated as a magnetosphere-ionosphere (MI) coupling region where physical processes that couple the magnetosphere to the ionosphere-thermosphere system are parameterized using simple models and relationships. Important for this investigation, we use MHD plasma parameters at the inner boundary to compute electron precipitation parameters, such as the electron energy flux and mean energy associated with diffuse and the discrete aurora. Details of these calculations can be found in *Raeder et al.* [1998, 2001b]; *Raeder* [2003]; *Raeder et al.* [2008b].

The ionosphere-thermosphere system is modeled using the NOAA CTIM (Coupled Thermosphere Ionosphere Model) [*Fuller-Rowell et al.*, 1996]. The original coupling between the OpenGGCM and CTIM is described in *Raeder et al.* [2001a].

The inner magnetosphere is modeled using the Rice Convection Model [*Toffoletto et al.*, 1996]. Some early results of the coupling between OpenGGCM and RCM were presented by *Hu et al.* [2010]. The coupling of these models has since been improved. Since the domains of the OpenGGCM and the RCM overlap, there exist two computed states of the plasma in that region. Because the RCM description is considered to be more adequate for the physics in this region, i.e., based on the energy dependent drifts, the RCM plasma density and pressure is used

Journal of Physics: Conference Series **767** (2016) 012021 doi:10.1088/1742-6596/767/1/012021

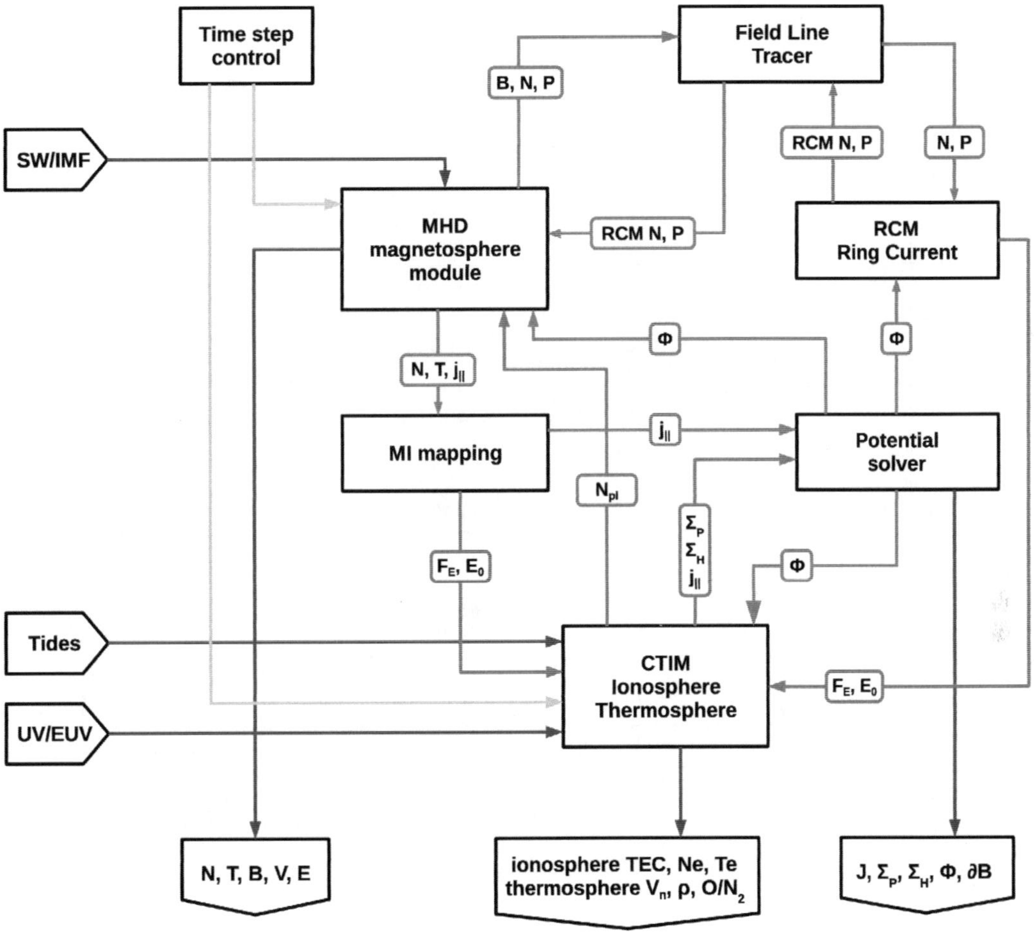

Figure 1. Block diagram of the OpenGGCM with its models and with data/control flow. Blue lines denote model input and output. Red lines denote data flow with strong coupling. Green lines denote data flow with weak or slow coupling. Orange lines denote control flow. B, N, and T are the magnetospheric magnetic field, plasma density, and temperature, respectively. The field aligned current is $j_{\|}$, Φ is the ionosphere potential, F_E and E_0 the energy flux and mean energy of precipitating electrons, Σ_H and Σ_P the ionosphere Hall and Pederesen conductances, and ∂B is the ground magnetic perturbation.

to override the MHD values. However, a simple overwrite feature will not work properly and is thus replaced by a blending algorithm that on a reasonable time scale guarantees that the MHD quantities follow the RCM quantities closely. In the region of overlap, the RCM also provides the electron precipitation for the IT system, which is otherwise, i.e., at high latitudes, provided by the MHD solution. In return, the MHD part of the code provides plasma boundary conditions for the RCM, and the electric potential that the RCM uses to calculate the $\mathbf{E}\times\mathbf{B}$ drift. The models are thus tightly coupled, and it can be demonstrated that the coupled model produces much more realistic R2 currents and inner magnetosphere shielding. OpenGGCM has in the past used for numerous studies, such as magnetopause reconnection [*Berchem et al.*, 1995a, b; *Raeder*, 2006], the plasma depletion layer [*Wang et al.*, 2003, 2004], plasma entry due to double lobe reconnection [*Li et al.*, 2005, 2008, 2009], ballooning modes [*Zhu et al.*, 2009; *Raeder et al.*, 2010, 2012], interplanetary shock impacts [*Oliveira and Raeder*, 2014; *Oliveira et al.*, 2016], and

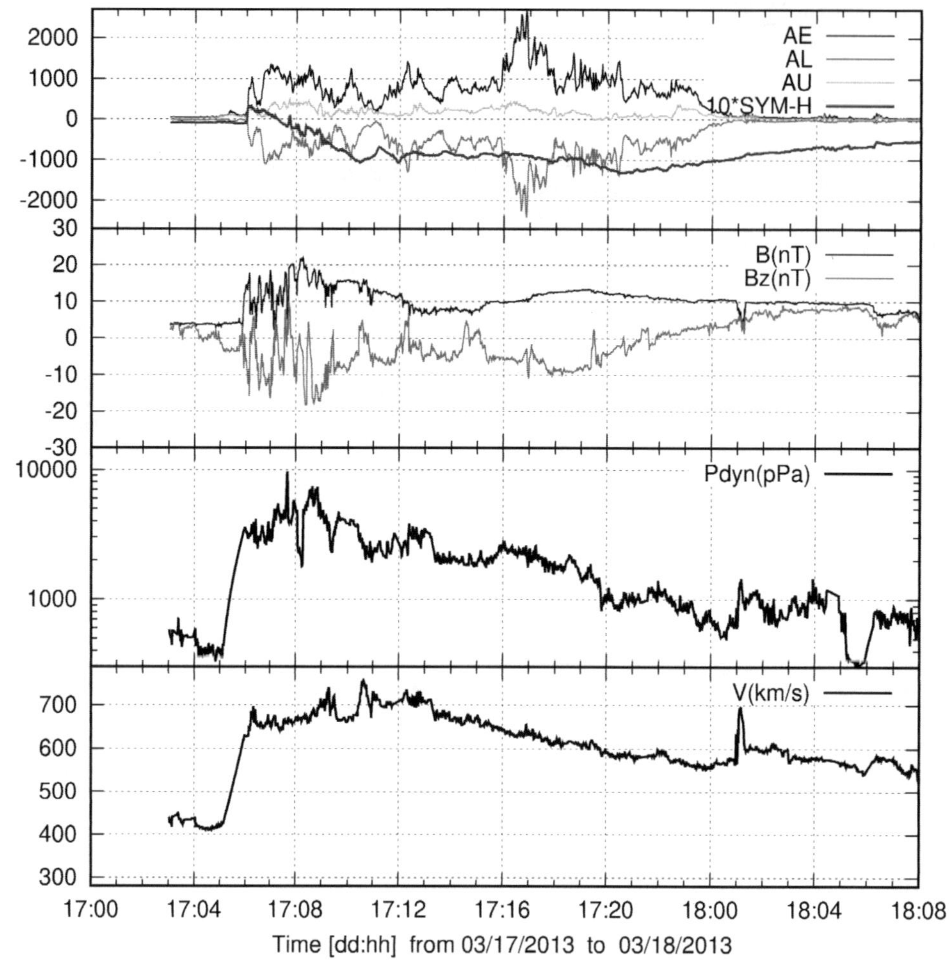

Figure 2. Solar wind, IMF, and geomagnetic conditions on March 17, 2013. The figure shows, from top to bottom: the geomagnetic indices AE, AU, AL, and SYM-H, derived from SuperMAG data; the IMF B_z component, and the field magnitude; the solar wind dynamic pressure; and the solar wind velocity. Note that SYM-H has been multiplied by 10 to fit on the same scale.

substorms [*Raeder*, 1995; *Raeder et al.*, 2001b; *Ge et al.*, 2011]. A more detailed description of the code and the methods used can be found in *Raeder* [2003]; *Raeder et al.* [2008a]. The OpenGGCM is available as a community model at the Community Coordinated Modeling Center (CCMC, `ccmc.gsfc.nasa.gov`), where it also has undergone a number of metrics evaluation studies [*Pulkkinen et al.*, 2010, 2011, 2013; *Rastätter et al.*, 2013]. Here, we demonstrate that the coupled model also produces SAPS.

3. SAPS event

The March 17, 2013 (St. Patrick's Day) storm event was chosen by the GEM and CEDAR communities for coordinated studies. It has been shown that during this event intense SAPS occurred [*Foster et al.*, 2014], which also led to the large-scale circulation of plasma to high latitudes and the convection throat. This leads to the formation of a Storm-Enhanced Density (SED) feature, and the so-called Tongue Of Ionization (TOI). However, in this paper, we will not address these phenomena and focus on SAPS.

The March 17 storm was caused by an Interplanetary Coronal Mass Ejection (ICME) that impinged on the magnetosphere. Figure 2 shows solar wind and Interplanetary Magnetic Field (IMF) data for that event. The storm commenced around 0600 UT on March 17, with a sudden increase in solar wind speed, dynamic pressure, and IMF magnitude. Initially, the IMF B_z fluctuated between southern and northern directions, until it turned mostly southward around 0900 UT. It stayed mostly southward until ~2000 UT. In the later phase of the storm B_z slowly and smoothly turned northward, indicating that the field represented an ICME flux rope. The SYM-H index fell gradually from the commencement of the storm at ~0600 UT to -100 nT at 0900 UT, and then stayed essentially constant until the beginning of the recovery at ~2100 UT. The storm is thus quite typical, and of weak to moderate strength. AE values during the storm are typically around 1000 nT, except for a burst of activity starting at ~1600 UT, where AE reaches values of up to 2000 nT

4. SAPS development

This storm was previously modeled using the Space Weather Modeling Framework [*Yu et al.*, 2015]. Their model also includes the outer magnetosphere, inner magnetosphere, and ionosphere, but lacks a dynamic and self-consistent ionosphere-thermosphere model. It uses an empirical model for ionosphere conductance, which may lead to missing the critical relation between R2 currents and conductance. Furthermore, there is no feedback from the inner magnetosphere model to the MHD solution. That model thus produces at best hints of SAPS with drift speeds of a few 100 m/s, whereas the observations show speeds of the order of 2000 km/s. (see Figure 8 in *Yu et al.* [2015].)

The OpenGGCM results are presented in Figures 3-5. Each figure shows a polar view of the northern hemisphere. There are four panels in each figure. Panel (A) shows the northward component of the electric field, i.e., E_ϑ in polar solar magnetic (SM) coordinates. Comparison to drift speeds is facilitated by noting that 50 mV/m correspond to ~1000 m/s drift speed. Panel (B) shows the diffuse electron precipitation energy flux. This flux is computed from the thermal energy flow from magnetospheric electrons, assuming a full loss cone. It essentially represents the auroral oval. Panel (C) shows the vertical Total Electron Content (TEC), as computed in CTIM. Panel (D) shows the ionosphere Pedersen conductance, computed self-consistently in CTIM from electron density, neutral densities, and collision frequencies. Obviously, the nightside conductance is closely related to precipitation.

Figure 3 shows the ionosphere state at 0736 UT, shortly after storm commencement. At this time, the dawn electric field has already intensified. The corresponding flow channel stretches from noon MLT to past midnight. The strong electric field is clearly separated, and equatorward of, the precipitation shown in panel (B). However, part of the channel is still on open field lines (the open-closed boundary is shown by the red line), thus, the afternoon portion of the flow should probably not be labeled SAPS. However, in the evening and night side, the flows are clearly separated from the polar cap. Furthermore, past 2000 MLT the flows split into two channels, a feature that is often observed with SAPS. Such separation is visible in the data as shown in the *Yu* paper (see Figure 8 in *Yu et al.* [2015].) Panel (C) shows that at this time no trough has formed yet, except maybe for a hint of depletion around dusk. It appears that at this time TEC in that region is still dominated by auroral electron precipitation.

Figure 4 shows, in the same format as Figure 3, the northern polar ionosphere at 1024 UT. By this time the SAPS is fully developed. Specifically, the flow channel now extends from ~1400 UT to past midnight and is well separated from the polar cap and from the precipitation region.

Journal of Physics: Conference Series **767** (2016) 012021 doi:10.1088/1742-6596/767/1/012021

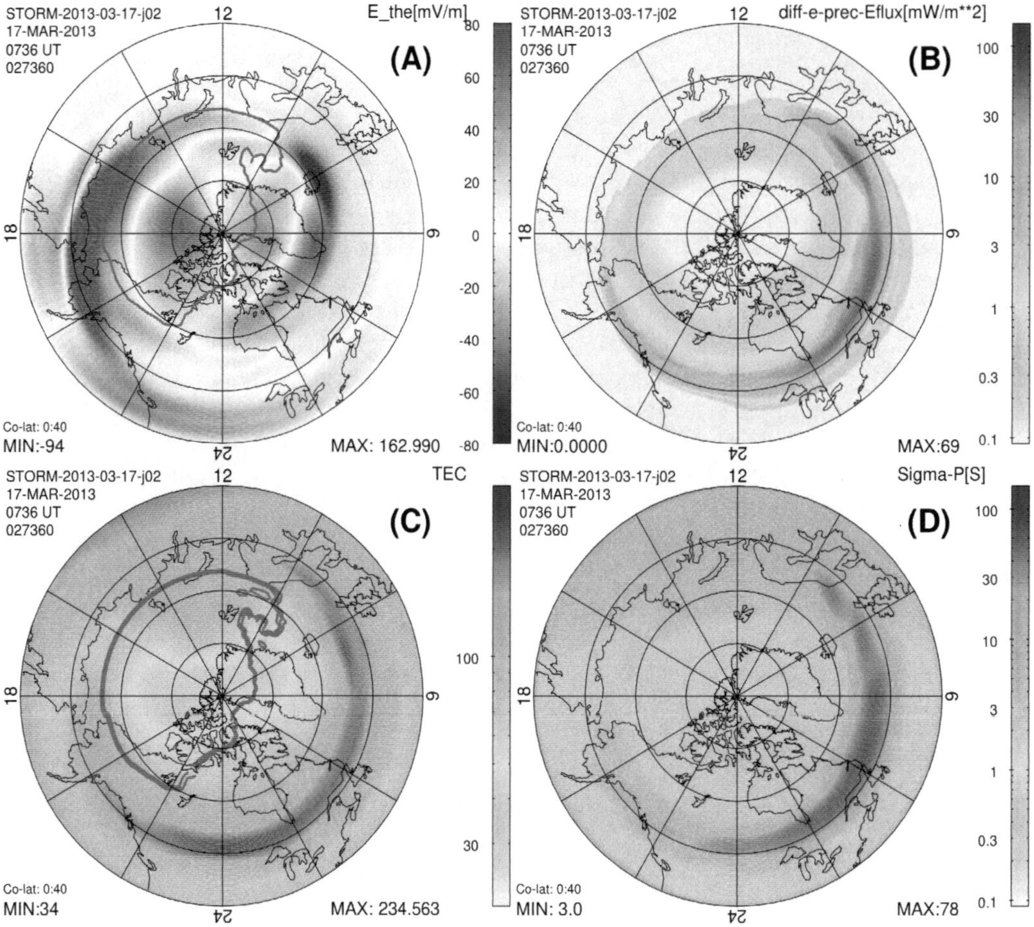

Figure 3. State of the ionosphere at 0736 UT. Polar views of (A) northward component of the northward component (B_ϑ in polar SM coordinates) ionosphere electric field, (B) diffuse electron precipitation, (C) the vertical Total Electron Content (TEC), and (D) the Pederen conductance. Each figure covers 50° latitude to the pole, and circles are drawn every 10 degrees.

Nearly coincident, a trough has developed and deepened in TEC. The trough is rotated some 2-3 hours local time past the flows, because the flow channel is stationary in the SM frame, whereas the electrons tend to co-rotate with the Earth, i.e., they are more stationary in the geographic frame. The trough thus extends substantially past midnight. A trough-like feature also begins to show in the conductance. It is also noteworthy that the latitude of the flow channel has a MLT dependence, such that the channel is at lower latitude as MLT increases. This has previously been found in statistical studies [*Foster and Vo*, 2002]. The simulation shows that such dependence can also exist at any given time.

Figure 5 shows the ionosphere at 1528 UT. This is still during the main phase of the storm; however, as Figure 2 shows, just after a brief phase of northward IMF. The SAPS now show some signs of recovery. The electric field is weaker, and the trough starts to disappear on the dayside, but still keeps co-rotating. Also, the flow channel progresses to lower latitudes in the night side. The SYM-H index (essentially the same as Dst) does not correlate well with SAPS development, since it reacts slowly to IMF changes. Likewise, the A-indices only seem to correlate weakly with SAPS as they primarily describe the high-latitude disturbances. Apparently, the SAPS electric

Journal of Physics: Conference Series **767** (2016) 012021 doi:10.1088/1742-6596/767/1/012021

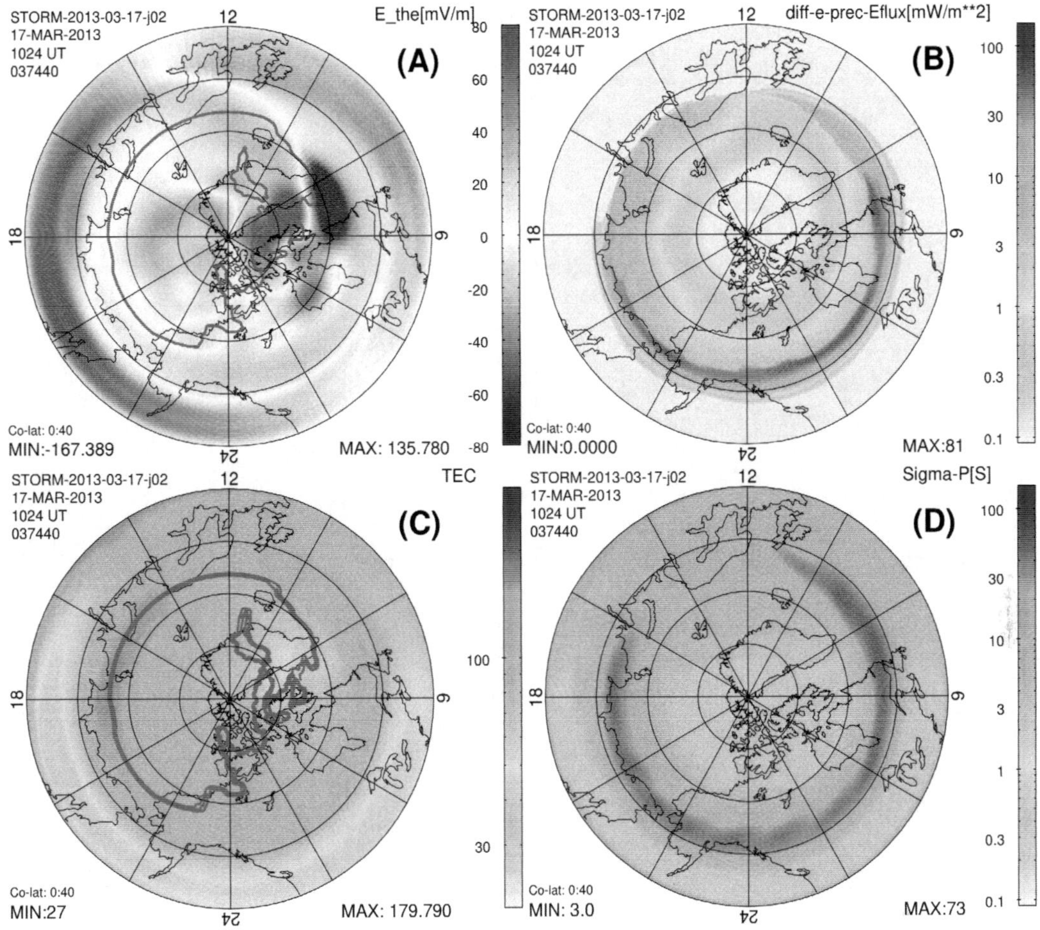

Figure 4. State of the ionosphere at 1024 UT, in the same format as Figure 3.

field reacts much more quickly to IMF changes, although the driver, i.e., the inner magnetosphere pressure distribution that drives the R2 currents is related to SYM-H. This warrants further investigation.

5. Summary and conclusions

We have presented a new model, consisting of the OpenGGCM for the outer magnetosphere, the RCM for the inner magnetosphere, and CTIM for the ionosphere-thermosphere system. We used the model to simulate the March 17, 2013 geomagnetic storm, which had previously been investigated and shown to develop significant SAPS.

We show that the model reproduces the salient features of SAPS, namely the strong northward fields and associated ion drifts, and the coincident trough in electron density. These features follow largely the characteristics established by statistical studies, i.e., the magnitude of the flows, and their latitude dependence on MLT. Also, as established in previous work, the model shows that the SAPS fields are well separated from the convection pattern itself, and that they lie equatorward of regions of electron precipitation.

Although fairly comprehensive data are available for this event, we forgo a detailed comparison, which is beyond the scope of this paper. We note, however, that a simple eye-

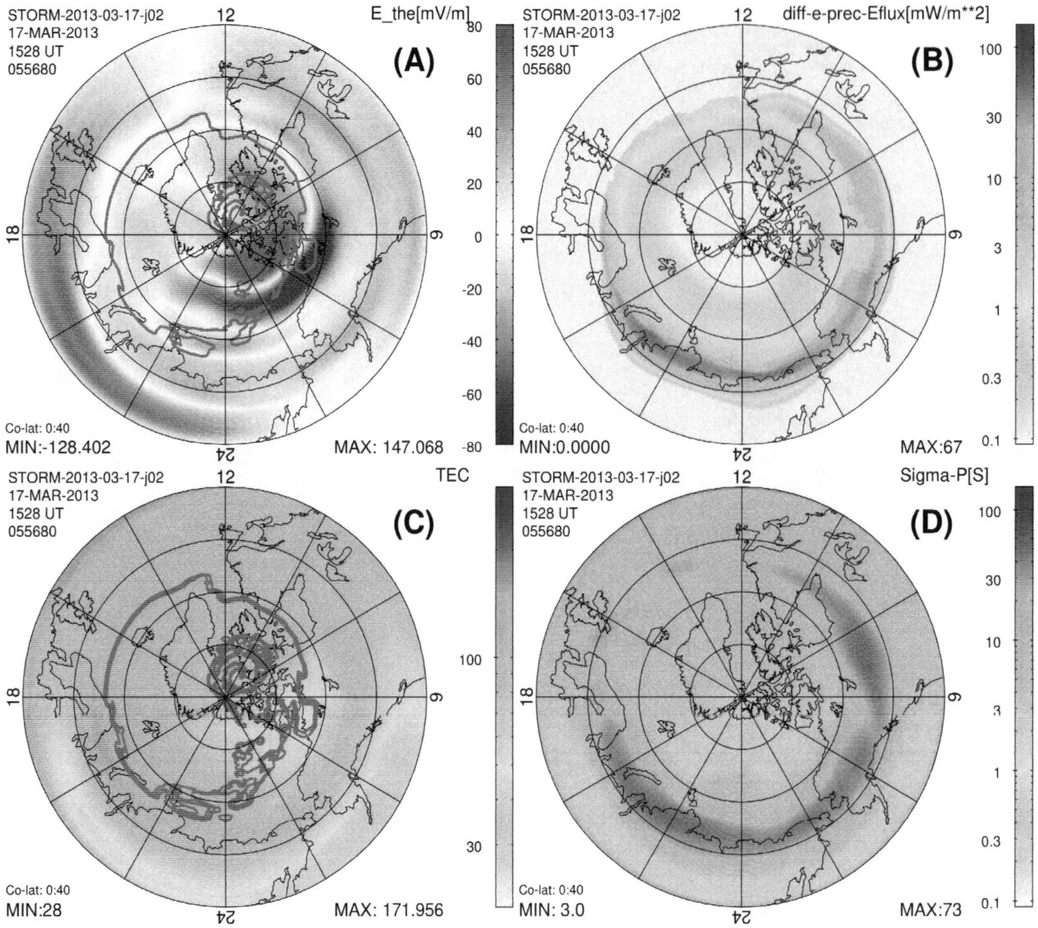

Figure 5. State of the ionosphere at 1538 UT, in the same format as Figure 3.

balling comparison with the DMSP passes presented in the *Yu et al.* [2015] paper reveals that the SAPS electric field and flow speed values in our model, which reach some 100 mV/m and 2000 km/s, respectively, compare well with the data.

We contrast this with the results of the *Yu* paper, which shows much weaker flows. The reason for this may be the fact that their model does not include a self-consistent ionosphere-thermosphere model to produce the ionosphere conductance, whereas ours does. Indeed, we observe in our results the development of a deep trough in the electron density, which lowers the conductance. Thus, our results indicate that the positive feedback between the flows, electron density, and conductance may be critical for development of SAPS, as indicated in the *Rodger* [2008] review, for example. We will investigate this in forthcoming studies by comparing with model runs that quench the feedback, and by comparing with data.

Acknowledgments
JR would like to thank J. Foster, P. Erickson, and P. Anderson for enlightening discussions. For the ground magnetometer data we gratefully acknowledge the SuperMAG contributors and PI Jesper Gjerloev, funded under NSF ATM-0646323, NSF AGS-1003580, and NASA NNX08AM32G S03. The work at UNH was funded through grant AGS-1143895 from the

National Science Foundation, and NASA/LWS grant NNX13AK31G. Computations were performed on Trillian, a Cray XE6m-200 supercomputer at UNH supported by the NSF MRI program under grant PHY-1229408.

References

Berchem, J., J. Raeder, and M. Ashour-Abdalla, Reconnection at the magnetospheric boundary: Results from global MHD simulations, in *Physics of the Magnetopause*, edited by B. U. Sonnerup and P. Song, vol. 90 of *AGU Geophysical Monograph*, p. 205, 1995a.

Berchem, J., J. Raeder, and M. Ashour-Abdalla, Magnetic flux ropes at the high-latitude magnetopause, *Geophys. Res. Lett.*, *22*, 1189, 1995b.

Fok, M.-C., R. A. Wolf, R. W. Spiro, and T. E. Moore, Comprehensive computational model of the Earth's ring current, *J. Geophys. Res.*, *106*, 8417, 2001.

Foster, J. C., and H. B. Vo, Average characteristics and activity dependence of the sub auroral polarization stream, *J. Geophys. Res.*, *107*, doi:10.1029/2002JA009,409, 2002.

Foster, J. C., W. Rideout, B. Sandel, W. T. Forrester, and F. J. Rich, On the relationship of SAPS to storm-enhanced densitys, *J. Atm. Sol. Terr. Phys.*, *69*, 303, 2007.

Foster, J. C., P. J. Erickson, A. J. Coster, S. Thaller, J. Tao, J. R. Wygant, and J. W. Bonnell, Storm time observations of plasmasphere erosion flux in the magnetosphere and ionosphere, *Geophys. Res. Lett.*, *41*, 762–768, 2014.

Fuller-Rowell, T. J., D. Rees, S. Quegan, R. J. Moffett, M. V. Codrescu, and G. H. Millward, A coupled thermosphere-ionosphere model (CTIM), in *STEP Report*, edited by R. W. Schunk, p. 217, Scientific Committee on Solar Terrestrial Physics (SCOSTEP), NOAA/NGDC, Boulder, Colorado, 1996.

Ge, Y. S., J. Raeder, V. Angelopoulos, M. L. Gilson, and A. Runov, Interaction of dipolarization fronts within multiple bursty bulk flows in global MHD simulations of a substorm on 27 February 2009, *J. Geophys. Res.*, *116*, A00I23, 2011.

Hu, B., F. Toffoletto, S. Sazykin, J. Raeder, D. Larson, and A. Vapirev, RCM simulation of the March 23rd 2007 substorm event using the OpenGGCM, *J. Geophys. Res.*, *115*, A12,205 doi:10.1029/2010JA015,360, 2010.

Jordanova, V. K., New insights of geomagnetic storms from model simulations using multi-spacecraft data, *Space Science Reviews*, *107*, 157, 2003.

Kozyra, J. U., V. K. Jordanova, J. E. Borovsky, M. F. Thomsen, D. J. Knipp, D. S. Evans, D. J. McComas, and T. E. Cayton, Effects of a high-density plasma sheet on ring current development during the november 2-6, 1993, magnetic storm, *Journal of Geophysical Research*, *103*, 26,285–305, 1998.

Li, W., J. Raeder, J. Dorelli, M. Øieroset, and T. D. Phan, Plasma sheet formation during long period of northward IMF, *Geophys. Res. Lett.*, *32*, L12S08, doi: 10.1029/2004GL021,524, 2005.

Li, W., J. Raeder, J. C. Dorelli, M. Thomsen, and B. Lavraud, Solar wind entry into the magnetosphere under northward IMF conditions, *J. Geophys. Res.*, *113*, A04,204, 2008.

Li, W., J. Raeder, M. Øieroset, and T. D. Phan, Cold dense magnetopause boundary layer under northward IMF: Results from THEMIS and MHD simulations, *J. Geophys. Res.*, *114*, A00C15, doi:10.1029/2008JA013,497, 2009.

Lyon, J. G., J. A. Fedder, and C. M. Mobarry, The Lyon-Fedder-Mobarry (LFM) global MHD magnetospheric simulation code, *J. Atm. Sol. Terr. Phys.*, *66*, 1333, 2004.

Maruyama, N., et al., Modeling storm-time electrodynamics of the low-latitude ionosphere–thermosphere system: Can long lasting disturbance electric fields be accounted for?, *J. Atm. Sol. Terr. Phys.*, *69*, 1182–1199, 2007.

Oliveira, D. M., and J. Raeder, Impact angle control of interplanetary shock geoeffectiveness, *J. Geophys. Res.*, *119*, 8188, 2014.

Oliveira, D. M., J. Raeder, B. T. Tsurutani, and J. W. Gjerloev, Effects of interplanetary shock inclinations on nightside auroral power, *Braz. Jour. Phys.*, *46*, 97–104, 2016.

Pulkkinen, A., L. Rastätter, K. Kuznetsova, M. Hesse, A. Ridley, J. Raeder, H. J. Singer, and A. Chulaki, Systematic evaluation of ground and geostationary magnetic field predictions generated by global magnetohydrodynamic models, *J. Geophys. Res.*, *115*, A03,206, 2010.

Pulkkinen, A., et al., Geospace environment modeling 2008-2009 challenge: Ground magnetic field perturbations, *Space Weather*, *9*, S02,004, 2011.

Pulkkinen, A., et al., Community-wide validation of geospace model ground magnetic field perturbation predictions to support model transition to operations, *Space Weather*, *11*, 369 doi:10.1002/swe.20,056, 2013.

Raeder, J., Global MHD simulations of the dynamics of the magnetosphere: Weak and strong solar wind forcing, in *Proceedings of the Second International Conference on Substorms*, edited by J. R. Kan, J. D. Craven, and S.-I. Akasofu, p. 561, Geophysical Institute, Univ. of Alaska Fairbanks, 1995.

Raeder, J., Global Magnetohydrodynamics – A Tutorial, in *Space Plasma Simulation*, edited by J. Büchner, C. T. Dum, and M. Scholer, Springer Verlag, Berlin Heidelberg New York, 2003.

Raeder, J., Flux transfer events: 1. Generation mechanism for strong southward IMF, *Ann. Geophys.*, *24*, 381, 2006.

Raeder, J., J. Berchem, and M. Ashour-Abdalla, The Geospace Environment Modeling grand challenge: Results from a Global Geospace Circulation Model, *J. Geophys. Res.*, *103*, 14,787, 1998.

Raeder, J., Y. L. Wang, and T. J. Fuller-Rowell, Geomagnetic storm simulation with a coupled magnetosphere - ionosphere - thermosphere model, in *Space Weather, AGU Geophys. Monogr. Ser.*, edited by P. Song, G. Siscoe, and H. J. Singer, vol. 125, p. 377, American Geophysical Union, 2001a.

Raeder, J., D. Larson, W. Li, E. L. Kepko, and T. Fuller-Rowell, OpenGGCM simulations for the THEMIS mission, *Space Sci. Rev.*, *141*, 535, doi:10.1007/s11,214–008–9421–5, 2008a.

Raeder, J., D. Larson, W. Li, E. L. Kepko, and T. Fuller-Rowell, OpenGGCM simulations for the THEMIS mission, *Space Sci. Rev.*, *141*, 535, 2008b.

Raeder, J., P. Zhu, Y. Ge, and G. L. Siscoe, OpenGGCM simulation of a substorm: Axial tail instability and ballooning mode preceding substorm onset, *J. Geophys. Res.*, *115*, A00I16, 2010.

Raeder, J., P. Zhu, and Y. Ge, Auroral signatures of ballooning mode near substorm onset: OpenGGCM simulations, in *Auroral Phenomenology and Magnetospheric Processes: Earth and other Planets*, edited by A. Keiling, E. Donovan, F. Bagenal, and T. Karlsson, vol. 197 of *AGU Geophysical Monograph*, pp. 389–395, 2012.

Raeder, J., et al., Global simulation of the geospace environment modeling substorm challenge event, *J. Geophys. Res.*, *106*, 381, 2001b.

Rastätter, L., et al., Geospace environment modeling 2008 - 2009 challenge: Dst index, *Space Weather*, *11*, 187–205, 2013.

Rodger, A., The mid-latitude trough—revisited, in *Midlatitude Ionospheric Dynamics and Disturbances*, pp. 25–33, Wiley-Blackwell, 2008.

Schunk, R. W., P. M. Banks, and W. J. Raitt, Effects of electric fields and other processes upon the nighttime high-latitude F layer, *J. Geophys. Res.*, *81*, 3271–3282, 1976.

Toffoletto, F. R., R. W. Spiro, R. A. Wolf, M. Hesse, and J. Birn, Self-consistent modeling of inner magnetospheric convection, in *Third International Conference on Substorms (ICS-3)*, edited by E. J. Rolfe and B. Kaldeich, p. 223, ESA SP-389, ESA Publications Division, Noordwijk, The Netherlands, 1996.

Toffoletto, F. R., S. Sazykin, R. W. Spiro, and R. A. Wolf, Modeling the inner magnetosphere using the rice convection model, *Space Sci. Rev.*, *108*, 175, 2003.

Toffoletto, F. R., S. Sazykin, R. W. Spiro, R. A. Wolf, and J. G. Lyon, RCM meets LFM: initial results of one-way coupling, *J. Atm. Sol. Terr. Phys.*, *66*, 1361, 2004.

Toth, G., et al., Space weather modeling framework: A new tool for the space science community, *J. Geophys. Res.*, *110*, 21, doi:10.1029/2005JA011,126, 2005.

Wang, Y. L., J. Raeder, C. T. Russell, T. D. Phan, and M. Manapat, Plasma depletion layer: Event studies with a global model, *J. Geophys. Res.*, *108*, 1010, doi:10.1029/2002JA009,281, 2003.

Wang, Y. L., J. Raeder, and C. T. Russell, Plasma depletion layer: Structure and forces, *Ann. Geophys.*, *22*, 1001, 2004.

Yu, Y., V. Jordanova, S. Zou, R. Heelis, M. Ruohoniemi, and J. Wygant, Modeling subauroral polarization streams during the 17 march 2013 storm, *J. Geophys. Res.*, *120*, 1738–1750, 2015.

Zhu, P., J. Raeder, K. Germaschewski, and C. C.Hegna, Initiation of ballooning instability in the near-Earth plasma sheet prior to the 23 March 2007 THEMIS substorm expansion onset, *Ann. Geophys.*, *27*, 1129, 2009.

Journal of Physics: Conference Series **767** (2016) 012022 doi:10.1088/1742-6596/767/1/012022

The Local Interstellar Magnetic Field and the IBEX Ribbon

Romana Ratkiewicz*/ and Marek Strumik****

*Institute of Aviation, al. Krakowska 110/114, 02-256 Warsaw, Poland
**Space Research Centre PAS, Bartycka 18A, 00-716 Warsaw, Poland
roma.ratkiewicz@ilot.edu.pl

Abstract. It is well known that the Interstellar Boundary Explorer (IBEX) discovered a region of enhanced energetic neutral atom (ENA) emission seen in all-sky maps as a ribbon. The enhanced fluxes of ENAs were between 2 and 3 times greater than adjacent regions of the sky. The ribbon itself was not predicted by any models of the heliosphere interface. In the paper by Ratkiewicz, Strumik and Grygorczuk published in the Astrophysical Journal, 756:3, 2012 the authors presented some arguments to show that it was possible to predict the IBEX ribbon. In the current paper the numerical results obtained by three-dimensional MHD simulations of the interaction between the solar wind and interstellar medium by Ratkiewicz et al. [2012] are summarized. Some new aspects of this issue are discussed.

1. Introduction

The heliosphere, a cavity carved in the local interstellar medium (LISM) by the solar wind (SW) blowing out from the Sun is a region filled by the SW plasma emerged in the solar magnetic field and remains under the substantial influence of the LISM ionized and neutral components and of the local interstellar magnetic field (ISMF; see e.g. Linde et al. 1998; Pogorelov & Matsuda 1998; Ratkiewicz et al. 1998, 2000, 2008; Ratkiewicz & Ben-Jaffel 2002; Opher et al. 2007, 2009; Izmodenov et al. 2009; Pogorelov et al. 2009; Strumik et al. 2011; Ben-Jaffel & Ratkiewicz, 2012; Ben-Jaffel et al. 2013, Zank 2015 and references therein). The heliosphere is separated from the LISM by the heliopause (HP) around which the SW decelerated by the termination shock (TS) is transported at its inner side. The LISM flow decelerated by the bow shock (BS) is flowing around the HP at its outer side. The region between the TS and the BS forms the heliosphere interface, between TS and HP - the inner heliosheath (IHS), and between HP and BS - the outer heliosheath (OHS). Hydrogen energetic neutral atoms (ENAs) produced by the solar-interstellar interaction has been discovered to form a ribbon seen by the Interstellar Boundary EXplorer in the sky maps as enhanced emission sites (IBEX; Funsten et al. 2009; Fuselier et al. 2009; McComas et al. 2009; Schwadron et al. 2009). Such a phenomenon as the ribbon was not predicted by any model of the heliosphere interface. Immediately after the discovery, McComas et al. (2009) proposed an interpretation of the ribbon as resulting from ENA production in locations in space ordered by the ISMF. The so-called secondary ENA mechanism for ribbon formation was proposed (Heerikhuisen et al. 2010; Chalov et al. 2010), where the ribbon was shown to be possibly associated with a region outside the HP, where the local ISMF $\mathbf{B}$ is perpendicular to radial vectors $\mathbf{r}$ from the Sun ($\mathbf{B} \perp \mathbf{r}$, i.e., $\mathbf{B} \cdot \mathbf{r} = 0$), which was also a posteriori suggested by McComas et al. (2009).

Several asymmetries and distortions associated with exterior field-flow obliquity were discussed in detail by Ratkiewicz et al. [2000], who pointed out the important influence of the ISMF on the shape and structure of the heliosphere. They noticed that, due to the draping effects around the HP just beyond it, the strength of the ISMF along the field lines achieves its maximum in a direction quasi-perpendicular to the ISMF direction. This result was used by Chalov et al. [2010]. They found by numerical simulations the location of domains of increased and minimal magnitudes of the ISMF (see their Fig. 1). The authors also pointed out that the transport of PUIs in the OHS is considerably affected in the regions of the strong magnetic field, which can be considered as magnetic mirrors or stagnation regions (see their Fig. 2), where the motion of charged particles along field lines is decelerated and some portion of the particles is reflected. These regions in the OHS are ideal places for the production of ENAs forming the ribbon. Therefore, provided that the regions of strong magnetic compression are located in the all-sky maps in the proximity of locations where $\mathbf{B} \cdot \mathbf{r} \approx 0$, the mirror effect may significantly contribute to the increased ENA fluxes. These considerations have become a base for exploration of physical arguments for the direct connection between the ribbon and the strong ISMF compression regions by Ratkiewicz et al. [2012]. They presented some arguments to show that it was possible to predict the IBEX ribbon.

2. Summary of the method and results

For this purpose Ratkiewicz et al. [2012] investigated the magnetic field draping effects in the OHS for many different heliospheric geometries. They tested the Ratkiewicz et al. [2000] postulate by studying sky map locations of magnetic compression regions and their relation to positions of $\mathbf{B} \perp \mathbf{r}$ in the OHS. More specifically, the regions of strong magnetic compression were identified as the maximum strength $|\mathbf{B}^{fl}{}_{max}|$ along the field lines, consistent with the physics of the mirror effect. Numerical computations showed the physical process leading to correlations between the perpendicularity and strong magnetic compression regions and further on to the ISMF–ribbon relationship.

The results in original paper were presented in the form of all-sky maps using the Mollweide projection in the ecliptic coordinates. The relation $|\mathbf{B} \cdot \mathbf{r}| \leq 0.05$ (here **B** and **r** are versors) was used in all figures presented to show the regions in the maps, where the perpendicularity condition $\mathbf{B} \perp \mathbf{r}$ was approximately satisfied. Obtained results in Figures 2–4 (Ratkiewicz et al. 2012) were presented for a range of ISMF strengths equal to 1.8, 2.4, 3.0, 3.6, 4.0, and 4.2 μG (rows) and for inclination angles 5°, 30° ,40°, 50° , and 85° (columns) to study quasi-parallel (5°) and quasi-perpendicular (85°) magnetic field directions in addition to oblique orientations.

Figure presented in the current paper summarizes the results shown in the original paper in Figures 2-4 for the inclination angle 40° and ISMF strengths equal to 1.8, 2.4, 3.6, and 4.2 μG. Plots in the left column illustrate the influence of the magnetic field draping effects on the geometry of the perpendicularity regions in the all-sky maps, i.e. the comparison of regions $\mathbf{B} \perp \mathbf{r}$ computed for an unperturbed (stripe between red lines) and perturbed (stripe between green lines) ISMF, respectively. Plots in the middle column show the comparison of regions $\mathbf{B}^{r}{}_{max} \perp \mathbf{r}$ (in gray, where $|\mathbf{B}^{r}{}_{max}|$ is the maximum strength along the given **r**) with locations of the maximum of the ISMF strength $|\mathbf{B}^{fl}{}_{max}|$ along the field lines (crosses). Plots in the right column show the comparison of the observed ribbon (black circles) with the regions shown in the middle column.
Full set of results in the original paper showed that the ribbon was reproduced best for the magnitude 3.6 ± 0.6 μG and orientation 40° ± 10° of the ISMF.

The systematic study made by Ratkiewicz et al. [2012] suggested the existence of a strong physical relationship between the ribbon and the ISMF. Hence, the conclusion was that by analyzing ENA detection from the perspective of the ISMF's influence on the heliospheric asymmetries (Ratkiewicz et al. 2000) it might have been possible to predict the IBEX ribbon. The open question, however, which was left, concerned the physical mechanism supporting a large enough number of energetic ionized particles necessary for ENA production in the near vicinity of the HP in regions of maximum strength of the draped ISMF. We expected that a direct leakage of high energy ions from the inner to the outer heliosheath through the HP could ensure sufficient plasma particles in the OHS (Schwadron et al. 2011). This expectation to some extent is confirmed by the recent paper by Zirnstein et al. [2016]. The authors conclude: "Our results suggest a significant number of primary ENAs from the inner heliosheath may contribute to the pickup ion source population outside the heliopause, depending on the ENA energy and SW speed."

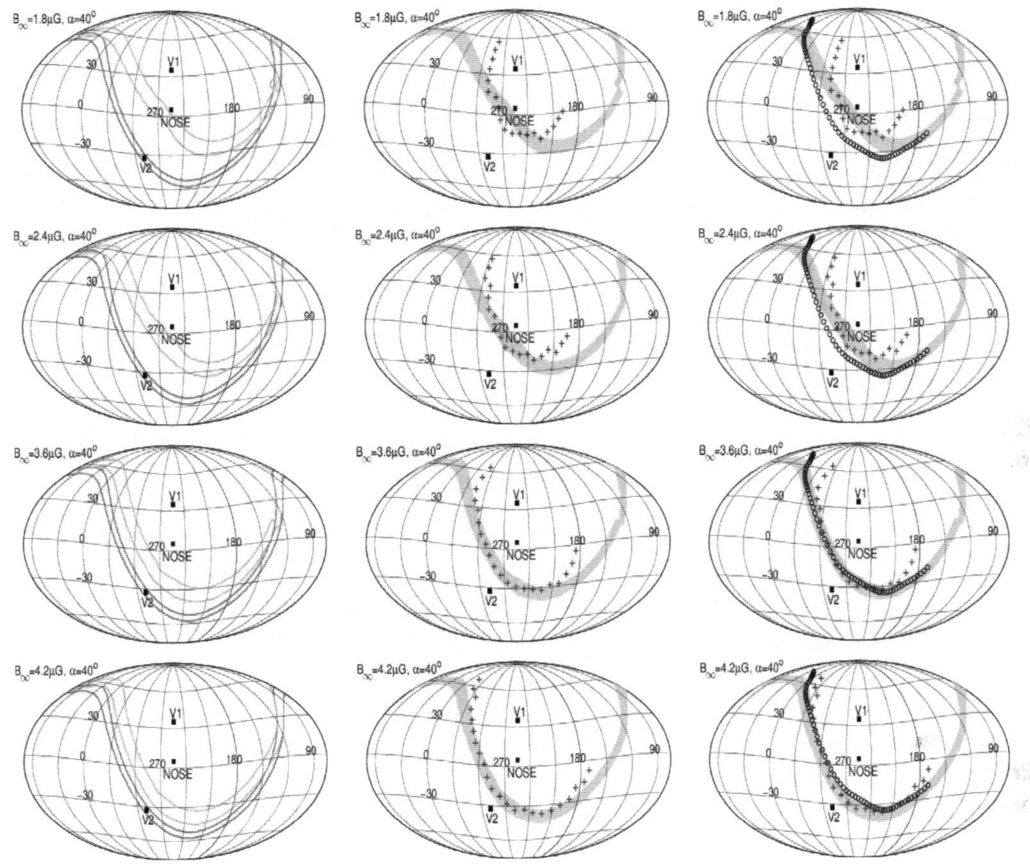

Figure. The influence of the magnetic field draping effects on geometry of the perpendicularity regions in the all-sky maps, i.e. the comparison of regions $\mathbf{B} \perp \mathbf{r}$ computed for an unperturbed (stripe between red lines) and perturbed (stripe between green lines) ISMF, respectively (left column). A comparison of regions $\mathbf{B}^r{}_{max} \perp \mathbf{r}$ (in grey, $\mathbf{B}^r{}_{max}$ is maximum of ISMF strength along lines of sight) with locations of the maximum of the ISMF strength $|\mathbf{B}^{fl}{}_{max}|$ along the field lines (crosses) (middle column). A comparison of regions $\mathbf{B}^r{}_{max} \perp \mathbf{r}$ (in grey) with locations of the maximum of the ISMF strength $|\mathbf{B}^{fl}{}_{max}|$ along the field lines (crosses) with the observed ribbon (black circles) (right column).

Acknowledgments

R.R. acknowledges the support from the Warsaw Institute of Aviation.

References

Ben-Jaffel, L. & Ratkiewicz, R. 2012, A&A,546, A78
Ben-Jaffel, L., Strumik, M., Ratkiewicz, R., & Grygorczuk, J. 2013, ApJ, 779, 130
Chalov, S. V., Alexashov, D. B., McComas, D. J., et al. 2010, ApJ, 716, L99
Funsten, H. O., Allegrini, F., Crew, G. B., et al. 2009, Science, 326, 964
Fuselier, S. A., Allegrini, F., Funsten, H. O., et al. 2009, Science, 326, 962
Heerikhuisen, J., Pogorelov, N. V., Zank, G. P., et al. 2010, ApJ, 708, L126
Izmodenov, V., Malama, Y. G., & Rudermann, M. S. 2009, Space Sci. Rev., 146,329
Linde, T. J., Gombosi, T. I., Roe, P. L., Powell, K. G., & Dezeeuw, D. L. 1998, J. Geophys. Res., 103, 1889
McComas, D. J., Allegrini, F., Bochsler, P., et al. 2009, Science, 326, 959
Opher, M., Stone, E. C., & Gombosi, T. I. 2007, Science, 316, 875
Opher, M., Alouani Bibi, F., Toth, G., et al. 2009, Nature, 462, L24
Pogorelov, N. V., & Matsuda, T. 1998, J. Geophys. Res., 103, 237
Pogorelov, N. V., Heerikhuisen, J., Mitchell, J. J., Cairns, I. H., & Zank, G. P. 2009, ApJ, 695, L31
Ratkiewicz, R., Barnes, A., Spreiter, J. R., et al. 1998, A&A, 335, 363
Ratkiewicz, R., Barnes, A., & Spreiter, J. R. 2000, J. Geophys. Res., 105, 25021
Ratkiewicz, R., & Ben-Jaffel, L. 2002, J. Geophys. Res., 107, 1007
Ratkiewicz, R., Ben-Jaffel, L., & Grygorczuk, J. 2008, in ASP Conf. Ser. 385, Numerical Modeling of Space Plasma Flows, ed. N. V. Pogorelov, E. Audit, & G. P. Zank (San Francisco, CA: ASP), 189
Ratkiewicz, R., Strumik, M., & Grygorczuk, J., 2012, ApJ 756:3(6pp)
Schwadron, N. A., Bzowski, M., Crew, G. B., et al. 2009, Science, 326, 966
Schwadron, N. A., Allegrini, F., Bzowski, M., et al. 2011, ApJ, 731, 56
Strumik, M., Ben-Jaffel, L., Ratkiewicz, R., & Grygorczuk, J. 2011, ApJ, 741,L6
Zank, G. P., 2015, Annual Rev. Astron. Astrophys., 53, 449
Zirnstein, E. J., Funsten, H. O., Heerichuisen, J., & McComas, D. J. 2016, A&A, 586, 13 pp., 2016

Element Abundances and Source Plasma Temperatures of Solar Energetic Particles

Donald V Reames

IPST, Univ. of Maryland, College Park, MD 20742-2431 USA
dvreames@umd.edu

Abstract. Thirty years ago Breneman and Stone [1] observed that the enhancement or suppression of element abundances in large solar energetic-particle (SEP) events varies as a power of the mass-to-charge ratio, *A/Q*, of the elements. Since *Q* during acceleration or transport may depend upon the source plasma temperature *T*, the pattern of element enhancements can provide a best-fit measure of *T*. The small SEP events we call ^{3}He-rich or "impulsive" show average enhancements, relative to coronal abundances, rising as the 3.6 power of *A*/Q to a factor of ~1000 for ($76 \leq Z \leq 82$)/O and temperature in the range 2-4 MK. This acceleration is believed to occur in islands of magnetic reconnection on open field lines in solar flares and jets. It has been recently found that the large shock-accelerated "gradual" SEP events have a broad range of source plasma temperatures; 69% have coronal temperatures of $T < 1.6$ MK, while 24% have $T \sim 3$ MK, the latter suggesting a seed population containing residual impulsive suprathermal ions. Most of the large event-to-event abundance variations and their time variation are largely explained by variations in *T* magnified by *A/Q*-dependent fractionation during transport. However, the non-thermal variance of impulsive SEP events (~30%) exceeds that of the ~3 MK gradual events (~10%) so that several small impulsive events must be averaged together with the ambient plasma to form the seed population for shock acceleration in these events.

1. Introduction

The relative abundances of the chemical elements vary greatly among the energetic particles from the Sun and the study of those abundances has revealed many underlying physical processes involved in their acceleration and transport. Most of these processes can only be studied using direct samples of the particles themselves, since photon production by accelerated ions is extremely limited and is poorly observed.

Logically the primary distinction was between SEP events we call "gradual" and "impulsive" [2, 3, 4, 5, 6, 7]. Historically, gradual events, being the most intense, were observed first. They are now known to be accelerated at shock waves driven out from the Sun by coronal mass ejections (CMEs). Most of the acceleration begins at 2 to 3 solar radii and samples ambient coronal plasma. It was recognized by Meyer [8] that the abundances, averaged over large SEP events and divided by the corresponding photospheric abundances shows a characteristic two-level pattern when plotted *vs.* the first ionization potential (FIP) of the element, as shown in the left panel of Figure 1. This occurs because elements with low FIP are ionized in the photosphere while those with high FIP are neutral so the former are much more easily transported up into the corona. Gradual SEP events, on average, measure coronal abundances. However a second important aspect of gradual event abundances, also

Journal of Physics: Conference Series **767** (2016) 012023 doi:10.1088/1742-6596/767/1/012023

shown in Figure 1 was that in individual events, abundances relative to the mean, varied as a power law in the charge-to-mass ratio (*Q/M* or *Q/A*) of the ions.

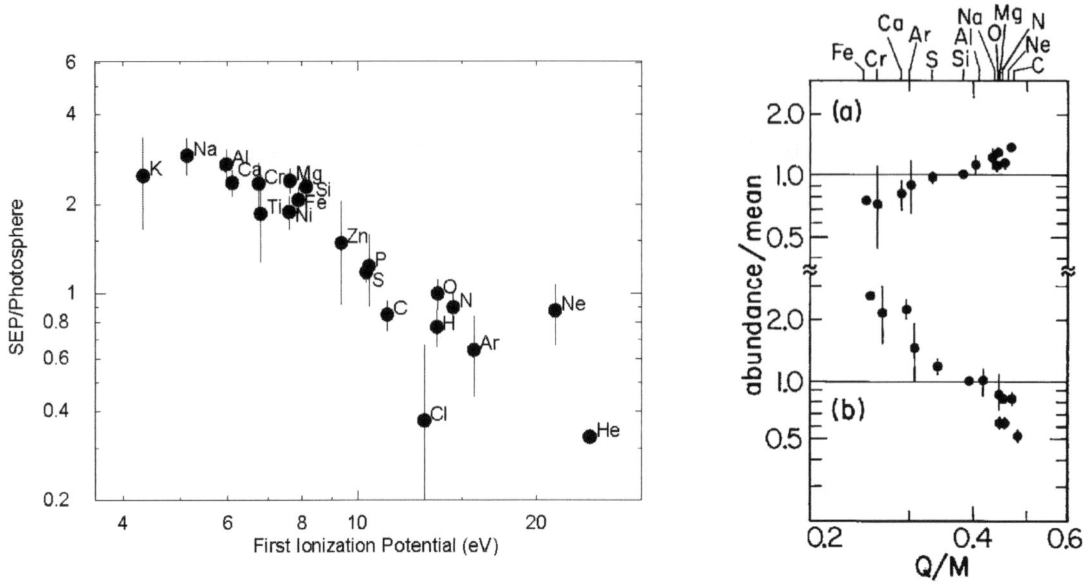

Figure 1. The left panel shows the average SEP abundances [9] relative to corresponding photospheric abundances [10] *vs.* FIP. The right panel shows *Q/M* dependence for two large events observed by Breneman and Stone [1].

It might be said that impulsive events were first distinguished by Wild, Smerd, and Weiss [11] as radio type III bursts produced by 10–100 keV electrons streaming out from the Sun. However the associated ions, found much later, turned out to have strange abundances, beginning with greatly enhanced ^{3}He/^{4}He ratios [12], exceeding that of the corona or solar wind by factors up to 10^4 [13]. These ^{3}He-rich events were eventually associated with the electron-rich type III bursts [13] and the heavier ions were found to have *A/Q*-dependent enhancements of a factor of ~10 at Fe [14]. These abundances led to the first estimate of source plasma temperature *T*, since He, C, N, and O appeared to be fully ionized, Ne, Mg, and Si had two orbital electrons, and Fe had a higher *A/Q*, consistent with *T* = 3–5 MK. Impulsive SEP events have been associated with magnetic reconnection on open field lines as occur in solar jets [15]. Particle-in-cell simulations support strong power-laws in *A/Q* [16] and the ^{3}He enhancements are believed to result from preferential absorption by ^{3}He of electromagnetic ion-cyclotron waves generated by the streaming electrons [17].

Our ability to cleanly distinguish impulsive and gradual SEP events using element abundances alone is prevented by the fact that shock waves frequently encounter residual suprathermal ions left over from impulsive SEP events among the seed population that becomes accelerated. Enhanced abundances of ^{3}He were found in large shock events [18] and are seen before and after shock acceleration at *in situ* observations [19]. Impulsive suprathermal ions can alter the energy dependence of abundance ratios such as Fe/O [20].

2. Element abundances and source plasma temperatures in impulsive SEP events

Recent measurements of ions in impulsive SEP events using two different instrument techniques [21] have extended the abundance measurements to elements throughout the periodic table (although not with single-element resolution). Figure 2 shows the abundance enhancement, relative to the SEP coronal abundances, averaged over 111 impulsive SEP events as a function of *A/Q* of the ions.

Journal of Physics: Conference Series **767** (2016) 012023 doi:10.1088/1742-6596/767/1/012023

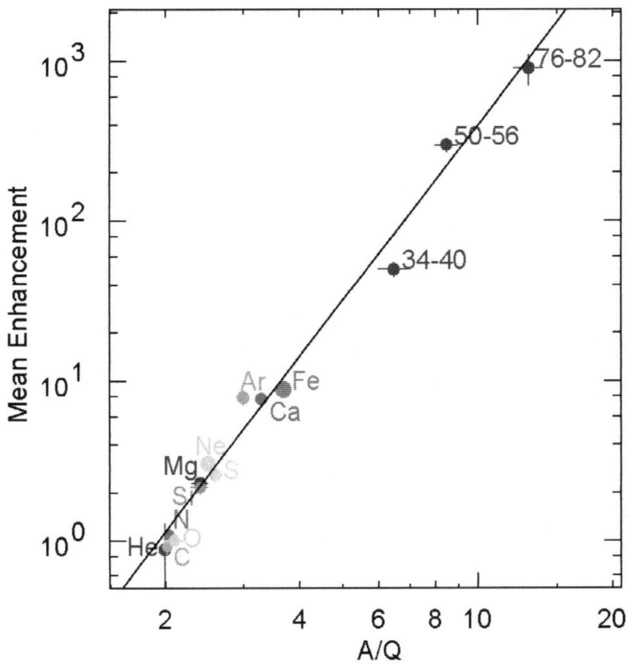

Figure 2. The mean element enhancement in impulsive SEP events (relative to coronal abundances) is shown *vs.* *A/Q* at 2.5–3.2 MK where values are taken from the shaded band in Figure 3. The power of the fitted line shown is 3.64±0.15 [22].

Figure 3 shows *A/Q* *vs.* temperature *T* using theoretical values of *Q* *vs.* *T* [23] for typical elements.

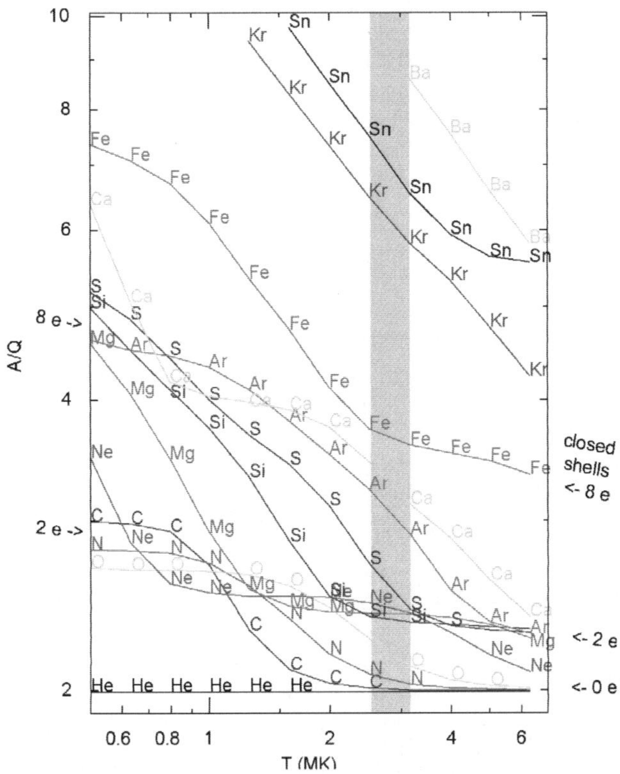

Figure 3. *A/Q* is plotted *vs.* the theoretical equilibrium temperature for elements that are named along each curve [23]. Points are spaced every 0.1 unit of log *T* from 5.7 to 6.8. Bands produced by closed electron shells with 0, 2, and 8 electrons are indicated in the margins, He having no electrons. Elements tend to move from one of these groups to another as the temperature changes. The red strip selects values used for impulsive SEP events in Figure 2 [23, 26].

Early observations [24] were ascribed to the range of 3–5 MK where He, C, N, and O are fully ionized and Ne, Mg, and Si have 2 orbital electrons. However, the lower range better fits the heavy elements, explains why Ne is enhanced more than Mg or Si, and explains the existence of "C-poor" and "He-poor" events where O has been enhanced so that He/O and C/O decrease [22]. The strong power-law dependence on *A/Q* agrees with theoretical particle-in-cell simulations of collapsing islands of magnetic reconnection [16].

It is also possible to estimate *T* for individual impulsive SEP events by fitting the observed enhancements *vs.* *A/Q(T)* for each of several temperatures of interest [25]. The fit with the lowest value of χ^2 is selected. Nearly all of the impulsive events have minimum values of χ^2 at either 2.5 or 3.2 MK. Most (69%) of the events associate with CMEs, most *slow* and *narrow*, suggesting solar jets, but, otherwise, the SEP parameters are not strongly correlated with CME or flare parameters [22, 25].

3. Element abundances and source plasma temperatures in gradual SEP events

Breneman and Stone [1] showed that abundance enhancements in gradual event also behave as a power law in *A/Q*. It can be shown from diffusion theory that the scattering of ions during transport can produce an enhancement (or suppression) of the abundance ratios that is a power law in *A/Q* with a power that varies with time *t* during the event as $a/t + b$ [26]. Here the parameter *b* is negative unless previously enhanced impulsive suprathermal ions are accelerated.

Figure 4 shows the analysis of several energy intervals in a large gradual event. Fits at each temperature yield values of χ^2 and the temperature of the minimum χ^2 is selected.

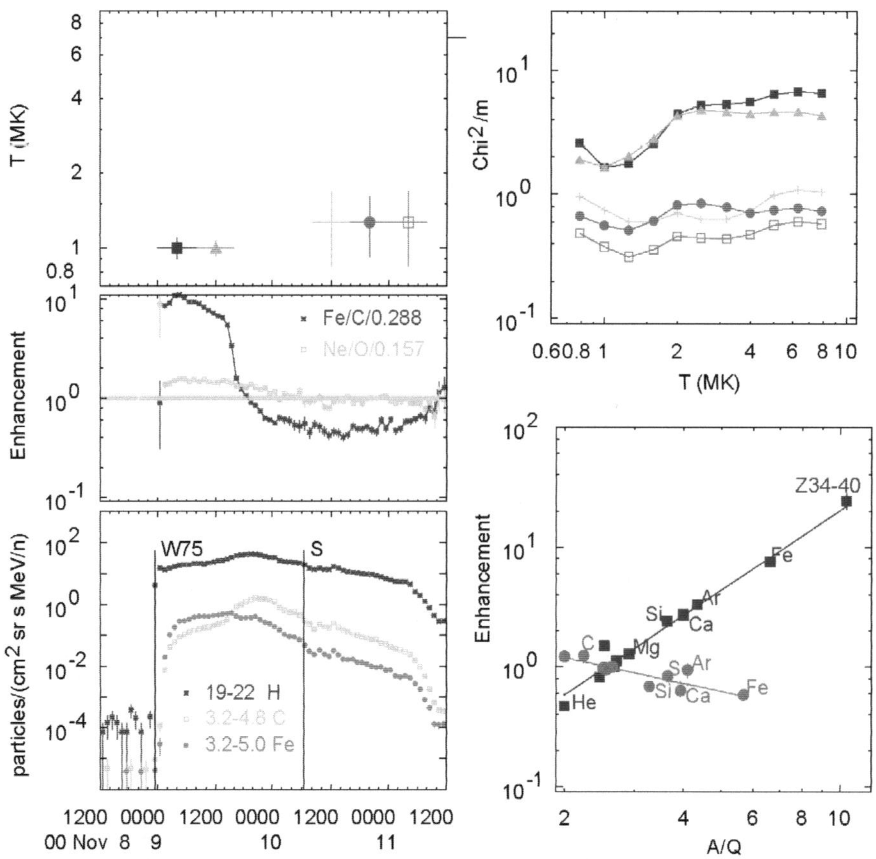

Figure 4. Clockwise from the lower left panel are the intensities of H, C, and Fe during the 8 November 2000 SEP event, the enhancements in Fe and Ne during the event, the best-fit temperatures in color-coded 8-hr intervals, values of χ^2/m *vs.* *T* for each time interval, and best power-law fits of the observed enhancements *vs.* *A/Q* at two times [26].

Journal of Physics: Conference Series **767** (2016) 012023 doi:10.1088/1742-6596/767/1/012023

The determination of temperatures is easily understood by returning to Figure 3. Proceeding downward in temperature, one sees the elements O, then N, then C cross from stripped to the two-electron shell. Meanwhile, the elements Ca, Ar, S, Si, then Mg cross from the 2- to the 8-electron shells. Since the observed enhancements mirror *A/Q*, the pattern of enhancement and the grouping of the elements shows the temperature. Of course, we could scale Fe/O to fit *A/Q* at any *T*, but the intermediate elements group poorly and raise χ^2 except at the best-fit temperature.

Figure 5 compares the pattern of enhancements in the first 8-hr interval of the 2000 November 8 event seen in Figure 3. In the figure the observed enhancements of C, N, and O have moved up above He to be near to Ne and Mg. Meanwhile Si and S have moved up near Ca, Fe is considerably enhanced and we even have a measure of $34 \leq Z \leq 40$, represented by Kr.

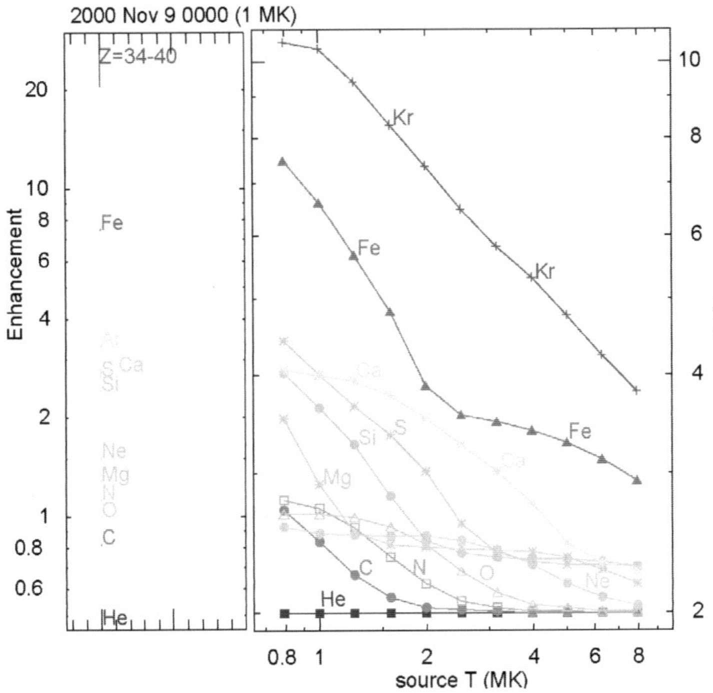

Figure 5. The left panel shows enhancements in element abundances during the interval 0000–0800 UT 9 November 2000. The right panel shows A/Q vs. T for various elements, as in Figure 3. The groupings of enhancements match those in A/Q near 1 MK [26].

Out of 45 gradual events that have been studied [26]:

- 11 events (24%) showed source plasma temperatures of 2.5–3.2 MK similar to impulsive SEPs,
- 31 events (69%) had source plasma temperatures of ≤ 1.6 MK,

For about 10% of the latter events, the values of χ^2/m turn down again at high temperatures. These events are surely not hot plasma but may be a component that has been stripped of electrons.

Generally correlations between source plasma temperatures and properties of the associated CME are rather weak. Figure 6 shows *T vs.* CME speed which has an un-weighted correlation coefficient of -0.49. Also, several events in the sample are ground level events (GLEs) in which GeV protons cause a nuclear cascade in the atmosphere that is measurable in instruments at ground level. These GLEs are identified in Figure 6. However, there seems to be nothing special about them.

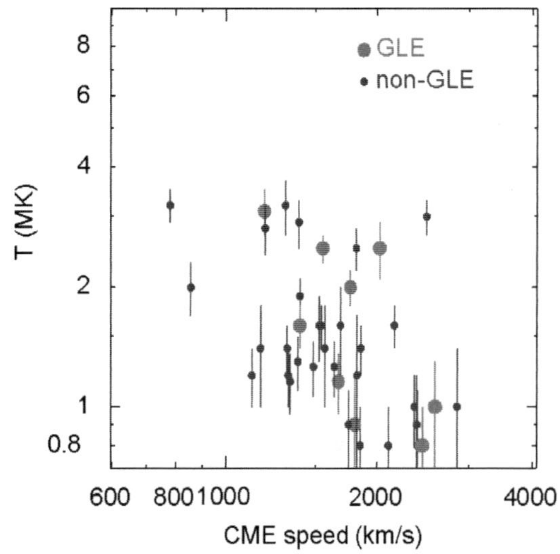

Figure 6. Source plasma temperatures of the SEP events are shown as a function of speed of the associated CME. GLEs among the sample, indicated as large red circles, have no special distribution [26].

Temperature differences explain many of the abundance variations between gradual SEP events that were previously unexplained. However, interesting new differences between impulsive and gradual events with the same source temperatures have also emerged as seen in Figure 7.

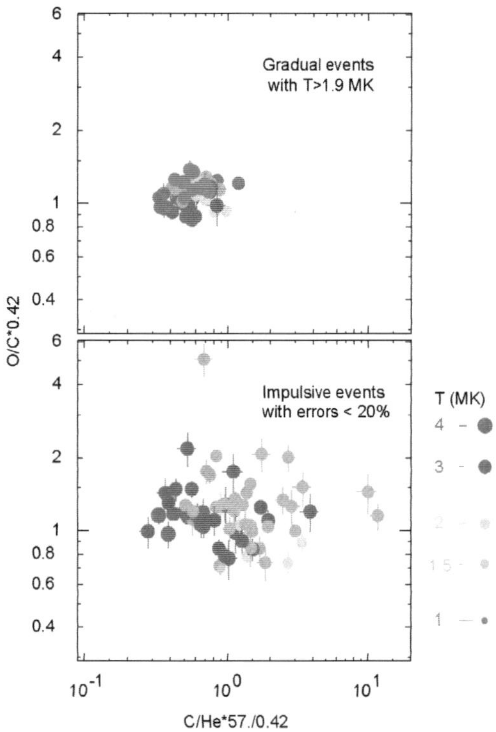

Figure 7. The figure compares normalized O/C *vs.* C/He for gradual SEP events with $T \geq 2$ MK (upper panel) and for well-measured impulsive SEP events (lower panel) at the same scale. The impulsive events show large non-thermal errors that must be reduced by averaging in the gradual events.

At the temperatures of impulsive SEPs, ~3 MK, He, C, and possibly even O, should be almost fully ionized so that no relative enhancements are possible. Thus the scatter of the impulsive events in this figure must come from *non-thermal* variations, possibly abundance variations in the underlying local plasma before acceleration. The comparison in Figure 7 suggests that *no single impulsive event can provide all seed particles for a gradual event*; if it did, the gradual events would have a spread like the impulsive events. However, averaging over many impulsive events would reduce the spread.

Averaging 10 impulsive events with 30% errors would reduce the error in the mean to 10%. Accelerating ambient active region plasma along with impulsive seed particles is also likely and would reduce the errors, but would also reduce Fe/O and Ne /O, possibly more than observed.

Where are the 10 impulsive events that contribute suprathermal seeds to each gradual SEP event? In 1984, Lin *et al.* [27] extended the size distribution of solar X-ray flares to smaller sizes using a large balloon-bourn detector; as the flare energy decreased the number of flares per day increased as a power law. This observation led Parker to suggest nanoflares [28] to heat the corona. Now flares involve magnetic reconnection on closed loops, capturing the energy as heat; we are interested in jets with reconnection on open field lines. Whether or not nanoflares are sufficiently numerous to heat the corona, we suspect that small *nanojets*, erupting frequently in active regions can supply a steady multi-jet seed population of impulsive suprathermal ions that would be available to the shock wave of a gradual event passing through. If the power of the size distribution were 1.72 [27], there would be over 50 times as many events, each contributing 10% of the suprathermal ions accelerated in the impulsive SEP events we see. The nanojets would sample different points in an active region.

Thus we see two situations for the production of a gradual SEP event:

- If an observer is magnetically connected to the *ambient corona* as a big shock goes through, SEPs come from seed ions at the local coronal temperature.
- If an observer is magnetically connected to an *active region* as a big shock goes through, SEP seeds average over the current multi-jet population of impulsive suprathermal ions as well as the sampling some of the ambient corona.

Ko *et al.* [29] found that Fe-rich gradual SEP events were commonly connected to active regions and observations of persistent and long-lived ^{3}He-rich sources have been observed [30].

4. Summary

Element abundances averaged over many of the large gradual events provide a measure of coronal abundances that differ from photospheric abundances by their well-known dependence on FIP. These SEP coronal abundances serve as a reference for both impulsive and gradual events.

Enhancements of the average abundances in impulsive events, relative to the reference corona, show a strong power-law in *A/Q* to the 3.6 power, determined at a temperature near 3 MK, extending from He to Pb. Such a power-law is obtained theoretically from simulations of islands of magnetic reconnection. These reconnections occur on open magnetic field lines in solar jets and are associated with slow, narrow CME.

We can find a temperature for individual impulsive events by fitting the observed enhancement vs. *A/Q(T)* at several values of *T* and selecting the fit and *T* with the minimum χ^2. Nearly all impulsive events fall at 2.5 or 3.2 MK.

Element abundances in gradual SEP events, relative to the reference corona, show either increasing or decreasing power laws in *A/Q*, as found originally by Breneman and Stone [1] and are expected from diffusive transport theory [26]. Values of *T* can again be found by fitting the observed enhancement vs. *A/Q(T)* at many values of *T* and selecting the fit and *T* with the minimum χ^2. This works because the grouping of *A/Q* values around closed shells is quite clear and varies strongly with *T*. Most (69%) of the gradual events have temperatures < 1.6 MK representing shock acceleration of ambient coronal material, 24% have *T* > 2 MK, probably from shock reacceleration of impulsive suprathermal ions contributed by multiple small nanojets as well as active-region plasma which dilutes the enhancements.

References

[1] Breneman H H and Stone E C 1985 *Astrophys. J. Lett.* **299** L57.
[2] Gosling J T 1993 *J. Geophys. Res.* **98**, 18937.

[3] Reames D V 1999 *Space Sci. Rev.,* **90**, 413.
[4] Mason G M 2007 *Space Sci. Rev.* **130**, 231.
[5] Kahler S W 2001 *J. Geophys. Res.* **106**, 20947.
Kahler S W 2007 *Space Sci. Rev.* **129**, 359.
[6] Cliver E W and Ling A G 2007, *Astrophys. J.* **658**, 1349.
Cliver E W and Ling AG 2009, *Astrophys. J.* **690**, 598.
[7] Reames D V 2013 Space Sci. Rev. 175, 53.
Reames D V 2015 *Space Sci. Rev.,* **194**: 303 arXiv: 1510.03449
[8] Meyer, J. P.: 1985, *Astrophys. J. Suppl.* **57**, 151.
[9] Reames, D.V.:2014, *Solar Phys.* **289**, 977, DOI: 10.1007/s11207-013-0350-4
[10] Asplund M Grevesse N Sauval A J and Scott P 2009 *Ann. Rev. Astron. Astrophys.* **47**, 481
[11] Wild J P Smerd S F and Weiss A A 1963, *Ann. Rev. Astron. Astrophys.* **1**, 291.
[12] Hsieh K C and Simpson J A 1970, *Astrophys., J. Lett.* **162**, L191.
[13] Reames D V von Rosenvinge T T and Lin R P 1985 *Astrophys. J.* **292**, 716.
Reames DV and Stone R G.1986 *Astrophys. J.* **308**, 902.
[14] Mason G M. Reames D V Klecker B Hovestadt D and von Rosenvinge T T 1986 *Astrophys. J.* **303**, 849.
[15] Shimojo M and Shibata K 2000 *Astrophys. J.* **542**, 1100.
Kahler S W Reames D V and Cliver EW 2015 *34th Intl. Cosmic Ray Conf.,* in press (arXiv: 1509.09260)
[16] Drake J F Cassak P A Shay M A Swisdak M Quataert E 2009 *Astrophys. J. Lett.* **700**, L16.
[17] Temerin M Roth I 1992 *Astrophys. J. Lett.* **391**, L105.
Roth I Temerin M 1997 *Astrophys. J.* **477**, 940.
[18] Mason G M Mazur J E Dwyer J R 1999 *Astrophys. J. Lett.* **525**, L133.
[19] Desai M I Mason G M Dwyer J R Mazur J E Gold R E Krimigis S M Smith C W and Skoug R M 2003 *Astrophys. J.* **588**, 1149.
[20] Tylka A J Cohen C M S Dietrich W F Lee M A Maclennan C G Mewaldt R A Ng C K and Reames D V 2005 *Astrophys. J.* **625**, 474.
Tylka A J and Lee M A.2006 *Astrophys. J.* **646**, 1319.
[21] Reames D V 2000 *Astrophys. J. Lett.* **540**, L111.
Mason G M Mazur J E Dwyer J R Jokipii J R Gold R E and Krimigis S M 2004 Astrophys. J. 606, 555.
Reames D V and Ng C K 2004 *Astrophys. J.* **610**, 510.
[22] Reames D V Cliver E W and Kahler S W 2014 *Solar Phys.* **289**, 3817 (arXiv: 1404.3322)
[23] Mazzotta P Mazzitelli G Colafrancesco S and Vittorio N 1998 *Astron. Astrophys Suppl.* **133**, 403.
Post D E Jensen R V Tarter C B Grasberger W H and Lokke W A 1977 *At. Data Nucl. Data Tables*, **20**, 397
[24] Reames D V Meyer J P and von Rosenvinge T T 1994 *Astrophys. J. Suppl.*, **90**, 649.
[25] Reames D V Cliver E W and Kahler S W 2014 *Solar Phys.* **289**, 4675 (arXiv:1407.7838)
[26] Reames D V 2016 *Solar Phys.* **291**, 911, DOI: 10.1007/s11207-016-0854-9 (arXiv: 1509.08948)
Reames D V 2016 *Solar Phys.*, submitted (arXiv: 1603.06233)
[27] Lin R P Schwarz, R A Kane S R Pelling R M Hurley K C 1984 *Astrophys. J.* **283** 421.
Aschwanden M J 2011 *Solar Phys.* **274** 99.
[28] Parker E N 1988 *Astrophys J.* **330** 474.
[29] Ko Y-K Tylka, AJ Ng C K Wang Y-M Dietrich W F 2013 *Astrophys. J.* **776**, 92.
[30] Wiedenbeck M E Cohen C M S Cummings A C de Nolfo G A Leske R A Mewaldt R A Stone E C von Rosenvinge T T 2008 *Proc. 30th Int. Cosmic Ray Conf.* (Mérida) **1**, 91.
[31] Bučík R Innes D E Mall U Korth A Mason G M Gómez-Herrero R 2014 *Astrophys. J.* **786**,71.

Comparison of astrospheres around cool and hot stars

K. Scherer[1,2], D.J. Bomans[3,2], S.E.S. Ferreira[4], H. Fichtner[1,2], J. Kleimann[1], R.D. Strauss[4], K. Weis[3]

[1] Institut für Theoretische Physik IV: Weltraum- und Astrophysik, Ruhr-Universität Bochum, Germany

[2]Research Department, Plasmas with Complex Interactions, Ruhr-Universität Bochum, Germany

[3]Astronomische Institute, Ruhr-Universität Bochum, Germany

[4]Center for Space Research, North-West University, 2520 Potchefstroom, South Africa

E-mail: `kls@tp4.rub.de`

Abstract. Simulations of astrospheres around hot and cool stars have recently move into the focus of scientific research. We describe here the differences between the astrospheres around hot and cool stars. While those around the former are huge (on pc scales) because of their high stellar wind momentum flow, those around cool stars are on a AU scale and the heliosphere can be used as a standard. Here, we will compare the differences and commonalities between such astrospheres.

1. Introduction

Every supersonic wind-driving star moving supersonically (superalvfénically) through the interstellar medium exhibits a bow shock, an astropause and a terminations shock. The mass flux of stellar winds around cool stars (mainly F,G,K stars) is usually small compared of those around hot stars (O and B stars). The most detailed observations of an astrosphere around a cool star are available for the heliosphere, the astrosphere around the Sun. Especially due to the Voyager 1 spacecraft we have nowadays in situ observations from the region beyond the heliopause (astropause). Some astrospheres of nearby cool stars could be observed in the Lyman-α line [1, 2], the observations of the astrospheres of hot stars are mainly based on Hα light (e.g., [3]). To understand these other astrospheres one has to transfer our detailed knowledge of the heliosphere. But this has to be done with some care, because radiative cooling plays an important role in the physics of astrospheres around hot stars, for which the characteristic length scales are parsec, compared to hundreds of AU for those around cool stars.

In the following we use a model of the heliosphere as example for an astrosphere around a cool star, which we compare to that of the astrosphere around the hot O star λ Cephei.

2. Astrosphere models

To model astrospheres we use the set of ideal Euler equations described in [4], in case of hot stars including cooling and heating functions in the energy equation. While the ideal hydrodynamic (HD) and and magneto-hydrodynamic (MHD) Euler equations are scale-invariant, this property is lost when adding either cooling and heating functions or a neutral component. Nevertheless, it is in principle possible to use scale-invariance to compare astrospheres around cool and hot stars, keeping in mind that this is only approximately true.

Thus neglecting the influence of neutrals and using the fact that cooling and heating functions are not effective for cool star astrospheres, they can be modeled using ideal MHD, and the heliosphere is used as the standard. For the astrospheres around hot stars we use the O star λ Cephei with cooling and heating, also neglecting neutrals because of the high flux of ionizing photons.

2.1. The heliosphere

The heliospheric large-scale structures is described with HD or MHD models of increasing complexity (e.g., [5, 6, 7, 8, 9, 10, 11, 12]). We base our simulations on the Cronos code with ideal (M)HD [13, 14, 15, 11, 4] using the following parameters [8] at 1 AU: $v_{\rm sw} = 375$ km/s, $n_{\rm sw} = 7$ cm^{-3}, $T_{\rm sw} = 7.364 \times 10^4$ K, whereas those for the ISM are $v_{\rm ism} = 26.4$ km/s, $n_{\rm ism} = 0.06$ cm^{-3}, and $T_{\rm ism} = 6.53 \times 10^3$ K. As usual, an axisymmetric configuration forms, which is characterized by the termination shock (TS) terminating the supersonic solar/stellar wind, a bow shock (BS) on the upwind side, where the interstellar flow (as seen in the rest frame of the Sun/star) changes from supersonic to subsonic speed, and a tangential discontinuity, the helio-/astropause (HAP) in between. The inner helio-/astrosheath (IHAS) is the region between the TS and the HAP, and the outer helio-/astrosheath (OHAS) that between the HAP and the BS (see Fig. 1). The bow shock may not exist, when the fast magnetosonic wave speeds are higher than the inflow speed [16, 17], but see also [18], who pointed out that heavier elements can lower the fast magnetosonic speed.

To be comparable with the O-star astrosphere, we refer here to a heliosphere without neutral interaction, that is the local ISM is completly ionized. It is very propable that the heliosphere has or will pass through Such an environement during its passage around the galactic center [19]. Because the heliosphere is small compared to O-star astrospheres heating or coolling (described below) does not play a role.

2.2. λ Cephei

In [20] we discussed in detail the hydrodynamic structure of the astrosphere of λ Cephei. In Fig. 1 we now present a MHD model of λ Cephei: the stellar magnetic field is assumed to be 100 times large than the solar field [21], and a hypothetical spiral field at 1 AU is then also 100 times stronger. This hypothetical field is then transported as a spiral magnetic field after [22] to the inner boundary at 0.03 pc. For the undisturbed interstellar magnetic field (IMF) we assumed a strength of 10 μG [23], and a field orientation deviating from the inflow direction (along the x-axis) by 45° in both φ and ϑ. Because the ram pressure of the interstellar medium (ISM) $\rho_{ism} v_{ism}^2$ (with the ISM density $\rho/m_p = 11\,{\rm cm}^{-3}$, $v_{ism} = 80$ km/s and m_p is the proton mass) is higher than that of the IMF:

$$\beta = \frac{8\pi \rho_{ism} v_{ism}^2}{B_{ism}^2} \approx 280 \tag{1}$$

there is only negligible dynamical influence of the magnetic field and the λ Cephei astrosphere is quite similar to a single-fluid hydrodynamic one. During the modeling it turned out that the cooling function by [24] leads to unstable results, thus we used for the recent calculation that by [25]. The model is shown in Fig. 1.

2.3. Comparison of the cool and hot models

2.4. The HD structure

For ideal HD shocks the structure of the heliosphere is well known (see [4] for an overview). On

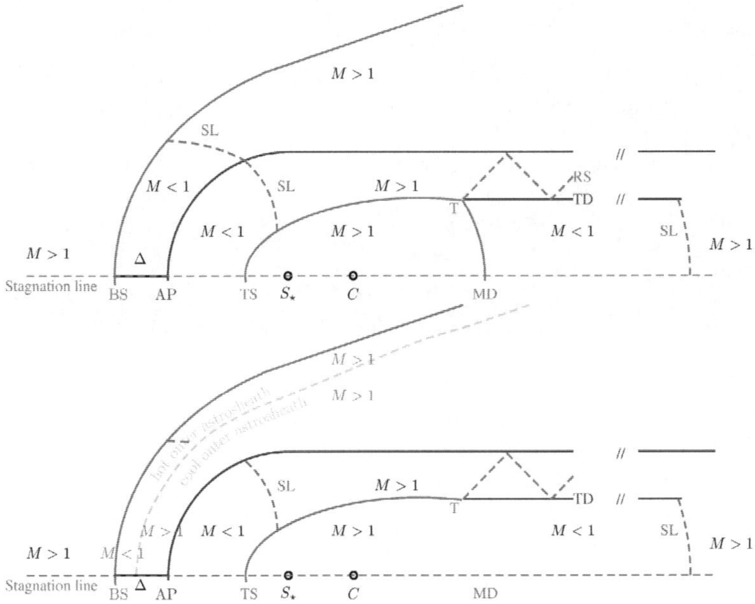

Figure 1. The standard HD model without cooling is shown in the upper panel. The star $S_\star$ is at the origin, SL denotes the sonic line, BS, AP, and TS are the bow shock, astropause and termination shock, MD is the spherical Mach disk, with origin at C, TD and RS are the triple point and an additional tangential discontinuity. M is the Mach number and Δ the distance between BS and AP (see [4] for details). Astrospheres including cooling (lower panel) show an additional feature: the outer astrosheath is split in a cool and a hot one.

the contrary, astrospheres around hot stars with cooling functions show an additional feature, a cool outer region between the bow shock and the astropause (the cool outer astrosheath, COA) together with a hot region directly after the bow shock (the hot outer astrosheath, HOA) [26, 4]. The reason is that the cooling length is much shorter than the distance Δ between the bow shock and the astropause. This cooling results also in an huge increase in the number density close to the astropause in the outer astrosheath. The cooling is also responsible for a shrinking of the bow shock-astropause distance, see e.g., [4]

2.5. The MHD structure

In Fig. 2 we show a the model of the cool astrosphere in the left panel and that of the hot astrosphere in the right panel. It can be clearly seen that the heliosphere is much more asymmetric than the λ Cephei astrosphere. The reason is that in the ISM of the latter the ram pressure dominates the magnetic field pressure. The ratio of these pressure, the plasma β is usually used to describe the plasma properties. The ratio of the thermal pressure to that of the magnetic field is valid in subsonic flows, but extending the fraction with the ram pressure, we can use it also in the supersonic case:

$$\beta = \frac{8\pi P}{B^2} = \frac{2}{\gamma}\frac{\gamma P}{v^2\rho}\frac{4\pi\rho v^2}{B^2} = \frac{2}{\gamma}\frac{M_A^2}{M_c^2} = \frac{2}{\gamma}\frac{v_c^2}{v_A^2} \tag{2}$$

with the Alfvén speed v_A, the sound speed v_c, and the respective Mach numbers M_A, M_c. The values are displayed in Table 1. It can be seen that for the ISM around the heliosphere the plasma β is small, while for that around λ Cephei is larger than one. Thus the ram pressure is dominated by the magnetic field pressure in the case of the heliosphere while it dominates that of the magnetic field in the case of λ Cephei. Therefore, the heliosphere is asymmetric because of the asymmetric magnetic field pressure, while the λ Cephei astrosphere is symmetric because here the symmetric ram pressure dominates.

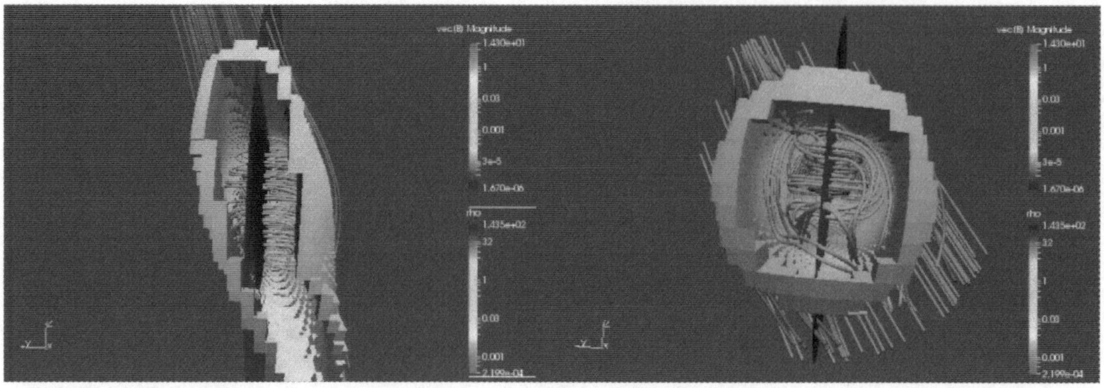

Figure 2. MHD model of the heliosphere (left panel) and λ Cephei (right panel). The star is in the origin. The solid (yellow and green) walls are the region where the number density is between 0.01 and 0.5 cm^{-3}, which is approximately the astropause in both cases. The tubes are the magnetic field lines, and the plane is perpendicular to the inflow. The axis orientation is given in the lower left corner in both panels. In the left figure one sees nicely the inner part with Parker type field lines (the closed ones) and the wrapping of field lines around the heliopause.

Table 1. Comparision of the charateristic parameters between the helio- and an astrosphere.

	heliosphere	astrosphere	
V_A	27.3	6.7	km/s
c_s	9.5	11.4	km/s
M_A	0.97	11.9	
M_c	2.8	7.0	
β	0.14	3.5	

Also in the MHD model the distance between the bow shock and the astropause decreases when cooling is effective (not shown here).

3. Recombination rate

We are here only interested in the projection of a model astrosphere onto the sky, thus we do not discuss the details of the λ Cephei model. But take the number density n and temperature T of the model to calculate the Hα volume recombination rate R:

$$R = 2 \cdot 10^{-10} T^{-4/3} n^2 dr [cm^3 s^{-1}] \tag{3}$$

while dr is the path element chosen to be constant. In principle, one has to calculate the path dr throughout each cell, but the error assuming a constant dr is sufficiently small for the approach described below, because we are only interested in the principle effects.

Because of the low β the compression ratio of the ISM is about 20 similar to the hydrodynamic model, see [20]. The number densities between the bow shock and the astropause range from 44 to 240 cm^{-3}. In this region, the COA, the temperature is similar to that in the ISM, and thus from this region, the highest flux of Hα radiation is emitted. This is easily explained by Eq. 3 which is proportional to the square of the density and decays with the temperature to power -4/3. Thus in the region between the astropause and the termination shock, the region of the shocked stellar wind, the recombination rate can be neglected, because of number densities below $10^{-2}\,\mathrm{cm}^{-3}$ and temperatures above 10^7 K. Because of the high photon flux close to the star we can neglect the recombination rate inside the termination shock as well. The recombination rate in the ISM is at least a factor 16 lower than in the COA. From our line-of-sight (LOS) integrations we found, that the contribution of he other regions is orders of magnitudes below that of the COA. Thus, we can safely neglect them, and in the following we present only the recombination rate in the COA.

4. Rotations

We shifted the model data to its position in the sky (distance 600 pc, $b = 2.61°$, $l = 103.83°$) and rotated it around the y- and z-axis in steps of 45° each. Because the model is approximately symmetric around the x-axis a rotation about this axis does not give new information and thus is left out here. Nevertheless, if the axisymmetry around the x-axis is broken, these rotations needed to be discussed also (see for example heliospheric models by [27], [28]) Due to the small

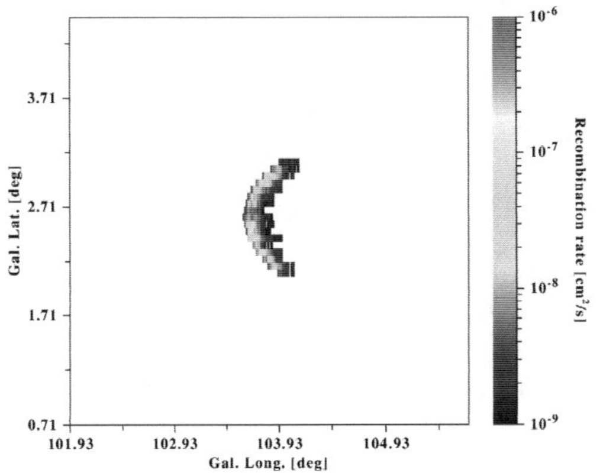

Figure 3. The projection of the λ Cephei MHD model onto the sky. The x- and y-axes denote the galactic longitude and latitude. The color bar gives the logarithmic column recombination rate in cm^2/s. This shape is not rotated.

projection area of $\approx 3° \times 3°$ on the sky, perspective distortions can be neglected, allowing for the use of a simple orthographic projection. In Fig. 3 we show the projection of λ Cephei with no rotation.

In Fig. 4 we show the rotated astrosphere, but without any notation on the axes, because we are only interested in the shape. The axes notation are the same as in Fig 2. The rotation is denoted by the triple (α, δ, γ), where α, δ, and γ are the rotation angles around the x-,y-, and z-axis, respectively.

In Fig. 4 we show the rotation around the z-axis in steps of 45°

A more detailed discussion for these projection features will be given in an upcoming publication.

With these simple rotations, we can get structures, like [29] described for the Herschel infrared observations as "fermata" and "irregular". The "eyes" and "rings" we do not see, because for the eyes we have to include an inner region, and especially the star itself, and the rings are astrospheres with no relative motion with respect to the ISM.

Some of the projections presented above do even look like a bow shock structure.

5. Cosmic rays

In [30] we discuss the propagation of high energetic cosmic rays (above TeV) through an λ Cephei like astrosphere 600 pc away from the observer. We found that such an obstacle modulates the cosmic ray flux (CRF) at the observer. In Fig. 5 a variation along the azimuthal angle is shown where the astrosphere is located at 45° and the above mentioned distance. It can be seen

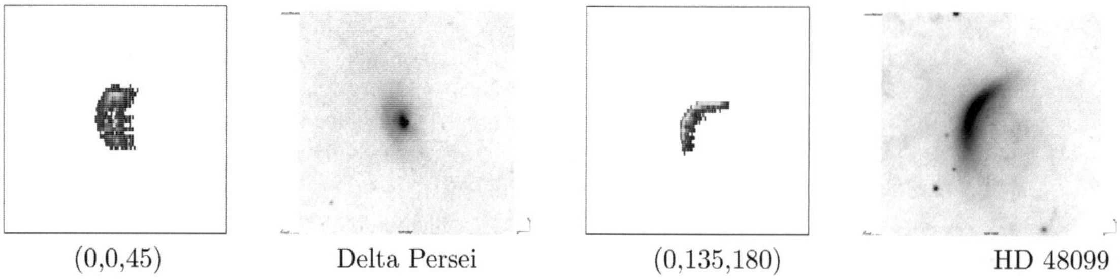

Figure 4. Rotation around the z-axis and observed bow shocks. The rotation angles are given in the brackets **below the panels as well as the identifier for the stars. Both images are obtained in the WISE 22μ infrared band and closely resemble the basic structure of one projected line emission image. The large scale morphology of the WISE 22μ images correlate reasonable well with the morphology observed in ionized gas (Bomans et al., in prep.) and can therefore be used as proxy for ionized gas.**

that there is a variation in the 0.1% range as observed by the large area telescopes e.g., [31]. Moreover, the minimum of the CRF is not at the position of the obstacle. The reason is that

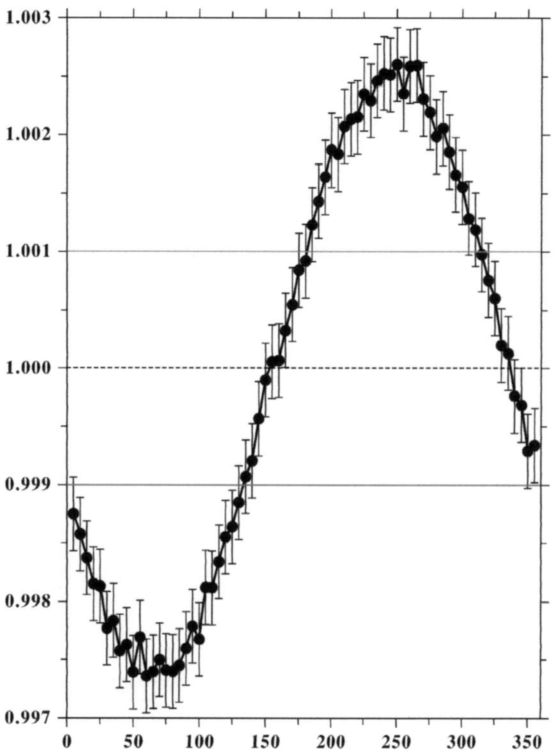

Figure 5. Variation of the cosmic ray flux at Earth for an obstacle with a radius of 10 pc at 600 pc and an angular offset of 45°. On the x-axis the observer angle ϕ is given, while on the y-axis the relative flux variation is shonw. To guide the eye the two red lines denote the 0.1% level.

the perpendicular diffusion cannot be detected due to the magnetic field being aligned along the y-axis. Due to the latter the position of the minimum CRF varies with the position of the obstacle, for details see [30].

6. Conclusion

We have used a model of λ Cephei to study the projection effects of astrospheres onto the sky. We find that with a single model the principle shapes of a lot of astrospheres can be explained including asymmetric and irregular ones. We have integrated the Hα recombination rate, which gives a nice representation of the cool outer astrosheath in the Hα light. It may be possible to use other processes, like bremsstrahlung in the hot inner astrosheath, a region, where the stellar wind is subsonic but the temperatures are above 10^7 K.

The astrospheres around hot stars can also act as sinks for the high energy cosmic ray flux.

Acknowledgments
This work was carried out within the framework of the bilateral BMBF-NRF-project 'Astrohel' (01DG15009) funded by the Bundesministerium für Bildung und Forschung. The responsibility of the contents of this work is with the authors. KS, HF, and JK are grateful to the *Deutsche Forschungsgemeinschaft (DFG)* funding the projects SCHE334/9-1, SCHE334/9-2, and FI706/15-1.

References
[1] Wood B E, Izmodenov V V, Linsky J L and Alexashov D 2007 *Astrophys. J.* **659** 1784–1791 (*Preprint* arXiv:astro-ph/0701274)
[2] Linsky J L and Wood B E 2014 *ASTRA Proceedings* **1** 43–49 (*Preprint* 1408.5934)
[3] Brown D and Bomans D J 2005 *Astron. Astrophys.* **439** 183–194
[4] Scherer K, Fichtner H, Kleimann J, Wiengarten T, Bomans D J and Weis K 2016 *Astron. Astrophys.* **586** A111 URL http://dx.doi.org/10.1051/0004-6361/201526137
[5] Pauls H L and Zank G P 1996 *J. Geophys. Res.* **101** 17081–17092
[6] Fahr H J, Kausch T and Scherer H 2000 *Astron. Astrophys.* **357** 268–282
[7] Borrmann T and Fichtner H 2005 *Adv. Space Res.* **35** 2091–2101
[8] Müller H R, Florinski V, Heerikhuisen J, Izmodenov V V, Scherer K, Alexashov D and Fahr H J 2008 *Astron. Astrophys.* **491** 43–51 (*Preprint* 0804.0125)
[9] Arthur S 2007 *Rev. Mex. Astron. Astrophys.* **30** 64–71
[10] Izmodenov V V, Alexashov D B and Ruderman M S 2014 *Astrophys. J. Lett.* **795** L7 (*Preprint* 1409.8128)
[11] Scherer K, van der Schyff A, Bomans D J, Ferreira S E S, Fichtner H, Kleimann J, Strauss R D, Weis K, Wiengarten T and Wodzinski T 2015 *Astron. Astrophys.* **576** A97 (*Preprint* 1502.04277)
[12] Zank G P 2015 *Ann. Rev. Astron. Astrophys.* **53** 449–500
[13] Kissmann R, Kleimann J, Fichtner H and Grauer R 2008 *Mon. Not. R. Astron. Soc.* **391** 1577–1588 (*Preprint* 0901.4495)
[14] Kleimann J, Kopp A, Fichtner H and Grauer R 2009 *Annales Geophysicae* **27** 989–1004
[15] Wiengarten T, Fichtner H, Kleimann J and Kissmann R 2015 *Astrophys. J.* **805** 155 (*Preprint* 1504.01858)
[16] McComas D J, Alexashov D, Bzowski M, Fahr H, Heerikhuisen J, Izmodenov V, Lee M A, Möbius E, Pogorelov N, Schwadron N A and Zank G P 2012 *Science* **336** 1291–
[17] Zank G P, Heerikhuisen J, Wood B E, Pogorelov N V, Zirnstein E and McComas D J 2013 *Astrophys. J.* **763** 20
[18] Scherer K and Fichtner H 2014 *Astrophys. J.* **782** 25 (*Preprint* 1312.1197)
[19] Frisch P C 2006 *Solar Journey: The Significance of our Galactic Environment for the Heliosphere and Earth* (Edited by Priscilla C. Frisch, Astrophysics and Space Science Library, Vol. 338. Springer Dordrecht)
[20] Scherer K, Fichtner H, Fahr H J and Röken C 2015 *Submitted to Astrophys. J.*
[21] Walder R, Folini D and Meynet G 2012 *Space Sci. Rev.* **166** 145–185 (*Preprint* 1103.3777)
[22] Parker E N 1963 *Interplanetary dynamical processes.* (New York, Interscience Publishers, 1963.)
[23] Beck R 2012 *Space Sci. Rev.* **166** 215–230
[24] Siewert M, Pohl M and Schlickeiser R 2004 *Astron. Astrophys.* **425** 405–416 (*Preprint* astro-ph/0409075)
[25] Schure K M, Kosenko D, Kaastra J S, Keppens R and Vink J 2009 *Astron. Astrophys.* **508** 751–757 (*Preprint* 0909.5204)
[26] Mackey J, Gvaramadze V V, Mohamed S and Langer N 2015 *Astron. Astrophys.* **573** A10 (*Preprint* 1410.0019)
[27] Pogorelov N V, Borovikov S N, Zank G P and Ogino T 2009 *Astrophys. J.* **696** 1478–1490
[28] Opher M 2012 *The Astronomical Review* **7** 010000–78
[29] Cox N L J, Kerschbaum F, van Marle A J, Decin L, Ladjal D, Mayer A, Groenewegen M A T, van Eck S, Royer P, Ottensamer R, Ueta T, Jorissen A, Mecina M, Meliani Z, Luntzer A, Blommaert J A D L, Posch T, Vandenbussche B and Waelkens C 2012 *Astron. Astrophys.* **543** C1

[30] Scherer K, Fichtner H, Kleimann J, Wiengarten T, Bomans D J and Weis K 2016 *Astron. Astrophys.* **586** A111 (*Preprint* 1512.01089)

[31] IceCube Collaboration, Aartsen M G, Abraham K, Ackermann M, Adams J, Aguilar J A, Ahlers M, Ahrens M, Altmann D, Anderson T and et al 2016 *ArXiv e-prints* (*Preprint* 1603.01227)

Interstellar Mapping and Acceleration Probe (IMAP)

N. A. Schwadron[1,2], M. Opher[3], J. Kasper[4], R. Mewaldt[5], E. Moebius[1], H. E. Spence[1] and T. H. Zurbuchen[4]

[1] University of New Hampshire, 8 College Road, Durham NH, 03824, USA
[2] Southwest Research Institute, 6220 Culebra Road, San Antonio TX, 78238, USA
[3] Boston University, 725 Commonwealth Ave, Boston, MA 02215 USA
[4] Dept. of Climate and Space Sciences and Engineering (CLaSP), University of Michigan, 2455 Hayward, Ann Arbor, MI 48109-2143
[5] Caltech, Pasadena, CA 91125, USA, 91125

Abstract. Our piece of cosmic real estate, the heliosphere, is the domain of all human existence -- an astrophysical case history of the successful evolution of life in a habitable system. By exploring our global heliosphere and its myriad interactions, we develop key physical knowledge of the interstellar interactions that influence exoplanetary habitability as well as the distant history and destiny of our solar system and world. IBEX is the first mission to explore the global heliosphere and in concert with Voyager 1 and Voyager 2 is discovering a fundamentally new and uncharted physical domain of the outer heliosphere. In parallel, Cassini/INCA maps the global heliosphere at energies (~5-55 keV) above those measured by IBEX. The enigmatic IBEX ribbon and the INCA belt were unanticipated discoveries demonstrating that much of what we know or think we understand about the outer heliosphere needs to be revised. This paper summarizes the next quantum leap enabled by IMAP that will open new windows on the frontier of Heliophysics at a time when the space environment is rapidly evolving. IMAP with 100 times the combined resolution and sensitivity of IBEX and INCA will discover the substructure of the IBEX ribbon and will reveal, with unprecedented resolution, global maps of our heliosphere. The remarkable synergy between IMAP, Voyager 1 and Voyager 2 will remain for at least the next decade as Voyager 1 pushes further into the interstellar domain and Voyager 2 moves through the heliosheath. Voyager 2 moves outward in the same region of sky covered by a portion of the IBEX ribbon. Voyager 2's plasma measurements will create singular opportunities for discovery in the context of IMAP's global measurements. IMAP, like ACE before, will be a keystone of the Heliophysics System Observatory by providing comprehensive measurements of interstellar neutral atoms and pickup ions, the solar wind distribution, composition, and magnetic field, as well as suprathermal ion, energetic particle, and cosmic ray distributions to diagnose the changing space environment and understand the fundamental origins of particle acceleration. This paper, the first citable reference for IMAP, is similar to an unpublished whitepaper that was presented to the National Academies of Sciences, Engineering and Medicine Committee for Solar and Space Physics. We provide the IMAP objectives and instrument straw man traced from the Solar and Space Physics Decadal Survey. It is fitting that our paper is published in the volume of papers that celebrates the 80th birthday of Ed Stone.

1. Introduction

In the 2012 Heliophysics Decadal Survey [1], IMAP (Figure 1) was rated the highest priority for implementation in the Solar Terrestrial Probe (STP) mission line based on its urgency in the context of recent Voyager observations, alignment with the objectives of the Heliophysics Decadal survey, and relevancy across the Heliophysics division. IMAP is urgently needed to understand the heliosphere's direct connection to the rapidly changing space environment as solar activity subsides while Voyager 1 and Voyager 2 directly probe the inner and outer heliosheath. IMAP is ready to be implemented and explores fundamental outstanding problems in Heliophysics concerning the outer boundaries of our solar system, the physics of interstellar interactions with the solar wind, the origin and physics of the IBEX ribbon, and the fundamental origins of particle acceleration throughout the heliosphere.

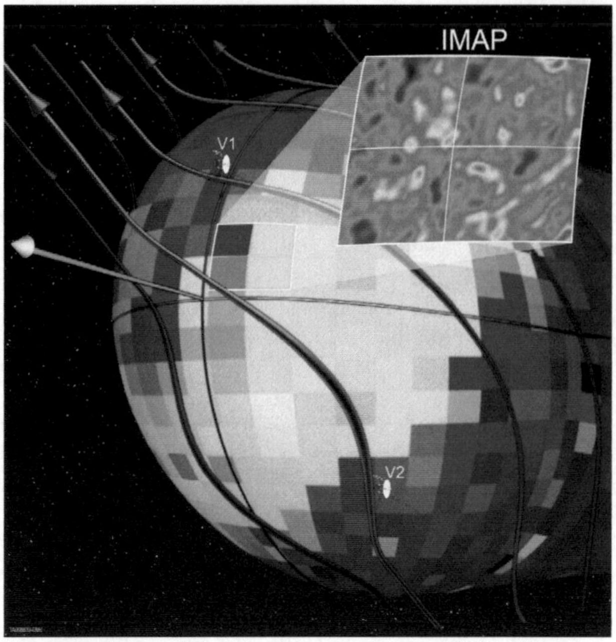

Figure 1. The Interstellar Mapping and Acceleration Probe (IMAP) will solve fundamental mysteries of our heliosphere's interaction with the interstellar medium and particle acceleration in the solar wind. Shown here projected onto the outer boundary of the solar system is the IBEX Ribbon, which was discovered by the Interstellar Boundary Explorer (IBEX) mission. The enigmatic Ribbon raises basic and profound questions related to its origin, the nature of the outer boundaries of our solar system, and the surrounding galactic medium. Most ideas involve a population of electrically charged matter existing near the boundaries of our solar system. These charged particle populations very likely originate from uncharged matter that streams out from the Sun (the neutral solar wind). Several new sources for the Ribbon have also been proposed, involving regions in the galaxy further out from the Sun. IMAP with more than twenty times the resolution of IBEX will probe the detailed source of the ribbon. Shown in the blowout is a depiction of the substructure that scientists have only so far been able to hypothesize. Image credit: D. McComas(SwRI) based on Adler Planetarium/SwRI/NASA image from [2] with mock IMAP data taken from WMAP (GSFC/Princeton/UofC/UCLA/UBC/Brown/NASA).

2. IMAP's Scientific Context and Motivation

As the Sun travels through interstellar space on its quarter billion year journey around the center of our galaxy, the solar wind—the supersonic outflow of magnetized plasma (or ionized gas) from the Sun's upper atmosphere—inflates an enormous bubble within the dilute plasma of the interstellar medium (Figure 2). Known as the heliosphere, this solar-wind-dominated cavity in the local galactic environment has been an object of speculation and study ever since its existence was first predicted in the 1950s.

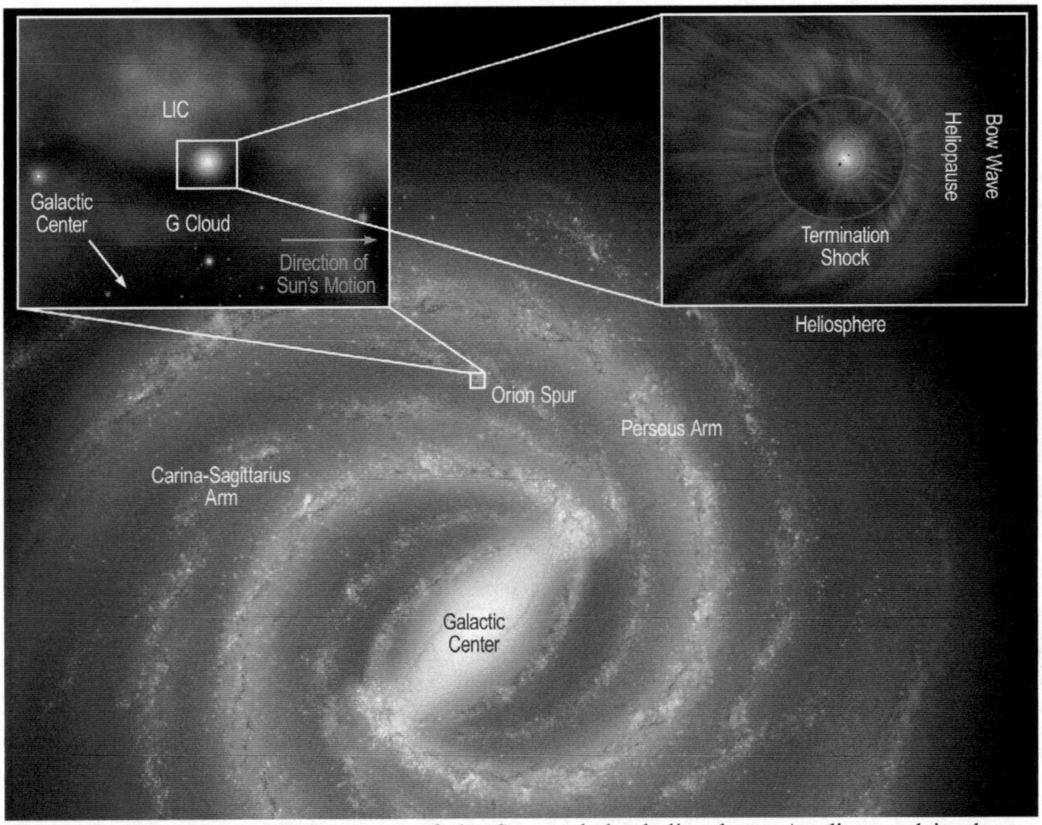

Figure 2. The galactic environment of the Sun and the heliosphere. As discussed in the text, the heliosphere is currently believed to be located near the edge of the Local Interstellar Cloud. Although it has been suggested that the heliosphere has passed into a transition region between the LIC and the G Cloud [3], IBEX measurements of interstellar helium indicate that the heliosphere remains inside the LIC, although "such a conclusion should await further refinement of analysis" [4]. Image credits: NASA/Adler/U. Chicago/Wesleyan/JPL-Caltech (Milky Way and LISM); NASA/IBEX/Adler (heliosphere)

Our heliosphere, its history and future in the Galaxy are key to understanding the conditions on our evolving planet and its habitability over time. By exploring our global heliosphere and its myriad interactions, we develop key physical knowledge of the heliospheric and interstellar interactions that influence our understanding of our home system in its current state, the distant history and destiny of our solar system, as well as the habitability of exoplanetary star systems.

During the last half century, analytic theory and increasingly sophisticated numerical simulations led to the development and refinement of a standard model of the heliosphere as a bullet-shaped obstacle in the local interstellar flow, with a blunt nose in the upstream direction, a long comet-like heliotail downstream, and complex boundaries separating the heliosphere from the interstellar environment. In 2004, Voyager 1 crossed the innermost of these boundaries, providing the first in-situ measurements of the termination shock and the shocked solar wind beyond; Voyager 2 crossed the termination shock three years later.

Although invaluable as direct samples of the outer heliosphere and interstellar medium, the Voyagers' single-point measurements along their trajectories cannot reveal the heliosphere's global structure. Thus, as the Voyagers continued their outward journey, work was under way to develop and implement NASA's Interstellar Boundary Explorer (IBEX) mission. From the vantage point of a highly elliptical Earth orbit, IBEX would generate global images of the heliosphere and its boundaries

Journal of Physics: Conference Series **767** (2016) 012025 doi:10.1088/1742-6596/767/1/012025

by detecting energetic neutral atoms (ENAs) created in the solar wind's interaction with the local interstellar medium.

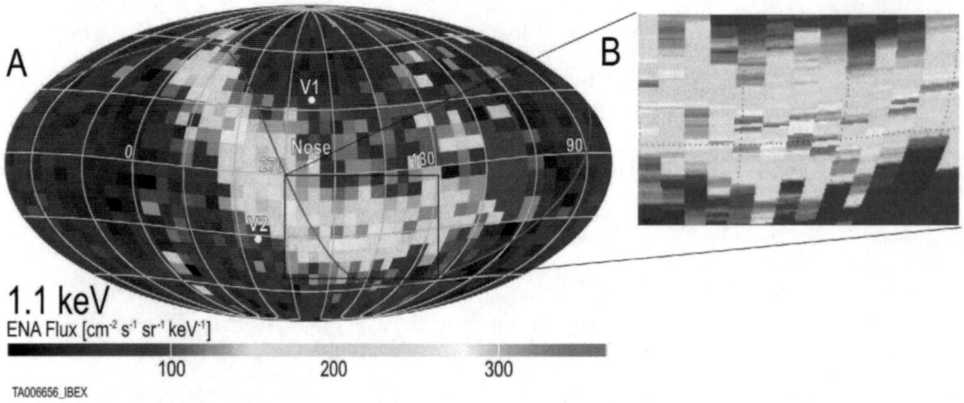

Figure 3. A. IBEX Ribbon at 1.1 keV in ecliptic coordinates using a Mollweide projection. The red line marks the galactic equator. The locations of the heliospheric nose and of the Voyager termination shock crossings are shown. B. Detail of a segment of the Ribbon showing apparent fine structure. From [2].

In October 2008, the tiny (<0.5 m^3 and ~100 kg) IBEX spacecraft, equipped with its own additional solid rocket motor, was launched from the Kwajalein Atoll on a Pegasus rocket. Two months later, IBEX began gathering ENA data for the first-ever maps of the global heliosphere. Creation of the first, energy resolved maps took six-months as the ENA cameras swept successive swaths of the sky, registering ENAs arriving from the outer boundaries of our solar system. As the data accumulated and the first detailed map emerged from the individual pixels, what the IBEX team saw was a feature fundamentally different from anything their pre-launch models of the "ENA sky" or any of the other models of the solar wind/LISM interaction had led them to expect: a bright "Ribbon" of intense emissions nearly encircling the sky and apparently aligned with the external interstellar magnetic field (Figure 3). The discovery of the Ribbon was reported in a series of papers in the November 13, 2009 issue of *Science*, which also featured the Ribbon on its cover. In the same issue of *Science,* Krimigis et al. [5] revealed the first ENA maps of the global heliosphere at energies greater than ~5 keV (above the IBEX energy range). INCA observed a belt of emissions (Figure 4) that is broader and offset from the IBEX ribbon. The white outline in Figure 5 shows the location of the IBEX ribbon. IBEX continues to complete a full scan of the sky every six months, providing better statistics and enabling the detection of time variations. The mapping by INCA also continues. However, the origin of the Ribbon and the Belt—where and how they are formed—remains a mystery and a serious challenge to the heliophysics and astrophysics communities. In addition to the surprising discovery of the Ribbon, the first in-situ sampling of the neutral interstellar H and O wind, along with He, was reported in this issue of Science [6].

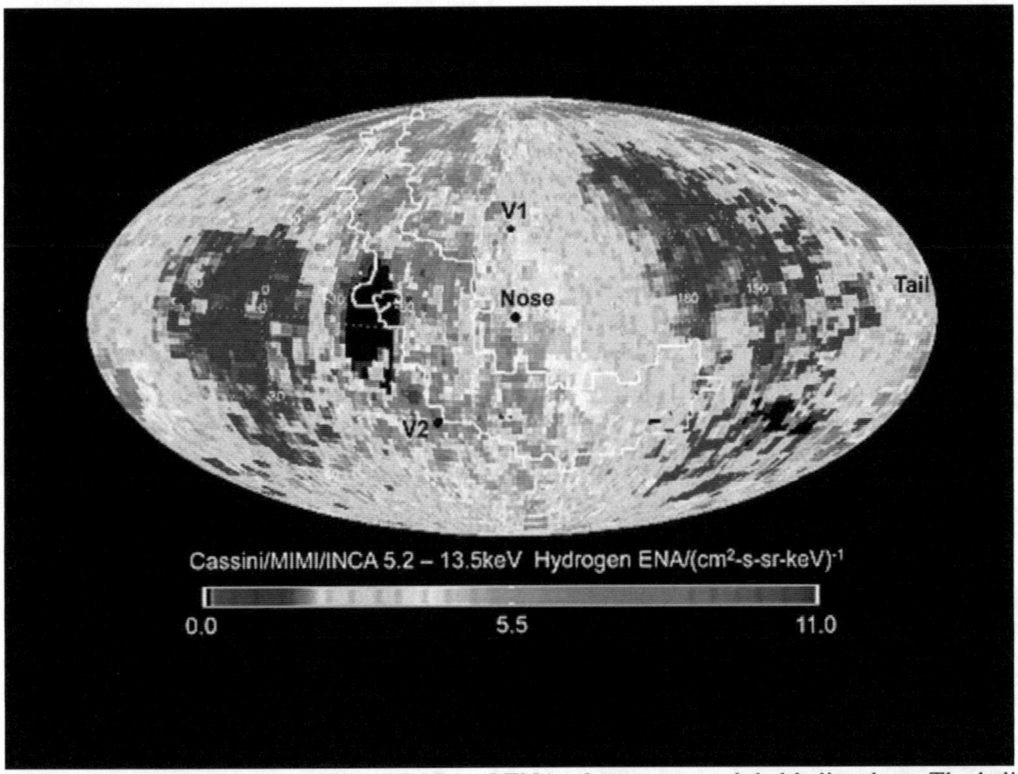

Figure 4. All-sky survey by Cassini/INCA of ENAs that map our global heliosphere. The heliosphere shape is formed as the Sun moves through the surrounding interstellar medium, with the front marked 'nose' and the rear marked 'tail'. The color-coding is by energy flux of ENAs measured by INCA. Also marked are the two Voyager space probes (V1 and V2), and the 'ribbon' of charged particles (white outline) discovered by the IBEX spacecraft. Shown here are energetic neutral atoms (ENA) in the 5.2 – 13.5 keV energy range.

In a similar vein, particle acceleration in the heliosphere is a fundamental problem with implications for the future and history of life within the solar system and beyond. The acceleration of energetic particles creates a population of atoms that penetrate all forms of matter, and deposit energy in Earth's atmosphere, planetary and exoplanetary atmospheres, and materials such as tissue, organic matter, and regolith. Because of the chemical changes induced by particle radiation, it is important to forming the building blocks of life. Yet, for astronauts on potentially long flights in deep space, high-energy radiation can induce cancer and at high levels of exposure leads to radiation sickness. Particle radiation thereby presents a formidable hazard. The many significant ramifications of particle radiation make understanding particle acceleration an imperative. While particle acceleration occurs throughout the cosmos, the details of the particle acceleration process remain difficult to discern because it is so difficult to observe directly. The heliosphere presents a unique environment in which we can study the problem locally. High-energy particle acceleration occurs not only at the termination shock, which surrounds the entire solar system, but also at the plethora of shocks and disturbances the travel past Earth with the solar wind.

3. Critical Next Steps with IMAP

The recent ground-breaking all-sky images of the heliospheric boundaries from IBEX mission and the Cassini INCA instrument, in concert with dual point *in situ* observations of the inner heliosheath from both Voyager spacecraft, as well the direct IBEX measurements of interstellar neutral H, He, O and Ne flow have made outer heliospheric science one of the most exciting and fastest developing areas in Heliophysics. As a next step, substantial improvements in spatial and temporal resolution along with sensitivity gains are required to resolve the substructure of the ENA ribbon and its evolution in time.

Understanding the substructure of the ribbon is critical for establishing the physical mechanisms at work in the creation of the IBEX ribbon. With a factor of x100 combined increase in resolution and sensitivity and a broader energy range of ENA observations, IMAP will provide these capabilities and thus will, in relation to IBEX, enable a "quantum leap" forward in understanding the heliosphere.

3a. What is the Physical Origin of the Ribbon and the Belt?

The Ribbon stretches across much of the sky and its origin remains an enigma despite its persistence for over five years and after almost more than a dozen theories that attempt to explain it. While each theory that has been posed has its strengths, each one also contradicts *IBEX* observations or demonstrates significant flaws in internal consistency [see review, 7].

An example of the more than a dozen theories proposed to explain the ribbon is that it is produced by a *spatial* region [8] in the local interstellar medium where newly ionized atoms are temporarily contained through increased rates of scattering by locally generated waves in the electromagnetic fields (Figure 5). The particles in the ribbon are created predominantly from neutralized solar wind and neutralized pickup ions from inside the solar wind termination shock and inside the heliopause.

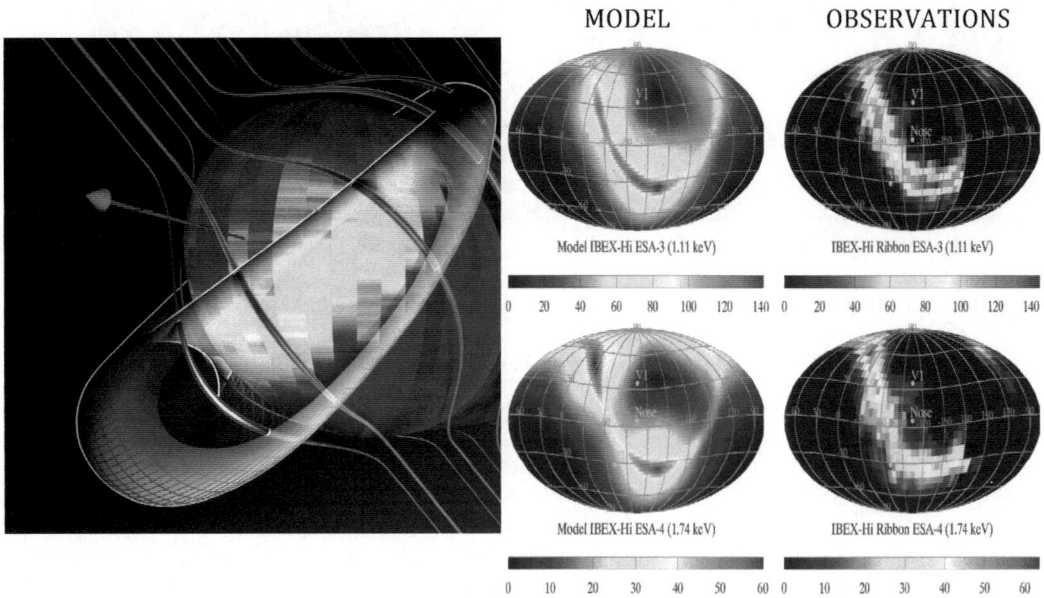

Figure 5. (Left) A new theory for the IBEX ribbon considers its source as a spatial region beyond the heliopause, which surrounds and protects the solar system from the harsh radiation environment of the local galactic medium. The heliopause is the inner surface pictured here with the IBEX ribbon ENA fluxes superimposed on the surface. The grey curves show the interstellar magnetic field lines warping around the heliopause and then stretching beyond it into the local galactic medium. The grey structure outside the heliopause that looks like an overinflated inner tube indicates the region that, according to the new theory, should hold higher concentrations of particles that form the IBEX ribbon. (Right) Left column panels show model results in comparison to observations of the ribbon on the right column Credit: NASA/IBEX/UNH.

The new theory and its competitors must be tested completely. However, the ~7° resolution of IBEX maps poses a major limitation. For example, the spatial retention concept [8] requires the existence of substructure to <1° to account for the observed fluxes of energetic neutral atoms. Until we achieve the x100 combined resolution and sensitivity observations from IMAP, the question of substructure in the ribbon will remain as a major question. Thus, the higher resolution observations of IMAP are critical to fully understanding the origin of the IBEX ribbon.

In contrast to the much narrower ribbon, the INCA belt has a width ~100° FWHM. The ribbon is inclined to the Belt in both ecliptic latitude (~25°) and longitude (~30°). The overlap in energy between Voyager ions and Cassini ENA intensities (averaged over the ENA line of sight) enables estimation of ion fluxes in the heliosheath, thus providing a continuous spectrum $5 \leq E \leq 4000$ keV. These measurements have been used to estimate the local partial pressure over this energy range (~ 0.1 pPa), suggesting the thickness of the heliosheath ~ 50 AU [9].

3b. What is the Global Structure of the Heliosphere?

IBEX measurements of energetic neutral atoms created by collisions at the solar system's boundaries have for the first time mapped out the structure of our solar systems' tail, which is shaped like a four-leaf clover (Figure 6). While telescopes have spotted such tails around other stars, it has been difficult to see whether our own star produced one. The particles found in the tail cannot be seen with conventional instruments, and so the shape and structure of its tail remained unknown. IBEX data have shown that, very much like a comet's tail, the heliotail is stretched out behind our solar system where the Sun's million mile per hour solar wind flows down and ultimately escapes the heliosphere, slowly evaporating because of charge exchange.

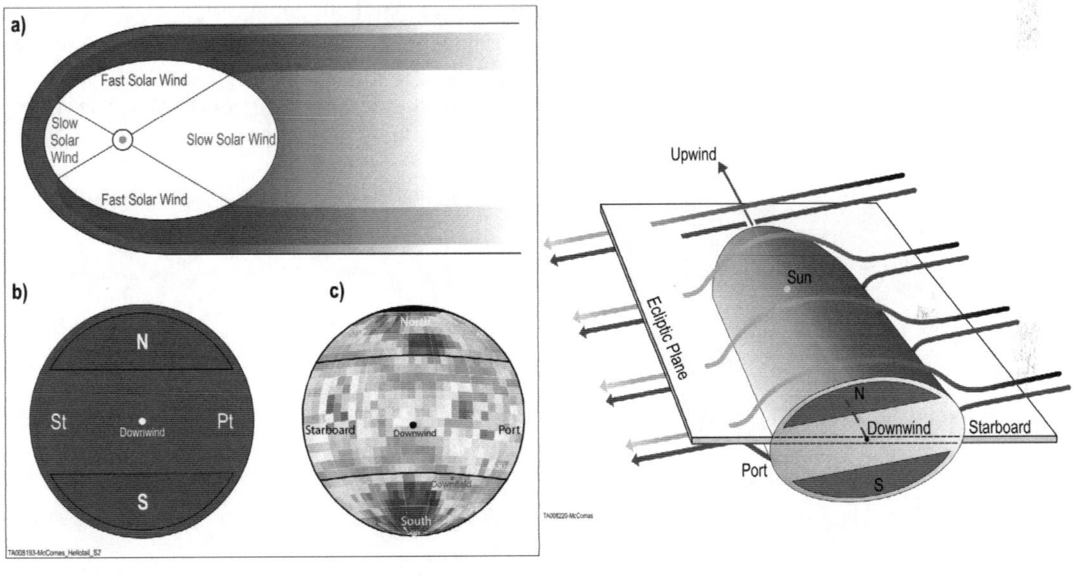

Figure 6. Data from NASA's IBEX (panel c) shows the spectral slope of the particles looking down the solar system's tail – the yellow and red colors represent areas of slow-moving particles, and the blue represents the fast-moving particles. Panel a) shows the latitude structure of typical solar wind, with fast wind at high latitudes and slower wind at low latitudes. Panel b) shows the corresponding solar wind structure looking down tail with faster wind in blue and slower wind in green. N & S refer to the North and South directions. St refers to the starboard direction and Pt refers to port (these are nautical terms). The panel on the right shows a cartoon of the structure of the heliotail. (Image: NASA/IBEX)

One of the fundamental quantities measured by IBEX is the line-of-sight integrated pressure of ions from the inner heliosheath (Figure 7). Determination of an absolute pressure requires some absolute distance scale, which is provided by the Voyager satellite boundary crossings of the termination shock and the heliopause. Once absolute distance scales at individual locations are established, the IBEX *global* pressure maps allow us to form a rough picture of the heliosphere (termination shock and heliopause boundary) near the ecliptic plane. The direction of the interstellar magnetic field is given by the center of the IBEX ribbon and is projected into the ecliptic plane. The draping configuration is illustrative.

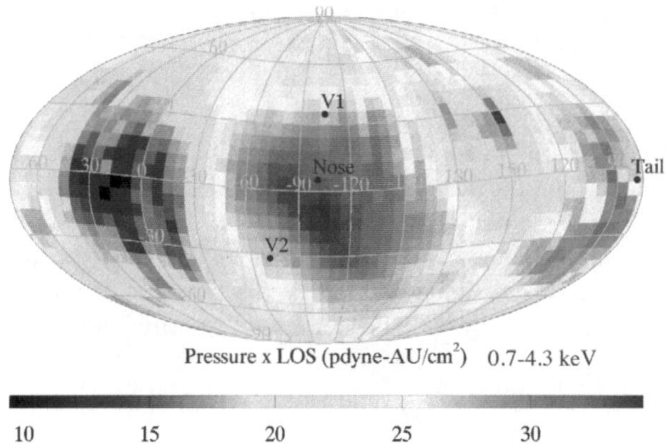

Figure 7. ENA imaging from IBEX and in the future from IMAP will resolve global line-of-sight integrated pressures, which reveal the global dynamic balance between the solar wind (the inner heliosheath plasma) and the interstellar plasma including the interstellar magnetic field. The extended energy range of IMAP compared to IBEX will allow more accurate estimation of pressure and of the line-of-sight integrations that change with energy. Shown here is the pressure of plasma protons from the inner heliosheath that form observed ENAs integrated over line-of-sight (LOS) as observed by IBEX from 0.7 to 4.3 keV.

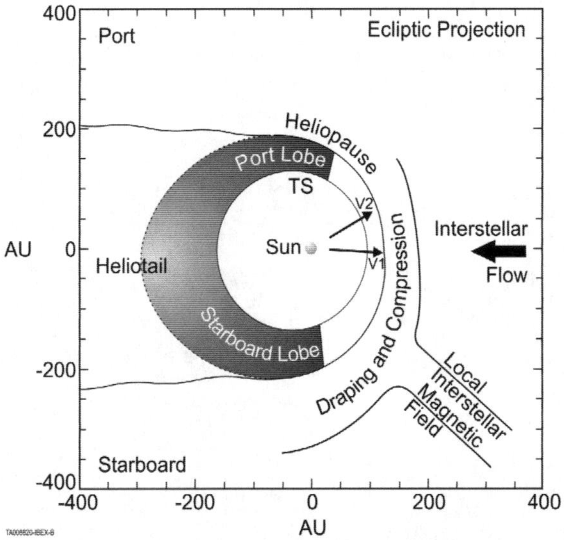

Figure 8. Analysis of IBEX data is used to form an approximate structure of the heliosphere projected into the ecliptic plane (viewed here from heliographic North). The complete energy range of distributions measured by IMAP will provide significant improvements over IBEX. In the formulation of such pictures of the global heliosphere, the distance scales provided by Voyager provide critical insight. The wavy lines extending from the lobes show regions beyond the line-of-sight sensitivity of IBEX and suggest outflow. The tail region and structure of heliosheath is asymmetric. The clearest signatures of asymmetry include the small starboard (~10°) offset of the core tail from the interstellar downwind direction, and the deeper reduction in ENA emissions from the port lobe of the heliosphere. A key question for IMAP is how these features extend down to low energies (<0.2 keV) and above 6 keV).

Figure 8 shows a rough picture of the structure of the heliosphere near the ecliptic plane. This picture is made possible only through the combined observations of IBEX and direct measurements by Voyager. Future measurements from IMAP will make fundamental use of Voyager observations to help identify distance scales, and thus, form a global picture of the heliosphere through complete energy integrations across the ENA distributions emanating from the inner heliosheath.

Sophisticated 3D MHD models (one example is shown in Figure 9) are being developed with the goal to produce ENA maps that can be compared with future observations from IMAP. The standard picture of the heliosphere is a comet-shape like structure with the tail extending for 1000's of AUs. This standard picture stems from the view that magnetic forces from the solar magnetic field are negligible and that the solar magnetic field is convected passively down the tail. Recent work [10, 11] shows that the magnetic tension of the solar magnetic field plays a crucial role on organizing the solar wind in the heliosheath (HS) into two jet-like structures. The two heliospheric jets are separated by the interstellar medium that flows between them. The heliosphere then has a "croissant"-like shape where the distance to the heliopause down tail is almost the same as towards the nose.

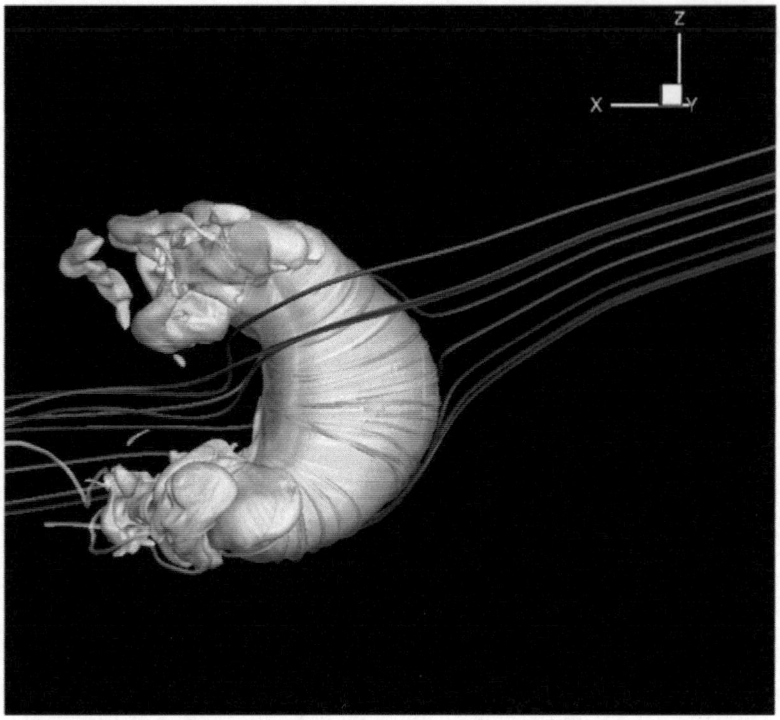

Figure 9. Croissant Heliosphere" / "Heliosphere with Jets": Two-lobed structure of the heliosphere [10]. Yellow surface shows the heliopause surface. Grey curves are solar magnetic field lines; red curves are interstellar magnetic field lines.

Several observational consequences of this new view are being explored. The heliospheric jets are very turbulent with large-scale turbulence ~ 100AU and with a turnover of years. One should expect that this turbulence will cascade to smaller scales and shorter time scales. These turbulent jets are the largest turbulent structures in the heliosphere. Furthermore we do expect that the heliospheric jets will vary in their structure and intensity with solar cycle. These signatures may manifest in ENA maps, particularly over large timescales. Initially signatures in variability should exist over mid-scale ENA energies (0.7-4keV) and some suggestions have been made that this variability may extend to higher energies (>10 keV) that will be measured by IMAP. Both IBEX as well as CASSINI/INCA show time variability, although IBEX measurements indicate very little variability in the region of the heliotail. The current baseline over which variability could be detected is half of a solar cycle, and a much larger timescale (more than a decade) is needed with IMAP, IBEX and INCA to test for the existence turbulent evolution of the heliotail. IMAP measurements should pin down the following questions: a) Why the ENA images on IBEX (0.7-4keV) of the tail reveal two lobes? Is it only related to the bimodal distribution of the solar wind (slow vs. fast wind) and will disappear as the cycle progresses, or is it a permanent feature of the tail associated with the heliospheric jets? (b) Why the ENA images on CASSINI of the tail reveal such strong time variability? Is this variability a solar cycle effect or is it related with the turbulent jets? IMAP ENA mapping will have high sensitivity and be able to detect temporal and spatial variations due to the turbulence generated by the heliospheric jets and their associated variations over the solar cycle.

3c. What are the Conditions of the Interstellar Medium?

After initial sampling of the neutral He ISN flow with Ulysses GAS [12, 13], observations of the interstellar neutral (ISN) flow with IBEX have provided the first direct multi-species measurements of the ISN flow parameters, with a strong indication that the local interstellar cloud (LIC) conditions around the heliosphere are largely isothermal [4,12]. The local determination of the LIC temperatures

Journal of Physics: Conference Series **767** (2016) 012025 doi:10.1088/1742-6596/767/1/012025

from the neutral gas velocity distribution for several species provides the most detailed independent measurement of this key quantity, with a handle on differential heating and cooling processes for the different interstellar species, provided that also the distributions of the minor species (O and Ne) can be obtained with excellent counting statistics and angular resolution. IBEX has also provided the first measurements of the ISN H velocity distribution and the effect of the Sun's radiation pressure, based on direct neutral gas observations [15].

Differences in radiation pressure have also enabled the first local detection of ISN D [16]. The interstellar D/H is a powerful probe of big bang nucleosynthesis and the chemical evolution of the Milky Way galaxy [e.g. 17, 18], yet observations in the interstellar medium are scarce and line-of-sight integrated [19]. IMAP will take advantage of the differential deflection of these flow distributions by radiation pressure [20], in combination with the varying angular aberration along the 1 AU orbit, to separate the D signal from the He ISN flow distribution. As with other ISN flow related observations IMAP provides the critically needed capability to track the ISN flow distributions in the sky along the orbit around the Sun. The remaining improvement provided by IMAP to obtain a D/H ratio with statistical uncertainties <10%, is achieved with a combined collecting power and observation time increase for D by >10x.

With IBEX ISN observations, the flow vector and temperature of ISN He have been constrained within a narrow tube in the 4-dimensional parameter space (ISN inflow longitude, latitude and speed, as well as temperature), where the latter three can be expressed as a function of inflow longitude [21-23], but with a larger uncertainty along the parameter tube. The allowable parameter region has recently been constrained further [24 – 28], and the ISN flow vector is now largely consistent with that obtained with Ulysses GAS [29, 30], but with a LIC temperature that is substantially higher than in previous determinations [13, 31].

The speed and direction of the ISN flow have profound implications for the LIC-heliosphere interaction, i.e. whether it is largely super- or sub-sonic [23] and, in conjunction with the determination of the interstellar magnetic field direction based on ENA ribbon observations [32, 33], the Voyager termination shock passages [34-36], and interstellar H flow direction [37, 38], the shape and structure of the heliosphere relative to the $V_{ISM} - B_{ISM}$ symmetry plane can be determined. In fact, rather small differences in the ISN flow direction have substantial leverage on the $V_{ISM} - B_{ISM}$ plane [21, 39] and thus on the large-scale heliosphere structure. Furthermore, the small original difference in the best fitting inflow longitude [21, 22] compared with previous observations [13, 31] had initiated the intriguing discussion whether we could potentially witness variations in the ISN flow vector over a few decades [40-42]. While the differences in the flow vector have been largely reconciled, some recent work has focused on the possibility of variations of the local ISN flow vector due to turbulence in the ISM over a large range of scales [42]. Previous studies recognized the changes of the global flow vector due to variations in individual interstellar clouds along the line-of-sight used in integrating across absorption spectra [43]. This line-of-sight integration poses issues in comparing local determinations from measurements of interstellar neutral atoms with remote line-of-sight integrated observations.

All the aforementioned topics require a precision determination of the ISN flow vector, preferably with contiguous data sets over decades. Imaging of the ISN flow and the evaluation of the speed cut-off of interstellar pickup ion distributions [44] provide highly complementary methods, which can independently validate each other, if performed over a range of ecliptic longitudes. ISN bulk flow and pickup ion distributions over a large portion of the orbit around the Sun are needed to obtain such precision measurements of the H, He, O, and Ne ISN flow vector. IMAP provides the needed next step by enabling tracking of the ISN flow, along with a pointing accuracy that is as good or better than IBEX and continuous observations of He, O, D and Ne (Figure 10) pickup ion distributions.

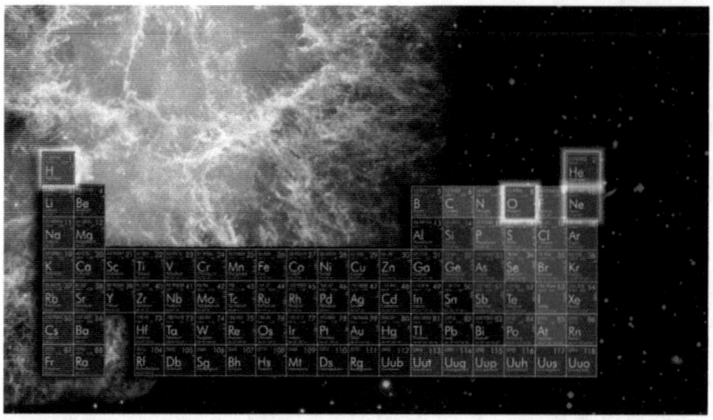

Figure 10. Most elements on Earth are produced by supernova explosions. About thirty elements in the periodic chart have been detected in the interstellar medium. Supernovae, such as the Crab, have produced the interstellar oxygen and neon that is observable by IMAP. In addition, IMAP will resolve Deuterium (D), which has major implications for big bang cosmology.

IMAP will also measure in detail the secondary components of the interstellar neutrals as pioneered with the IBEX-Lo maps for O [6, 45] and extracted as a secondary component from the He flow distributions [46, 47]. Definitive measurements of secondaries by IMAP provide powerful complementary tools to probe the LIC-heliosphere interaction. In combination with the ribbon and the precise ISN flow direction, these observations will put multiple strong constraints on heliospheric structure and the interstellar magnetic field. IMAP provides both images of these distributions over most of the orbit around the Sun and substantially improves collecting power for the heavy neutrals, such as O, over IBEX. In addition, for the H ISN flow distribution, IMAP observations over large fractions of a 1 AU orbit and over dramatically varying radiation pressure with solar activity will provide a tool to obtain more complete ISN H distributions and to clearly separate the primary ISN flow and the secondary neutrals, which for H are typically only observed as a combined distribution. With the superior neutral atom collecting power and its continuous coverage of the sky for low flux distributions, IMAP will be able to seriously test, whether the observed distributions have additional non-thermal components and, if so, will characterize them and connect them to physical processes in the ISM and the heliospheric boundary.

3d. What is the Direction and Strength of the Interstellar Magnetic Field?

Determining the direction of the local interstellar magnetic field (LISMF) is important for understanding the heliosphere's global structure, the properties of the interstellar medium, and the propagation of cosmic rays in the local galactic medium. Measurements of interstellar neutral atoms by Ulysses for He and by SOHO/SWAN for H provided some of the first observational insights into the LISMF direction. Because secondary neutral H is partially deflected by the interstellar flow in the outer heliosheath and this deflection is influenced by the LISMF, the relative deflection of H versus He provides a plane - the so-called B-V plane in which the LISMF direction should lie. IBEX subsequently discovered a ribbon, the center of which is conjectured to be the LISMF direction. The most recent He velocity measurements from IBEX and those from Ulysses yield a B-V plane (Figure 11) with uncertainty limits that contain the centers of the IBEX ribbon at 0.7-2.7 keV.

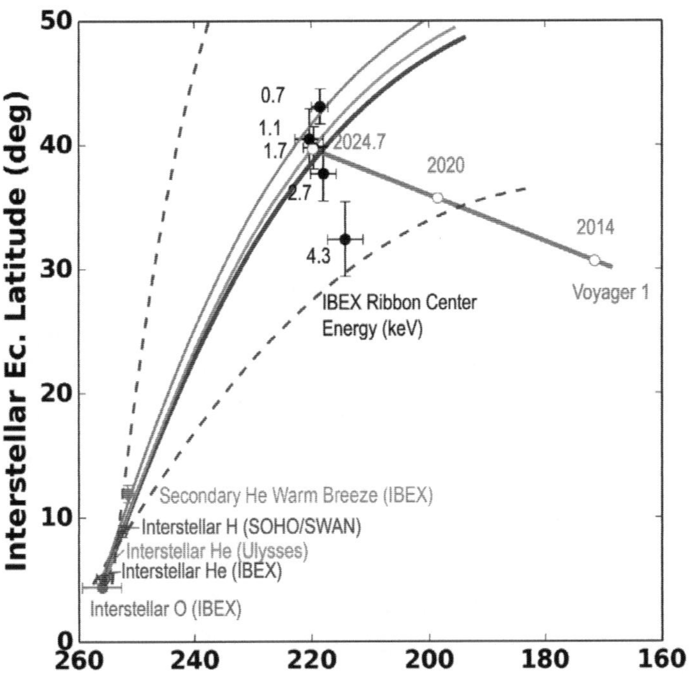

Figure 11. We combine seven different sets of observations to determine the direction of the interstellar magnetic field. The red line shows the linear fit to Voyager 1 observations of the interstellar magnetic field [50]. This linear fit is projected forward in time (the red circles surrounding a "V" show discrete points in time along the Voyager trajectory). The H flow direction from SOHO/SWAN [37, 49] is shown with the He flow direction derived by Schwadron et al. [28] in blue. The H inflow is more strongly deflected by secondary interactions in the heliosheath than the He inflow. Therefore, the BISM-VISM plane contains the deflection of H relative to He [37]. The region bounded by the dark blue dashed curves shows the limits of the BISM-VISM plane, which contains the orientation of the IBEX ribbon [32] from 0.7-2.7 keV. The purple closed circle shows the He inflow direction based on the most recent analysis of Ulysses ISN flow observations [30]. The purple line shows the corresponding B-V plane connecting the Ulysses He and SOHO/SWAN H observations. The center (closed black circles) of the IBEX ribbon is shown at separate energy steps observed by the Hi sensor on IBEX. The projected interstellar field direction from Voyager 1 converges with the 1.7 keV IBEX ribbon center on the date of 2024.7. The grey curve is the B-V provided by fitting secondary He in addition to H and primary He [47]. Also shown in green is the direction of interstellar O based on a recent determination from IBEX data [51].

Immediately outside and beyond the heliopause lies interstellar space, where Voyager 1 is currently making ground-breaking observations. The possibility that Voyager 1 has moved into the outer heliosheath now suggests that Voyager 1's direct observations provide another independent determination of the LISMF. The LISMF direction measured by Voyager 1 is > 40° off from the IBEX ribbon center and the B-V plane (Figure 11). However, taking into account the temporal gradient of the field direction measured by Voyager 1, we extrapolate to a field direction that passes directly through the IBEX ribbon center (0.7-2.7 keV) and the B-V plane (Figure 11), allowing us to triangulate the LISMF direction and estimate the gradient scale size of the magnetic field [50, 30]. The linear projection of the Voyager 1 data suggests that it could observe a field direction at the IBEX ribbon center by 2025 when the spacecraft is at 165 AU from the Sun. This also indicates a draping region of ~ 45 AU in radial extent near Voyager 1.

Because Voyager 1 is at a single location, it is difficult to develop a complete global picture from its measurements. Recently, Schwadron et al. [48] used near-Earth IBEX ENA data from the past five years in conjunction with observations of highly energetic cosmic ray particles with 99% the speed of light streaming in from elsewhere in the Milky Way galaxy to shed new light on our cosmic neighborhood, and propose an explanation for a decades-old mystery – why we measure more incoming high-energy cosmic rays on one side of the heliosphere than on the other. The IBEX ribbon appears to be ordered by the direction of the interstellar magnetic field direction that drapes around the heliosphere. The direction of the interstellar magnetic field revealed by the IBEX ENA ribbon observations is nearly perpendicular to the motion of our solar system through the galaxy. This information was then used to predict the distribution of high-energy (~10 TeV) cosmic rays that penetrate into the heliosphere. These predictions are shown on the right in Figure 12. The blue regions represent a lower intensity of cosmic rays whereas the red regions signify a higher intensity of cosmic rays. This uneven distribution looks similar to what is actually observed, shown on the left of Figure 12, thus supporting IBEX's findings.

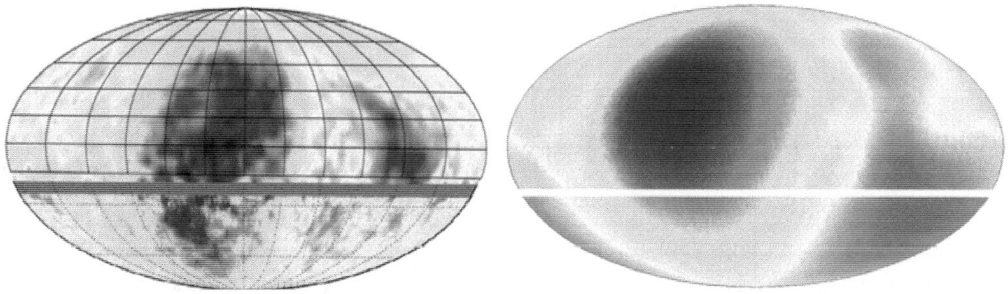

Figure 12. The magnetic fields in interstellar space inferred from IBEX lead to a prediction for the global anisotropies of TeV cosmic rays as shown on the right. Regions in blue have a decrease in the intensity of cosmic rays, while regions in red represent an increase in cosmic ray intensity. This looks similar to what is actually observed, shown on the left, thus supporting IBEX's findings. These global maps of TeV cosmic ray anisotropies are shown in Mollweide projection and in standard J2000 equatorial coordinates.

How does the ribbon change over ~10 year timescales? Stability of the ribbon over these timescales would support the idea that the ribbon is ordered by the interstellar magnetic field. However, if the ribbon changes significantly on these timescales and IMAP detects a very different structure than observed by IBEX, this would call into question not only the leading ideas to explain the ribbon, but also the concept that it is ordered by the interstellar magnetic field. Thus, IMAP is positioned to answer a critical question about the stability of the ribbon with major implications not only for the interstellar magnetic field, but also for the interstellar interactions with the magnetic field and the global structure of our heliosphere. Ultimately, these questions must be answered to understand how precisely our interplanetary plasma environment changes with time and how cosmic rays are controlled by the structure of the heliosphere.

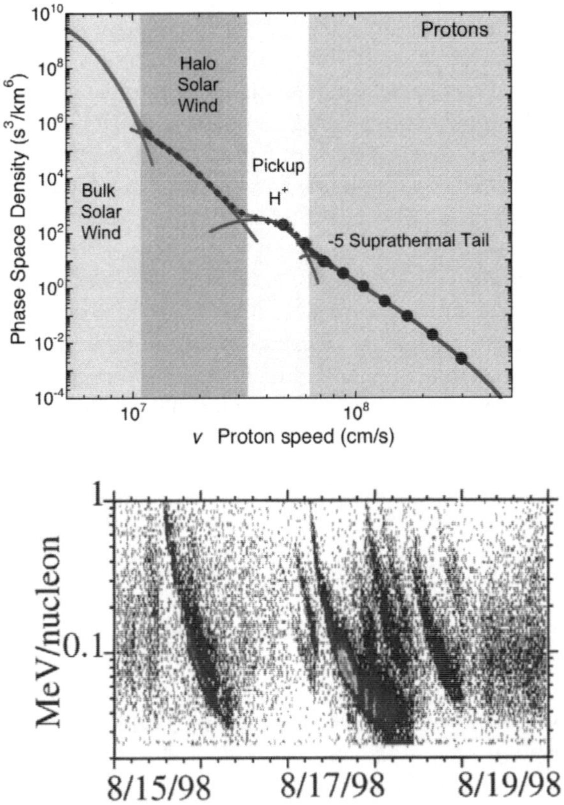

Figure 13. (Upper panel) Ion velocity distribution measured in the solar wind frame. Four components are clearly distinguishable: 1) Bulk solar wind studied since the beginning of the space age (red), 2) much hotter halo solar wind (blue), 3) interstellar pickup H^+, observable at 1 AU during the deep solar minimum (green), and 4) suprathermal tail (ST), just above the pickup ions. The ST spectrum is well approximated by a power law, with a gradual exponential rollover at ~$2.5 \cdot 10^8$ cm/s.

(Lower panel) Suprathermal particles injected at the Sun and in interplanetary shock events stay relatively close to their field lines, providing seed populations that vary by orders of magnitude in space and time. [52].

3e. What are the Suprathermal Seed Populations for Particle Acceleration in the Heliosphere?

Observations from many spacecraft in the Heliophysics System Observatory have contributed dramatically to our understanding of solar energetic particle events, the importance of suprathermal particles in interplanetary space for their effective acceleration, the source and evolution of solar wind, solar wind and energetic particle inputs into geospace, and the evolution of the coupled solar and heliospheric magnetic field. It is abundantly clear that a myriad of complex physical effects, variable with time and location, contribute to the properties of the solar wind, the heliospheric magnetic field, suprathermal and energetic particle populations at 1 AU (Figure 13), that are still poorly understood. With its combination of highly sensitive pickup and suprathermal ion sensors, IMAP will provide the species and spectral coverage as well as the temporal resolution to associate emerging suprathermal tails with interplanetary structures and physical processes.

IMAP, like ACE before it, will be a keystone of the Heliophysics System Observatory by providing comprehensive solar wind observations, measurements to diagnose the source and evolution of solar wind and suprathermal ions, provide solar wind and energetic particle inputs into the geospace environment, and track the evolution of cosmic rays and of the coupled solar and heliospheric magnetic field. IMAP's comprehensive interplanetary monitoring suite is critical to support on going geospace interaction studies and space weather observations at the ideal location of the Lagrangian point L1. The high societal relevance of comprehensive solar wind, suprathermal, magnetic field and cosmic ray observations from L1 makes the IMAP mission an imperative as a successor to ACE.

4. IMAP's Urgency in a New Paradigm of the Evolving Space Environment

The space environment is a complex system regulated by the solar wind's interaction with the Local Galactic Environment. Our local interstellar boundaries separate the solar wind plasma from the Local Interstellar Medium (LISM), which is composed of the galactic matter (neutral atoms, ionized matter, and dust) and galactic magnetic fields left over from supernovae, stellar winds, ultimately as a relic of the primordial field. The distant interstellar plasma boundaries surrounding our heliosphere and the outflowing solar wind partially protect our solar system by regulating the intensity of GCRs that enters our solar system. GCRs are charged particles with relativistic energies and they permeate our galaxy; because they are charged, their motions are governed by the magnetic fields they encounter. GCRs present one of the greatest hazards for long-term space exploration since these high energy (~GeV) particles are so difficult to shield against deep in interplanetary space, that is, at altitudes well above Earth's strong internal magnetic field. The most energetic GCRs penetrate even the powerful magnetic fields closest to Earth ultimately colliding with and producing complex interactions with Earth's massive atmosphere; the effects of GCRs on the Earth system, including the biosphere either directly or indirectly, remain poorly understood and oftentimes highly controversial [53]. Recent compilation of paleontological data into estimates of global diversity suggest there is a significant ~62 million year cycle to biodiversity that as of yet has no agreed upon cause and is possibly driven by extraplanetary processes [54].

GCRs not only present a hazard to life through the breakdown of DNA, but also may help to stimulate evolution by increasing the rate of cell mutation [57]. In other words, the radiation environment of the Earth and planets, which is largely defined by the intensity and composition of GCRs in the solar system, may play a fundamental role in the formation and evolution of life. Similarly, as we begin our search for life elsewhere in the cosmos, particularly on planets surrounding other stars, we must also investigate the interstellar boundaries surrounding these stellar systems, and the effects these interstellar boundaries have on the cosmic rays within these systems. GCRs may also affect life in indirect ways through climate variability [53], although this relationship remains highly controversial.

The deep solar minimum between cycles 23 and 24 and the activity in cycle 24 differed significantly from those of the prior cycle. During this period, the fast wind was slightly slower, was significantly less dense and cooler, had lower mass and momentum fluxes, and weaker heliospheric magnetic fields compared to earlier cycles [58]. During the rise of activity in cycle 24 the mass flux of solar wind remained low and the magnetic flux of the heliosphere remained at significantly lower levels than observed at previous solar maxima in the space age. Cycle 24 is the weakest solar maximum of the space age, which continues the highly anomalous trends observed in the deep cycle 23-24 minimum. Conditions during the cycle 23-24 minimum appear to be similar to conditions at the beginning of the 1800's at the start of the Dalton Minimum [59]. Taken together, these recent changes suggest that the next solar minimum may continue to show declining sunspot numbers, associated with declining values of magnetic flux and further reductions in solar wind particle flux.

Journal of Physics: Conference Series **767** (2016) 012025 doi:10.1088/1742-6596/767/1/012025

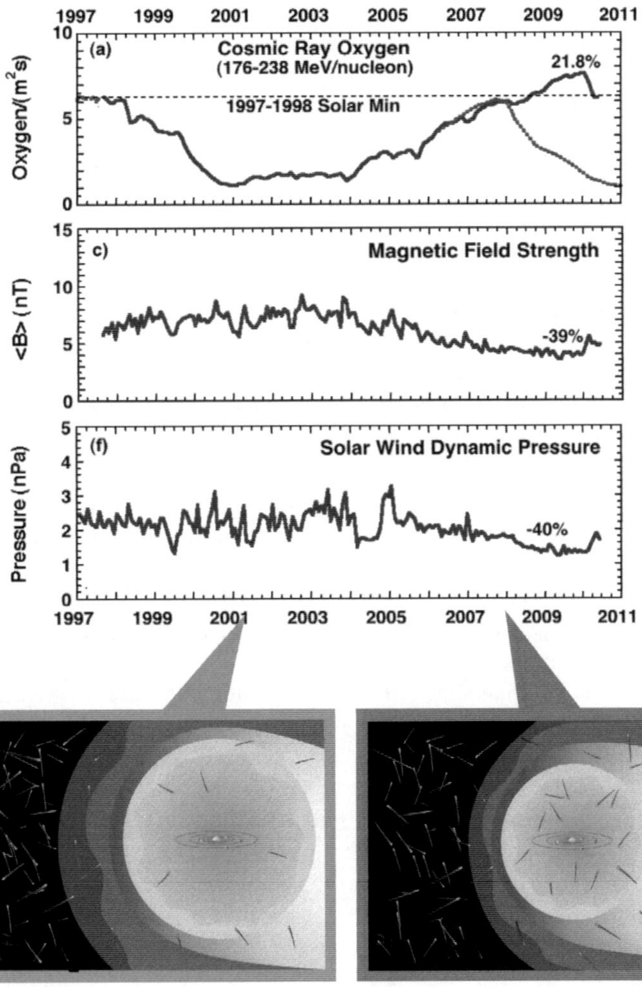

Figure 14. (Top Panels from 53) The recent anomalous rise in GCR flux (a) is associated with both a drop in the interplanetary magnetic field (b), solar wind ram pressure (c), and in the size of the heliospheric boundaries (bottom) where as much as 90% of low-energy GCRs (shown as blue streaks) are deflected in the strong magnetic field of the heliosheath [56].

The anomalously weak heliospheric magnetic field and low solar wind flux during the last solar minimum (Figure 14) have resulted in GCRs achieving the highest flux levels of the space age [55, 60], and fluxes continue to be unusually elevated through the cycle 24 maximum. It is unknown if the recent anomalous deep solar minimum is a harbinger of larger changes in the near future, or if the unusual changes in GCR fluxes and conditions on the Sun have an impact on Earth's atmosphere. These compelling questions provide fundamental motivation for IMAP.

The changes in the space radiation are strongly controlled by the changes in the global heliosphere. Understanding the interplay between changing solar conditions, the resulting changes in the global heliosphere, and the resulting changes in the radiation environment throughout the solar system remains a fundamental challenge. IBEX and INCA have opened a new window on the interaction between the Interstellar Medium and the Heliosphere. IMAP extends and expands this domain of discovery through:

- Improved resolution (higher sensitivity and suppression of background) of interstellar boundaries. While gross features of the interstellar boundaries such as the nose, tail and lobes have been identified by IBEX, these features have required years of observation and subsequently averaged over a changing medium.

- Improved time resolution of observations of interstellar boundaries to identify the connection between changes in the solar wind including and motion of interstellar boundaries

- Increased resolution of structures in both time and angular coverage allowing significant improvement in the connection between detailed observations of Voyager 1 and Voyager 2 and IMAP.

5. Overview of IMAP Scientific Objectives

IMAP builds on the highly successful first heliospheric ENA mapping mission, with its concurrent capability to provide in-situ observations of the interstellar gas flow for a variety of species, and on a broad array of in-situ pickup-ion (PUI) observations, to enable the discovery of the detailed processes and interaction between the heliosphere and the local interstellar medium (LISM). IMAP answers these fundamental questions:

1) What is the spatio-temporal evolution of heliospheric boundary interactions?
2) What is the nature of the heliopause and the interaction of the solar and interstellar magnetic fields?
3) What are the composition and physical properties of the surrounding interstellar medium?
4) How are particles injected into acceleration and what mechanisms energize them throughout the heliosphere and heliosheath?
5) What is the time-varying magnetic field, plasma, energetic particle, and galactic cosmic ray input at L1 into the Earth system and inner solar system?

The IBEX ribbon and the INCA belt provide fundamental motivation for IMAP. Since the first observations of the ribbon, the prospect of the existence of fine structure has remained a significant potential finding that strongly limits possible Ribbon mechanisms. However, the existence of this fine structure remains at the observing limit of IBEX both in terms of angular temporal resolution. IMAP's enhanced sensitivity and improved background suppression will allow a factor of almost 100 improvement in resolution compared to IBEX.

The mission's focus on understanding heliospheric boundaries makes it important to try to simultaneously take IMAP observations while also making direct *in situ* measurements via the two Voyager spacecraft in the heliospheric boundary region. IMAP will also enable observation of suprathermal ions of solar wind, interstellar, and inner-heliospheric origin with unprecedented collection power and time resolution. These measurements are essential for understanding particle acceleration. IMAP instruments will provide the environmental monitoring that is critical for effective background evaluation and removal in the ENA images and the interpretation of the pickup ion distributions. In addition, these instruments will serve as a comprehensive interplanetary monitoring suite in support of geospace interaction studies and space weather observations at the ideal location of the Lagrangian point L1.

Answering the questions motivating IMAP requires substantially advanced observations:

- High-resolution mapping and time evolution of heliospheric boundaries;
- Properties of interstellar neutral gas flow and its composition for H (including isotopes), He, O, and Ne (to also address Big Bang Cosmology with the first in situ D/H observations), and properties of the outer heliosheath;
- Pickup Ion Composition (implications for big bang cosmology and nucleosynthesis with a dedicated PUI instrument: He^3/He^4 and Ne^{22}/Ne^{20} with better than 5% accuracy);
- Seed populations of energetic particles with high time resolution (several minutes);

- Underlying time variations of ubiquitous suprathermal ions;
- Solar Energetic Particle composition, injection and acceleration;
- Suprathermal and energetic particle transport;
- ACR/GCR modulation and evolution with time; and
- L1 environmental monitoring and solar wind input for magnetospheric and ITM science

6. IMAP Mission and Instrument Implementation

The IMAP goals can be achieved with a mission concept that is largely drawn from ACE, with greatly-improved ENA imaging aspect infused from IBEX and INCA, with the first dedicated pickup ion instrument, and with a high-collecting power suprathermal ion sensor. IMAP is conceptualized as a Sun-pointed spinner, with re-adjustment of the spin axis every few days to provide all-sky maps every 6 months. To escape the ENA background environment of the magnetosphere and to allow continuous interplanetary observations, the spacecraft will be placed at L1. The mission goals can be achieved with a 2-year baseline mission, including the transit to L1, with extensions to a longer operation possible. To provide the necessary observations IMAP combines the following state-of-the-art measurement capabilities, for which no further development effort is required because the improvements over previous missions are based on experience with the instruments during calibration, testing, and operation.

High-Resolution ENA Maps: Two ENA cameras will produce critical new observations of ENAs from the heliospheric boundary region over an extended energy range and with significantly improved sensitivity, and spatial and energy resolution, compared to prior observations. They will cover the energy range 0.3-20 keV and 3-200 keV with 10x the angular resolution and ≈100x the combined sensitivity/duty cycle of the IBEX-Hi and CASSINI INCA sensors. With the sensitivity gain, these sensors will take advantage of oversampling the polar regions with high time resolution of a few days.

High-Resolution & Sensitivity ISN Flow Collection: An **Interstellar Neutral Atom Camera** and the first dedicated **Pickup Ion Sensor** will take coordinated high sensitivity observations of the interstellar gas flow through the inner solar system. The ISM Neutral camera will provide ISM flow observations of H, D, He, O, and Ne at 5-1000 eV with a pointing knowledge of 0.05° and >10x the combined sensitivity/duty cycle of IBEX-Lo, also extending the ENA maps below 0.3 keV. The pickup ion sensor provides pickup ion distributions of interstellar H, ^{3}He, ^{4}He, N, O, ^{20}Ne, ^{22}Ne, and Ar as well as inner source C, O, Mg, and Si over the energy range 100 eV – 100 keV/e with a combined sensitivity/duty cycle 100x that of SWICS, also providing SW heavy ion composition.

High-Cadence Suprathermal Ion Observations: Overlapping with the pickup ion sensor, a suprathermal ion sensor will provide composition (0.03-5 MeV/nuc) and charge state (0.03-1 MeV/e) for H through ultra-heavy ions (5 min cadence for H and He).

Solar Wind and Interplanetary Monitoring Suite will serve to understand and mitigate backgrounds from the local environment for high-sensitivity ENA observations and can also provide societally important real-time solar wind and cosmic ray monitoring. This suite measures SW ions (0.1-20 keV/e) and electrons (0.005-2 keV) every 15 s, the IMF to ≤1nT at 16 Hz, and SEP, ACR, and GCR electrons and ions (H-Fe) over 2-200 MeV/nuc.

7. Conclusion and Outlook

Our piece of cosmic real-estate, the heliosphere, is the domain of all human existence. Its history and future in our galaxy is key to understanding the conditions on our evolving planet and future expansion across the solar system. As we ask about the habitability of other planets surrounding other stars, we grapple with understanding the complex environments and interactions in the local parts of the galaxy where these stars exist. Our own heliosphere is an astrophysical case-history of the successful evolution of life in a habitable system. By exploring our global heliosphere and its myriad interactions, we develop key physical knowledge of the interstellar interactions that influence exoplanetary habitability as well as the distant history and destiny of our solar system and world. The interactions in the solar wind and the heliosphere produce highly energetic particles and help shield much of the galactic cosmic radiation that penetrates the heliosphere from the interstellar medium. Thus, the heliosphere presents a fundamental opportunity to study the basic processes that control particle radiation.

IBEX was the first mission to explore the global heliosphere and in concert with Voyager 1 and Voyager 2 is discovering a fundamentally new and uncharted physical domain of the outer heliosphere. The enigmatic ribbon is an unanticipated discovery demonstrating that much of what we know or think we understand about the outer heliosphere needs to be revised. The next quantum leap enabled by IMAP will open new windows on the frontier of Heliophysics at a time when the space environment is rapidly evolving and becoming increasingly hazardous due to rising levels of galactic cosmic ray fluxes.

The remarkable synergy between IMAP, Voyager 1 and Voyager 2 will remain for at least the next decade as Voyager 1 pushes further into the interstellar domain and Voyager 2 moves through the heliosheath. In fact, Voyager 2 moves outward in the direction of part of the ribbon and Voyager 2's plasma measurements will create singular opportunities for discovery in the context of IMAP's global measurements.

IMAP, like ACE before it, will be a keystone of the Heliophysics System Observatory by providing comprehensive solar wind observations, measurements to diagnose the source and evolution of solar wind and suprathermal ions, provide solar wind and energetic particle inputs into the geospace environment, evolution of cosmic rays and the evolution of the coupled solar and heliospheric magnetic field. IMAP's comprehensive interplanetary monitoring suite is critical to support on going geospace interaction studies and space weather observations at the ideal location of the Lagrangian point L1. *The high societal relevance of comprehensive solar wind, suprathermal, magnetic field and cosmic ray observations from L1 makes the IMAP mission an imperative as a successor to ACE.*

This paper is similar to an unpublished whitepaper that was presented to the National Academies of Sciences, Engineering and Medicine Committee for Solar and Space Physics (CSSP). The author list is identical to that of the white paper, and all authors were heavily involved in the IMAP deliberations as a part of the Solar and Space Physics (SSP) Decadal Survey. This paper provides a citable reference for IMAP objectives and the instrument straw man traceable to the SSP Decadal Survey. It is fitting that our paper appears in the volume that celebrates the 80th birthday of Ed Stone as part of the AIAC.

Acknowledgments
Authors wish to thank and acknowledge the many dedicated individuals that have made possible the Voyager missions, the IBEX mission, the Cassini/INCA project, the ACE mission, the individuals that have contributed SSP Decadal Survey, and the CSSP. We would like thank Ed Stone for his work in space science over many decades, which has lead to the remarkable successes of the ACE mission, the Voyagers, and countless other projects. Lastly, we would like to thank Gary Zank and Adele Corona, the organizers of AIAC, and the editors of this volume. The meeting and the volume of papers that has resulted is a great benefit to the field of Heliophysics and Heliospheric Physics.

8. References

[1] Committee on a Decadal Strategy for Solar and Space Physics (Heliophysics); Space Studies Board; Aeronautics and Space Engineering Board; Division on Engineering and Physical Sciences; National Research Council 2013 Science for a Technological Society: The 2013–2022 *Decadal Survey in Solar and Space Physics* DOI: 10.17226/13060

[2] McComas D J et al. 2009 Global Observations of the Interstellar Interaction from the Interstellar Boundary Explorer (IBEX) *Science* **326** 959 doi:10.1126/science.1180906.

[3] Redfield S and Linsky J 2008 *ApJ* **673** 283

[4] Möbius E et al. 2012 *ApJ Supp.* **198** 11 doi:10.1088/0067-0049/198/2/11

[5] Krimigis S M, Mitchell D G, Roelof E C, Hsieh K C and McComas D J 2009 Imaging the Interaction of the Heliosphere with the Interstellar Medium from Saturn with Cassini *Science* **326** 971

[6] Möbius E. et al. 2009 *Science* **326** 969

[7] McComas D J, Lewis W S and Schwadron N A 2014 IBEX's Enigmatic Ribbon in the sky and its many possible sources *Reviews of Geophysics* **52** 118

[8] Schwadron N A and McComas D J 2013 Spatial Retention of Ions Producing the IBEX Ribbon *ApJ* **764** 92

[9] Krimigis S, Roelof E, Mitchell D, Decker R and Dialynas K 2012 Relating the Heliosheath and Cassini Measurements: Voyagers' Ground Truth *39th COSPAR Scientific Assembly* **39** 982

[10] Opher M, Drake J F, Zieger, B and Gombosi T I 2015 Magnetized Jets Driven By the Sun: the Structure of the Heliosphere Revisited *ApJ* **800** L28 2015

[11] Drake J F, Swisdak M and Opher M 2015 A Model of the Heliosphere with Jets *ApJ* **808** L44

[12] Witte M et al 1996 *SSR* **78** 289

[13] Witte M 2004 *A&A* **426** 835

[14] Möbius E. et al. 2015 Proc. 13th Internat. Astrophys. Conf. *AIP Conf Proc.* **577** 012019

[15] Schwadron N A et al 2013 Solar Radiation Pressure and Local Interstellar Medium Flow Parameters from Interstellar Boundary Explorer Low Energy Hydrogen Measurements *ApJ* **775** 86

[16] Rodriguez Moreno D F et al 2013 Evidence of direct detection of interstellar deuterium in the local interstellar medium by IBEX *A&A* **557** A125

[17] Linsky J L 2003 Atomic Deuterium/Hydrogen in the Galaxy *SSR* **106** 49

[18] Linsky J L 2007 D/H and Nearby Interstellar Cloud Structures *SSR* **130** 367

[19] Linsky J L et al 2006 What Is the Total Deuterium Abundance in the Local Galactic Disk? *ApJ* **647** 1106

[20] Kubiak M A et al 2013 Assessment of detectability of neutral interstellar deuterium by IBEX observations *A&A* **556** A39

[21] Bzowski M 2012 Neutral Interstellar Helium Parameters Based on IBEX-Lo Observations and Test Particle Calculations *ApJ Supp* **198** 12

[22] Möbius E et al 2012 *ApJ Supp* **198** 11 doi:10.1088/0067-0049/198/2/11

[23] McComas D J et al 2012 The Heliosphere's Interstellar Interaction: No Bow Shock *Science* **336** 1291 doi:10.1126/science.1221054.

[24] Bzowski M et al 2015 *ApJ Supp* **220** 28

[25] McComas D J et al 2015 *ApJ Supp* **220** 22

[26] Leonard T. et al. 2015 *ApJ* **804** 42

[27] Möbius E. et al. 2015 *ApJ Supp* **220** 24

[28] Schwadron N A et al. 2015 *ApJ Supp* **220** 25

[29] Bzowski M et al 2014 *A&A* **569** A8

[30] Wood B, Mueller H-R and Witte M 2015 *ApJ* **801** 62

[31] Möbius E. et al. 2004 *A&A* **426** 897

[32] Funsten H O et al 2013 Circularity of the Interstellar Boundary Explorer Ribbon of Enhanced Energetic Neutral Atom (ENA) Flux *ApJ* **776** 30

[33] Schwadron N A et al 2014 *Science* **343** 988

[34] Opher M, Stone E C and Gombosi T I 2007 The Orientation of the Local Interstellar Magnetic

Field *Science* **316** 875

[35] Opher M et al 2009 A strong, highly-tilted interstellar magnetic field near the Solar System *Nature* **462** 1036 doi:10.1038/nature08567

[36] Pogorelov N V et al 2009 Heliospheric Asymmetries and 2-3 kHz Radio Emission Under Strong Interstellar Magnetic Field Conditions *ApJ* **695** L31

[37] Lallement R et al 2005 Deflection of the Interstellar Neutral Hydrogen Flow Across the Heliospheric Interface *Science* **307** 1447

[38] Izmodenov V, Alexashov D and Myasnikov A 2005 Direction of the interstellar H atom inflow in the heliosphere: Role of the interstellar magnetic field *A&A* **437** L35

[39] Möbius E et al 2015 *Proc. 13th Internat. Astrophys. Conf.* AIP Conf Proc **577** 012019

[40] Frisch P C et al 2013 *Science* **341** 1080

[41] Lallement R and Bertaux J-L 2014 *A&A* **565** A41

[42] Frisch P C et al. 2015 *ApJ* **801** 61

[43] Redfield S and Linsky J. 2008 *ApJ* **673** 283

[44] Möbius E et al 2015 *ApJ* **815** 20

[45] Park J et al 2015 *ApJS* **220** 34

[46] Kubiak M A et al 2014 *ApJ* **213** 29

[47] Kubiak M A et al 2016 Interstellar Neutral Helium in the Heliosphere from IBEX Observations. IV. Flow Vector, Mach Number, and Abundance of the Warm Breeze, *ApJ Supp* **223** 25

[48] Schwadron N A et al 2014 Global Anisotropies in TeV Cosmic Rays Related to the Sun's Local Galactic Environment from IBEX *Science* **343** 988

[49] Lallement R et al 2010 The Interstellar H Flow: Updated Analysis of SOHO/SWAN Data, *Twelfth International Solar Wind Conference* **1216** 555

[50] Schwadron N A et al 2015 Triangulation of the Interstellar Magnetic Field *ApJ* **813** L20

[51] Schwadron N A et al 2016 Determination of Interstellar O Parameters using the first 2 years of data from the Interstellar Boundary Explorer *ApJ* In Press

[52] Mason G M, Mazur J E and Dwyer J R 1999 3He Enhancements in Large Solar Energetic Particle Events *ApJ* ***525*** L133 doi:10.1086/312349

[53] Shaviv N and Veizer J 2003 Celestial driver of Phanerozoic climate 1340.

[54] Rohde R A and Muller R A 2005 Cycles in fossil diversity *Nature* **434** (7030) 208 doi:10.1038/nature03339

[55] Mewaldt R A et al 2010 Record-setting Cosmic-ray Intensities in 2009 and 2010 *ApJ* **723** L1 doi:10.1088/2041-8205/723/1/L1

[56] Florinski V, Zank G P and Axford W I 2003 The Solar System in a dense interstellar cloud: Implications for cosmic-ray fluxes at Earth and 10Be records *GRL* **30** 2206 doi:10.1029/2003GL017566

[57] Todd P 1994 Cosmic radiation and evolution of life on earth: Roles of environment, adaptation and selection *ASR* **14** 305 doi:10.1016/0273-1177(94)90483-9.

[58] McComas D J, Angold N, Elliott H A, Livadiotis G, Schwadron N A, Skoug R M and Smith C W 2013 Weakest Solar Wind of the Space Age and the Current "Mini" Solar Maximum, *ApJ* **779** 2 doi:10.1088/0004-637X/779/1/2

[59] Goelzer M L, Smith C W, Schwadron N A and McCracken K G 2013 An analysis of heliospheric magnetic field flux based on sunspot number from 1749 to today and prediction for the coming solar minimum *JGR* **118** 7525 doi:10.1002/2013JA019404

[60] Schwadron N A et al 2012 Lunar radiation environment and space weathering from the Cosmic Ray Telescope for the Effects of Radiation (CRaTER) *JGR* **117** doi:10.1029/2011JE003978.

[61] McCracken K G and J Beer 2007 Long-term changes in the cosmic ray intensity at Earth, 1428-2005 *JGR* **112** 10101 doi:10.1029/2006JA012117

[62] McCracken K G 2007 Heliomagnetic field near Earth, 1428-2005 *JGR* **112** 9106 doi:10.1029/2006JA012119

Journal of Physics: Conference Series **767** (2016) 012026 doi:10.1088/1742-6596/767/1/012026

Modification of Velocity Power Spectra by Thermal Plasma Instrumentation

P L Whittlesey[1,2,3], G P Zank[1,2], J W Cirtain[2,3], K H Wright, A W Case, and J C Kasper

[1]Department of Space Science, University of Alabama in Huntsville, Huntsville, AL 35899, USA

[2]Center for Space Plasma and Aeronomic Research (CSPAR), University of Alabama in Huntsville, Huntsville, AL 35899, USA

[3] Space Sciences Laboratory at University of California, 7 Gauss Way, Berkeley, CA 94720

[4] NASA Marshall Space Flight Center (NASA/MSFC) 320 Sparkman Dr NW, Huntsville, AL 35805

E-mail: pw0004@uah.edu

Abstract. The upcoming Solar Probe Plus mission (Launch 2018) will launch with the newest and fastest space plasma instrumentation to date. The Solar Wind Electrons, Alphas, and Protons (SWEAP) instrument suite, which measures thermal plasma, will make measurements faster than the local gyro-frequency and proton plasma frequency. By developing an end-to-end computer model of a SWEAP instrument, this work explores the specific instrumental effects of thermal space plasma measurement, particularly in the reproduction of velocity power spectra, or Power Spectral Densities (PSDs). This model reproduces the slowest measurement cadence of the Solar Probe Cup (SPC), a Faraday cup (FC) style instrument on that will measure thermal plasma density, velocity, and temperature on SPP. By using the calibrated model to model measurement of fully determined and synthetic turbulent time series data, a consistent underestimation of the velocity power spectral indices has been quantified, with possible implications for previous missions flying similar instrumentation.

1. Introduction

The spectral index is important in space physics because it describes the distribution of energy as a function of frequency. Typically magnetic field data are used to generate PSDs because magnetic field measurements are made at higher cadence, but in order to determine how Alfvénic a plasma is, comparison between the velocity PSDs and the magnetic PSDs is necessary. The amplitude of the PSD yields the overall power as a function of frequency, and the spectral index describes the relative distribution of power.

Power Spectra are often generated from magnetic field measurements, where the instrumentation often permits higher cadence measurements (Wüest et al., 2007). Power Spectra from velocity measurements often reveal a lower power spectral index of $P(f) \propto f^{-3/2}$, whereas magnetic field measurements usually reveal a $P(f) \propto f^{-5/3}$ (Alexandrova et al., 2013). This disparity between spectral indices indicates an unequal distribution of energy between magnetic field and velocity fluctuations δV and δB, violating usual assumptions of Alfvénicity and equipartition.

This work explores the influence that instrumental effects may have upon disagreement in spectral indices between velocity power spectra and magnetic power spectra. The upcoming Solar Probe Plus mission has instrumentation that will produce the fastest thermal plasma measurements to date. By using the calibrated model to model measurement of fully determined and synthetic turbulent time series data, a consistent underestimation of the velocity power spectral indices has been quantified, with possible implications for previous missions flying similar instrumentation.

This paper is split into six parts: Section one is this introduction. Section two is a primer on the traditional spectral analysis technique used in this paper. Section three describes the method of creating synthetic turbulent time series data. Section four includes details of the Solar Probe Cup and the computer model. Section five includes the results from feeding the SPC model the turbulent time series, and section six is a discussion of the results.

2. Spectral Analysis Background

The Power Spectrum, or more appropriately, the Power Spectral Density (PSD), is defined as the Fourier transform of the autocorrelation function R_{uu},

$$S_{uu}(\omega) = \frac{1}{2\pi} \int_{-\infty}^{\infty} e^{i\omega\tau} R_{uu}(\tau) d\tau, \tag{1}$$

where the autocorrelation function of a general function $u(t)$ is,

$$\begin{aligned} R_{uu}(\tau) &= \langle u(t)u(t+\tau)\rangle \\ &= \lim_{T\to\infty} \frac{1}{T} \int_{-T/2}^{T/2} u(t)u^*(t+\tau)dt, \end{aligned} \tag{2}$$

which is the time-averaged autocorrelation value (Zank, 2014, Blackman and Tukey, 1958).

Although $u(t)$ in Equation 2 is a continuous function in nature, it is discretized by instrumentation to a finite time series of point measurements. Thus, Equations 1 and 2 must be discretized. The autocorrelation function is rewritten for the case of discrete data with time Δt between measurements,

$$\begin{aligned} R_{uu}(\tau) &= \langle u(t)u(t+\tau)\rangle \\ R_{uu} &= \frac{1}{n-rh} \sum_{i=0}^{n-rh} u_i \cdot u_{i+rh}, \end{aligned} \tag{3}$$

where n is the number of velocity measurements, h is some integer greater than zero and $\Delta\tau = h\Delta t$, $r = 0, 1, 2, ...m$, where $mh < n$ and describes the longest lag $m\Delta t$ between measurements (Blackman and Tukey, 1958).

The PSD is now estimated via the real-valued Fourier transform of R_{uu}.

The domain of the PSD in frequency space is determined by two values: the total time duration of the data and the time between measurements Δt. The Nyquist, or folding, frequency determines the minimum detectable frequency, is directly related to the time-resolution of the data record,

$$f_N = \frac{1}{2\Delta t}, \tag{4}$$

where Δt is the sampling rate. For example, to detect a signal with a 2000Hz frequency, the sampling rate must be at least 1000Hz. Nyquist frequencies for different levels of SPC data are shown in Table 2 below.

Table 1: Nyquist Frequency for SPC Data Levels

Data Level	SPC Sampling Freq.	Minimum Detectable Freq.
Level 0	37500Hz	18750Hz
Level 1	146.5Hz	73.25Hz
Level 2	2.526Hz	1.263Hz

Table 2: Sampling Frequency and Detectable Frequency for different levels of SPC data sophistication. Level 0 is the least sophisticated data level, and requires Fourier processing before it can be interpreted physically. Level 1 data contains data for distinct, discrete sections of velocity space, and is in instrument units of current rather than density, velocity, or temperature. Level 1 data can be used to determine plasma flow angles only. Level 2 data is in physical units of density, velocity, and temperature, and is the product of fitting Level 1 data to the Faraday Cup instrument response function. See Section 4 for details of instrument operation.

3. Generation of Turbulent Velocity Fluctuations

To examine turbulent fluctuations in SPC measurement, synthetic time-series velocity measurements are generated. Here the estimated PSD, or autocorrelation tensor, is defined in terms of velocity fluctuation values,

$$P_{ij}(\vec{k}, t) = \langle \delta V_i(\vec{k}, t) \delta V_j^*(\vec{k}, t) \rangle. \tag{5}$$

Several simplifying assumptions are made in deriving an expression for δV as a function of solar distance. First, assume fast solar wind conditions. The idealized three-part turbulent spectrum consisting of the energy-containing, inertial, and dissipation ranges is most often and consistently associated with fast solar wind (Bruno and Carbone, 2005). Second, assume equipartition between magnetic fluctuations δB and velocity fluctuations δV. Assuming equipartition implies a purely Alfvénic plasma, a condition which is expected to become more frequent with increasing solar proximity as seen from *Helios* data (Bruno and Carbone, 2005, Bruno et al., 1985). Also assume that the magnetic field vector is well described by the Parker spiral configuration, thus the background magnetic field has a radial component pointing anti-sunward. Simplify coordinates by assuming the velocity fluctuations δV are antiparallel to the magnetic field fluctuations δB, as is often seen in spacecraft observations (Horbury et al., 2005). Finally, assume that the spacecraft is in the ecliptic plane.

The absolute direction of the velocity and magnetic field fluctuation vectors depends on whether the turbulence is slab or quasi-2D. In the case of slab turbulence, the magnetic field variation vector δB is parallel or antiparallel to the background magnetic field B_0, and the sense of δV is antiparallel to δB. For two-dimensional turbulence, the vector δB is in the plane perpendicular to the background magnetic field B_0, with the sense of δV again antiparallel to δB.

First assume a general expression for the spectrum energy contained in velocity fluctuations of form,

$$E_V(\vec{k},t) = \gamma k^a. \tag{6}$$

Now examine the form of the estimated PSD. The estimated PSD can be described in three regions: the energy-containing range, the inertial range, and the dissipation range (Figure 8). The transition between the energy-containing range is called the 'bendover scale' or 'bendover length', whereas the transition between the inertial range and the dissipation range is called the 'breakpoint frequency'. In this chapter's work the energy-containing range is neglected because the primary interest is in reproducing the inertial and dissipation ranges. Thus a range of frequencies significantly below the bendover scale are chosen. Typically the breakpoint frequency is on the order of either the ion cyclotron frequency,

$$\Omega_p = qB/m_p, \tag{7}$$

where m_p is the proton mass, or the ion-inertial length,

$$\begin{aligned} d_p =& V_A/\omega_{p,i} \\ =& \left(\frac{m_p}{\mu_0 q^2 n_p}\right)^{\frac{1}{2}} \end{aligned} \tag{8}$$

where n_p is the proton mass density and $\omega_{p,i}$ is the plasma ion frequency. The quantities B and n_p can be written as functions of radius. Here,

$$n_p = n_{p,0}\left(\frac{r_0}{r}\right)^2, \tag{9}$$

where the value $n_{p,0} = 5.6 \times 10^7 m^{-3}$ is the number density at 0.29AU, given in Adhikari et al. (2015) as calculated from Chandran and Hollweg (2009).

Likewise the radial and azimuthal magnetic field magnitude are expressed as functions of radius,

$$B_R(r) = B_{R,0}\left(\frac{r_0}{r}\right)^2, \tag{10}$$

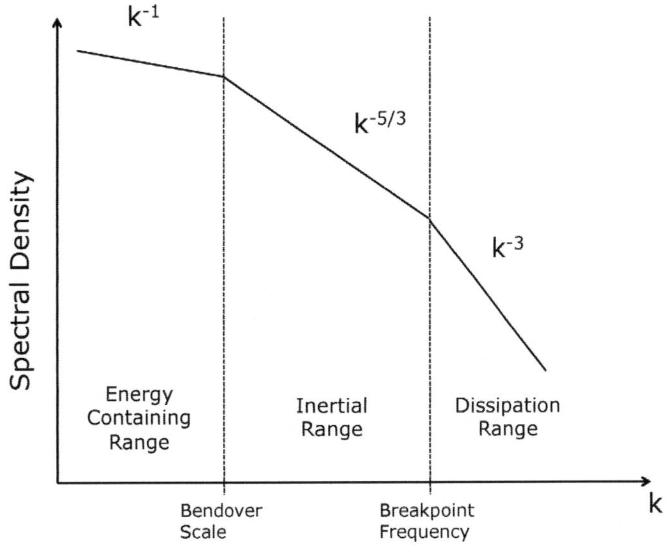

Figure 1: Shown is a cartoon diagram of what a full turbulent spectrum looks like. The Energy Containing Range usually has k^{-1} form, and the Dissipation range is thought to be of $\sim k^{-3}$ form. The Inertial Range above is shown with Kolmogorov $k^{-5/3}$ form.

$$B_\phi(r, \theta, \phi_0) = B_{R,0}(r_0, \theta, \phi_0)\frac{\Omega r_0^2 \sin\theta}{V_r r}, \quad (11)$$

where $B_{R,0}$ is the magnetic field at $r = r_0$, and θ is the colatitude angle (Owens and Forsyth, 2013). Making the simplifying assumption that all measurements occur in the ecliptic plane reduces $\sin\theta = 1$,

$$B_\phi(r, \theta = 0, \phi_0) = -B_{R,0}(r_0, 0, \phi_0)\frac{\Omega r_0^2}{V_r r}. \quad (12)$$

For completeness, assume a purely radial flow. Thus the θ component of the magnetic field,

$$B_\theta(r, \theta, \phi) = 0. \quad (13)$$

With the solar wind density and magnetic field defined as functions of radius, the remaining quantity V_r is left as a free parameter as long its value is within the range of acceptable solar wind radial speeds.

From Equation 7 the ion-cyclotron frequency is shown to be inversely proportional to the square of radial distance,

$$\Omega_p \sim \frac{1}{r^2} \quad (14)$$

and from Equations 8 and Equation 9 that the ion inertial length is proportional to radial distance,

$$d_p \sim \sqrt{\frac{1}{n_p}} \sim r. \quad (15)$$

Taking the inverse of the ion inertial length and multiplying by the phase velocity yields the equivalent frequency for use in the following analysis,

$$\frac{v}{d_p} \sim \sqrt{n_p} \sim \frac{1}{r}. \tag{16}$$

Thus, both the ion cyclotron frequency and the equivalent frequency for the ion-inertial length are expected to increase with growing solar proximity. More significantly, the separation between the two will also increase compared to their values at 1AU. At 1AU the frequencies are similar enough that there remains ambiguity as to which value is the true break-point frequency. One of the science goals of Solar Probe Plus is to distinguish which value is the actual breakpoint frequency, thereby determining the roles that turbulent heating and ion-cyclotron heating have in coronal temperatures.

To generate a synthetic time-series, a PSD is defined that describes the desired δV fluctuations. The PSD equation will differ depending on the turbulence model and dissipation mechanism. Once the PSD is defined by a continuous function (see Eq 6), the PSD's domain is discretized according to,

$$\begin{aligned} n_{PSD} &= \frac{n_{time}}{2} + 1 \\ n_{time} &= 2(n_{PSD} + 1) \end{aligned} \tag{17}$$

where n_{time} is the number of measurements in the desired time series result, and n_{PSD} is the number of discretized values from the estimated PSD (Blackman and Tukey, 1958). Once the number of discrete PSD values is known, the discrete Fourier transform (DFT) solves for frequencies at,

$$\frac{1}{T_0}, \frac{1}{2T_0}, \frac{1}{3T_0}, \ldots, \tag{18}$$

where T_0 is the total time length of the desired time series. These frequencies designate the values of the mean squared amplitudes $\langle A(k) \rangle^2$ in the PSD. The square root of the $\langle A(k) \rangle^2$ yields the average amplitude as a function of wave number or frequency.

Now construct a wave function for δV in wave number space,

$$\delta V(k) = \sum_{j=1}^{n_{PSD}} A(k_j) e^{-i\phi(k_j)}, \tag{19}$$

where $\phi(k_j)$ is a randomly-chosen phase angle on $[0, 2\pi)$ for each wavenumber or frequency j. The PSD by definition has no phase information, so a randomized phase must be substituted in the wave function. Thus, taking the inverse real DFT of the series in Equation 19 yields the synthetic time series $\delta V(t)$ fluctuations, which are then summed with the bulk velocity and used as input to the SPC model.

The upper and lower boundaries of the PSD are determined by 1) the length of the observation desired for analysis, and 2) the Nyquist frequency. The total length of the observation determines the slowest detectable frequency. The inverse of the length of the observation yields the slowest detectable frequency, since a signal of that frequency can be contained only once over the length of the observation (Blackman and Tukey, 1958).

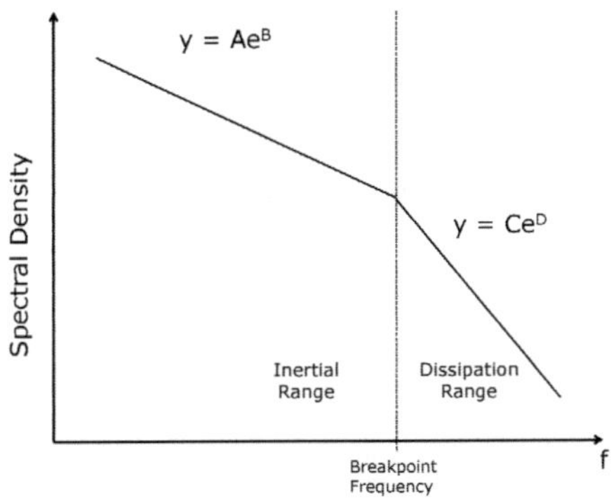

Figure 2: Cartoon diagram of the expected PSD for the range examined in this paper. The Inertial Range and Dissipation range are separately defined.

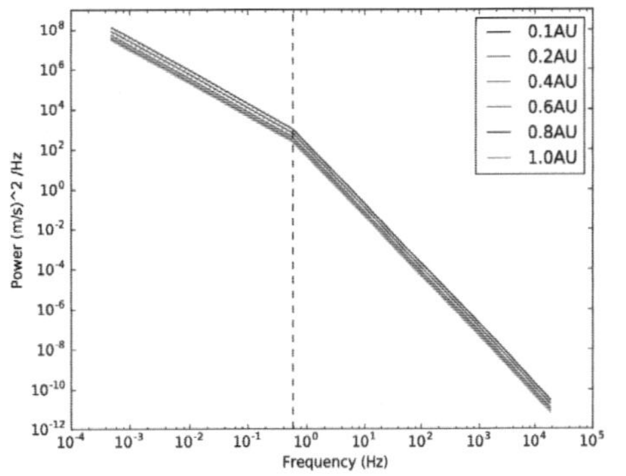

Figure 3: A comparison of Kolmogorov velocity fluctuation PSDs as a function of radius and frequency. The rate ϵ is the only free parameter; magnetic field, density, temperature, and bulk velocity are constant. The proton gyrofrequency Ω_p is indicated by a vertical dashed line, and is also constant. $B = -5/3$; $D = -3$

Here, 0.0005Hz is chosen as the slowest frequency; thus data is generated for a 2000 second period (33.3 minutes). Thus, the time series includes 75000000 measurements at 37500Hz. The chosen 0.0005 Hz frequency is sufficiently below the turbulence bendover scale separating the energy-containing range from the inertial range (Borovsky, 2012, Leamon et al., 1998).

To vary the PSD as a function of radial distance from the sun, first construct the PSD function using an isotropic Kolmogorov turbulent plasma as an example. The kinetic energy PSD is expressed as,

$$E(k) = C_K \epsilon^{2/3} k^{-5/3}, \tag{20}$$

where C_K is an experimentally derived constant as described in (Vasquez et al., 2007), and ϵ is the turbulent heating or dissipation rate. Determining the radial evolution of the heating rate ϵ allows us to study the radial evolution of the corresponding velocity fluctuations as measured by SPC.

To determine this dissipation or heating rate as a function of radius, the example set byVasquez et al. (2007) and developed in Verma et al. (1995) is followed,

$$\frac{dT_{pr}}{dr} + \frac{4}{3}\frac{T_{pr}}{r} = \frac{m_p}{(3/2)V_{SW}k_B}\epsilon, \tag{21}$$

in which the solar wind expansion experiences adiabatic cooling, in-situ heating, and all of the energy dissipation manifests as proton temperature T_{pr}. Equation 21 implies a radial proton temperature dependence of $T_{pr} = T_{pr,0}\left(\frac{r_0}{r}\right)^{\xi}$ where ξ has hitherto been determined by spacecraft measurement, and varies depending on the heliospheric location. For a system experiencing adiabatic cooling alone, $\xi = 4/3$. For radial distances

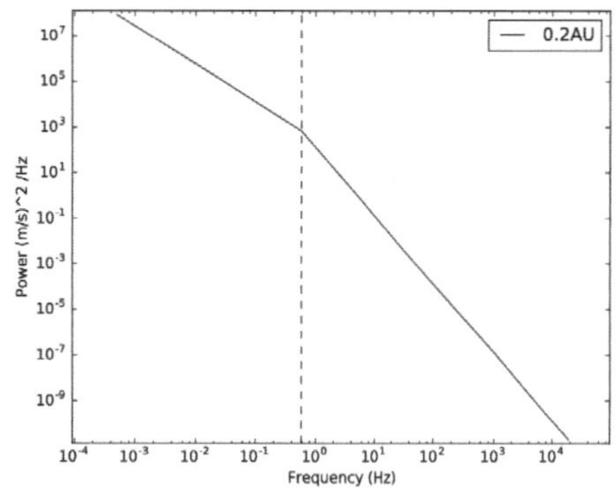

Figure 4: An example of the PSDs that are used to generate synthetic turbulent time series. In this example, the inertial range spectral index is $-5/3$, the dissipation range spectral index is -3, and the breakpoint frequency is the local cyclotron frequency.

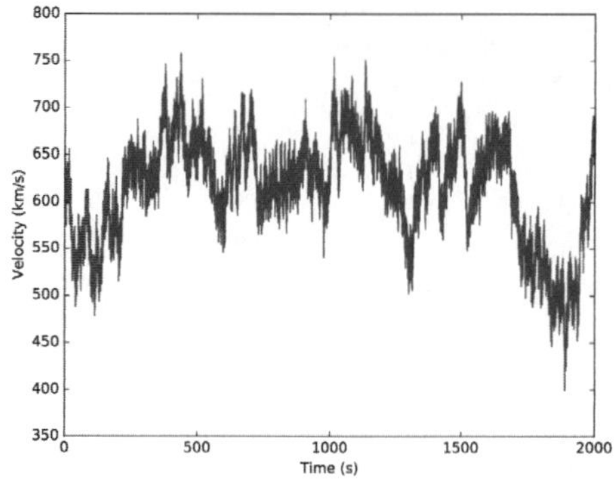

Figure 5: The corresponding synthetic velocity time series for 4, over 2000 seconds. The bulk velocity for the data above is $600km/s$.

between 0.3 AU and 1.0 AU however, $\xi = 0.9$, in support of in-situ heating (Totten et al., 1995, Vasquez et al., 2007). Solving for the heating rate,

$$\epsilon = \left(\frac{4}{3} - \xi\right) \frac{3}{2} \frac{V_{SW}}{r_0} \frac{k_B T_{pr,0}}{m_p}, \tag{22}$$

and substitute $\xi = 0.9$ for $r = (0.3, 1.0)\ AU$ to arrive at the radial expression for the heating rate,

$$\epsilon = 0.65 \frac{V_{SW}}{r_0} \frac{k_B T_{pr,0}}{m_p}, \tag{23}$$

where $T_{pr,0}$ is the proton temperature at r_0. An illustration of the PSD's variance with respect to radius is shown in Figure 3.

Table 3: Parameters for Generating Synthetic Turbulent Time Series

Type	Distance	Break Point	Model
Slab	1.0 AU	None	Kolmogorov
2D	0.5 AU	Cyclotron Freq.	Iroshnikov-Kraichnan
-	0.2 AU	Ion Inertial Freq.	-

The synthetic turbulent time-series are generated using all combinations of parameters shown in Table 3. Although the density and magnetic field were expressed as functions of radius previously to solve for radial expressions of the cyclotron frequency Ω_p and ion-plasma frequency $\omega_{p,i}$, we choose to keep them constant here. The density

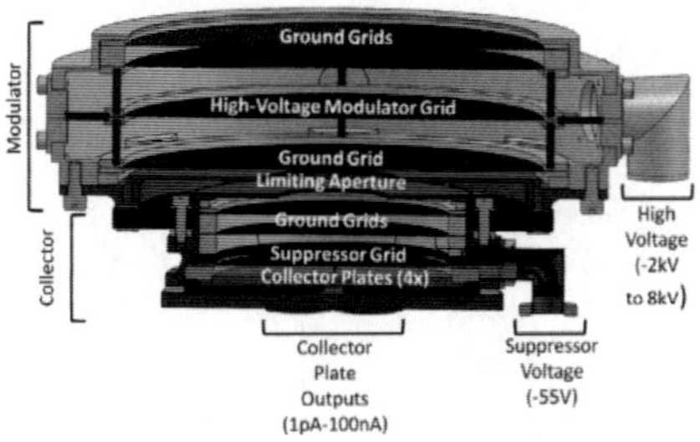

Figure 6: Cross section of the SPC instrument.

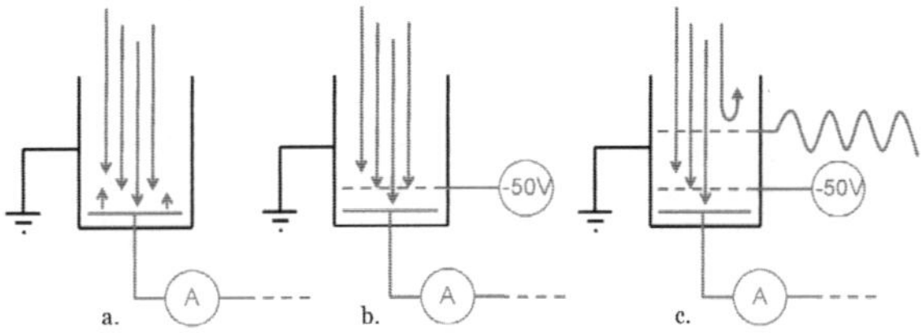

Figure 7: Cartoon Explaining Faraday cup operation: blue arrows indicate incoming charged particles. Charged particles impact the collector plate, inducing a small current. By inserting stacked grids (green) with a time-varying voltage bias, particles are sorted by energy.

is a constant $4cm^{-3}$, the thermal speed is $20km/s$, and the bulk speed is $600km/s$. Examples of typical PSDs and the resultant synthetic turbulent fluctuations are shown in Figures 4 and 5.

As a final step in validating the synthetic turbulent time series, the Fourier transform of the autocorrelation of the constructed velocity fluctuation time series is taken. The resulting PSD exactly reproduces the initial PSD, confirming that the time series accurately represent their originating PSD.

4. Faraday Cup Model

The Faraday cup's enduring success may be attributed to a simple design, robust against deterioration over time. An advantage that the Faraday cup has over other thermal plasma instrumentation is that it can make angular measurements simultaneously to energy measurements. Likewise, the instrument response can be modeled with an analytic response function. Successful science missions that have flown Faraday cup instruments include *Explorer 10*, *IMP-8*, *Voyager* 2, *Wind*, *DSCOVR* (formerly *Triana*), and others (Bonetti et al., 1963, Bellomo and Mavretic, 1978, Bridge et al., 1977, Ogilvie and Chornay, 1995, Valero et al., 1999). A summary of the differences between well known Faraday cups is shown in Table 4.

To produce a single ion spectrum, a Faraday cup scans incrementally through a preset number of energy windows, each window bounded by specific voltage boundaries $(V_i, V_i + \Delta V_i)$ induced on the modulator grid 7. A predetermined number of windows span SPC's full energy range without overlap. Usually, the window width increases for each successive window as SPC scans upward through the energy windows. Window spacing is defined by choosing a minimum measurable energy and a desired window width as a percentage of the bottom of each energy window,

$$\Delta E/E_{bottom} = (E_{top} - E_{bottom})/E_{bottom} = \left[(V_o + \Delta V)^2 - (V_o)^2\right]/V_o^2. \tag{24}$$

Hence, the width of the windows is a constant percentage $\Delta E/E$ of each E_{bottom}. SPC uses narrow windows according to Eq 24 at the bottom of its dynamic energy range, changing to fewer windows with broader energy widths at the upper end of its energy range.

Journal of Physics: Conference Series **767** (2016) 012026 doi:10.1088/1742-6596/767/1/012026

Table 4: Design Differences Between Famous Faraday Cups

Mission	Modulator Frequency	Collector #	Number of Cups	FOV°	e$^-$'s ?	Maximum Energy	Modulator Waveform
Voyager	400Hz	1	4	90°	Yes	5950eV	Square
DSCOVR	200Hz	3	1	30°	No	8000eV	Square
Wind	200Hz	2	2	60°	No	8000eV	Square
SPP	1171.875Hz	4	1	30°	Yes	8000eV	Sinusoid

Shown is a comparison between Faraday cup operation parameters. The number of collectors refers to the number of collector plate segments per individual cup. A few missions have multiple separate Faraday cups as indicated above. For Voyager, the FOV varies for each cup, but combined adds to a 90°half-cone field of view, not including the "D" cup. The maximum modulated energy is displayed in units of eV assuming that all incoming ions have a charge of $1q$. The SPC on SPP will be the first sinusoid modulated Faraday cup flown on a space mission.

Thus, the Faraday cup instrument response function is dependent on the size of, and rate of scanning through, successive energy windows. For a fully detailed derivation of the Faraday cup response function for a square wave modulated Faraday cup, see the dissertations of Kasper (2003) and Maruca (2012). In their derivation they begin by assuming the plasma is well described as a Maxwellian,

$$f(\vec{v}) = \frac{n}{\pi^{3/2}w^3}e^{-\frac{(\vec{v}-\vec{U})^2}{w^2}}. \tag{25}$$

where the total plasma velocity v has been rewritten in terms of thermal speed $\vec{w} = \sqrt{\frac{2k_BT}{m}}$, and the bulk speed $\vec{U}$, so that $\vec{v} - \vec{U} = \vec{w}$.

Defining a segment of phase space in terms of a current

$$I_{iso} = \frac{Aqn}{2}\left[\frac{w}{\sqrt{\pi}}\left(e^{-(V_o-U_z)^2/w^2} - e^{-(V_o+\Delta V-U_z)^2/w^2}\right) + U_z\left(\text{erf}\left(\frac{V_o+\Delta V-U_z}{w}\right) - \text{erf}\left(\frac{V_o-U_z}{w}\right)\right)\right]. \tag{26}$$

Equation (26) describes the instrument response in units of instrument current within one window of width ΔV between V_o and $V_o + \Delta V$ during square wave modulation, assuming the plasma sufficiently resembles a Maxwellian distributed plasma. However, to measure kinetic scale phenomena, which may have frequencies nearing 60Hz at SPP perihelion, SPC uses a sine wave modulation technique (Kasper et al., 2015). Thus, the SPC computer model analytically solves Equation 26 for each point measured at 37,500 Hz.

Because the range of each energy window is controlled only by the voltage bias on the modulator grid, the energy windows are highly configurable during the mission operation phase. For SPC, three main operational modes have been identified:

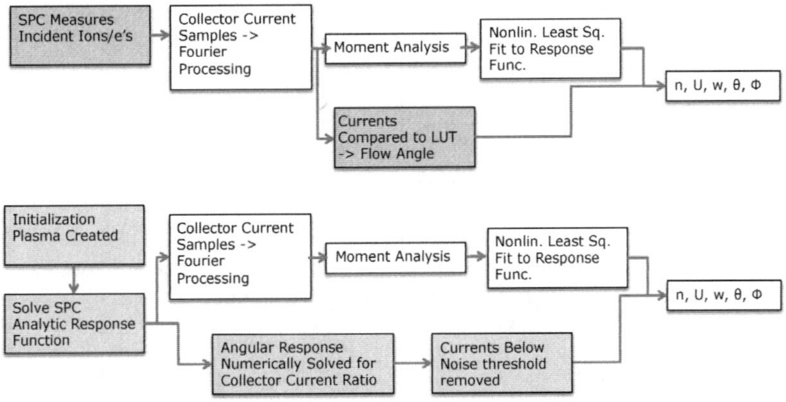

Figure 8: A flow chart describing differences between SPC measurement (Top) and the SPC model (Bottom). White background boxes indicate identical steps in each. For a full description of the SPC model, see Whittlesey (2016).

(i) Full Scan (FS) mode: SPC scans through all the energy windows incrementally from lowest to highest, and then back down to lowest before repeating the scan cycle.

(ii) Peak Tracking (PT) mode: SPC executes one Full Scan, identifies the energy window with the largest current response, and scans incrementally through a predefined number of windows centered about the peak of the distribution.

(iii) Flux or Flow Angle (FA) mode: SPC executes one Full or Peak Tracking scan, identifies the energy window with the largest current response, and samples the current in that window alone.

The time required for each of the scans outlined above is a function of the modulation frequency, the number of windows used, and the number of modulation cycles in each window. Typically each window consists of 8 integration cycles, the first two of which are discarded, leaving 6 cycles' worth of samples for each window. Therefore, the number of seconds for each ion spectrum to be taken is,

$$\frac{Seconds}{Spectrum} = \frac{NumberofCycles}{Window} * \frac{1}{ModulationFreq} * NumberofWindows \qquad (27)$$

SPC's sine wave modulation is sampled discretely, 32 times per cycle, yielding a total of 256 samples per 8-cycle energy window. Only the last 192 samples are used to create the measured ion spectrum.

For each energy window, these 192 measured points of Level 0 data are processed using Fourier analysis to create Level 1 data (Kasper et al., 2015).

Of the three operational modes, the Flux Angle (FA) mode contains the least information, followed by Peak Tracking (PT), while the Full Scan (FS) mode obtains the full electron and ion spectrum. Table 5 details the expected data output from each of these scans. In a worst case operational scenario where the PT mode does not function as expected, the FS mode is the default operational mode. For this work we consider only the FS mode.

Journal of Physics: Conference Series **767** (2016) 012026 doi:10.1088/1742-6596/767/1/012026

Table 5: SPC Products by Operational Mode

Mode	e^-'s?	Frequency	Density	p^+Window#	Bulk V	Thermal $\vec{v}$
Full Scan	Yes	TBD	Yes	All	Yes	Yes
Peak Tracking	No	18Hz	Yes	8	Yes	Yes
Flux Angle	No	146.5Hz	Yes	1	Yes	No

Shown above are characteristics of each SPC operational mode. Electrons are only measured during the Full Scan mode. Bulk velocity is measured in every mode, but thermal velocity is not measured during Flux Angle mode. These values are subject to change before launch.

5. Results

Using the method described in section 3, and the model in section 4, all possible combinations of the values in Table 3 are used to generate synthetic turbulent spectra.

The velocity fluctuations generated from Table 3 were used as input for the SPC model. The resultant Level 2 data (Figure 2) is used to generate PSDs, an example of which is shown in Figure 10.

Tables 7 and 8 show the results when the first third and last third of the data are fit according to a power law. The spectral index is indicated by k. The lower and upper thirds of the PSD frequencies are fit separately to prevent data points close to the breakpoint frequency contributing to the fit parameters. Spectral indices corresponding to the last third of the data roughly correspond to the dissipation range spectral indices; spectral indices corresponding to the first third of the data roughly correspond to the inertial range spectral indices.

Table 6: Fit Parameters - Kolmogorov, No Dissipation

Turbulence	R(AU)	k	σ_k
Slab	1.0	-1.5761	0.00032
Slab	0.5	-1.5875	0.00033
Slab	0.2	-1.5834	0.00034
2D	1.0	-0.1119	0.00066
2D	0.5	-0.0554	0.00060
2D	0.2	0.0101	0.00066

Table 6 lists fit values for simulated SPC data produced by exposure to a plasma where the spectral index of $-5/3$ is extended to the entire frequency range, *e.g.* without dissipation. From Table 6 it is shown that the SPC does not produce accurate fits for 2D turbulence, as expected since SPC is insensitive to fluctuations parallel to the instrument normal vector. Thus, 2D turbulence is consciously neglected in this chapter's remaining analysis. However, Table 6 shows that the SPC model produces data that underestimates the B parameter of the inertial range of the slab Kolmogorov spectrum, where $B = -1.66$ or $-5/3$ is expected. Reasons for this under-estimation of the PSD power are explored further in this section.

Journal of Physics: Conference Series **767** (2016) 012026 doi:10.1088/1742-6596/767/1/012026

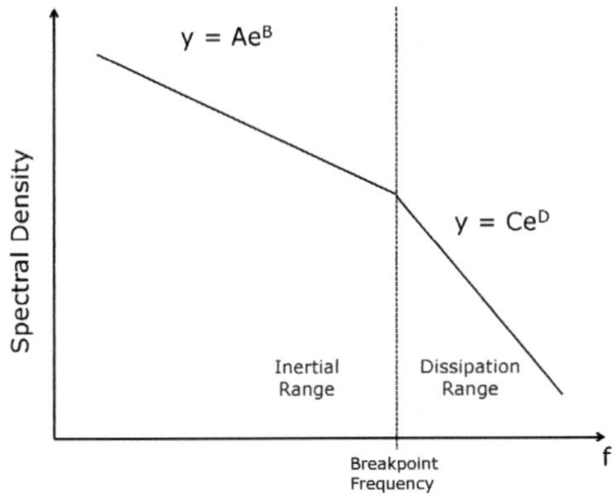

Figure 9: Idealized Cartoon Inertial and Spectral Ranges

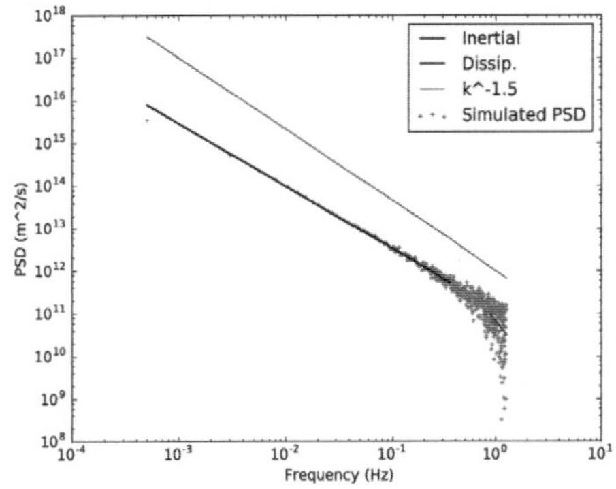

Figure 10: Example of the results of the SPC model's simulated measurement of turbulent plasma. The inertial range is fit in black with the first 1/3; the dissipation range in blue with the last 1/3.

Two breakpoint frequencies are used for the remaining analysis of slab turbulence in this chapter: the ion-inertial length, and the ion-cyclotron frequency. Table 7 contains the results from Kolmogorov turbulence in the inertial range, with a -3 spectral index imposed in the dissipation range. Again, the simulated SPC data consistently underestimates the magnitude of the spectral index in both the inertial and dissipation ranges. Note that the uncertainty is significantly lower in the inertial range than those for a straight Kolmogorov spectrum (Table 6), but that the uncertainty for the dissipation range is much higher.

Table 7: Fit Parameters - Slab Kolmogorov

R(AU)	Breakpoint Frequency	Ideal Spectral Indices	Inertial Spectral Index	Dissipative Spectral Index
1.0	Ω_p	-1.67/-3.0	-1.6313 ± 0.00002	-2.5048 ± 0.18583
1.0	$\omega_{p,i}$	-1.67/-3.0	-1.6340 ± 0.00002	-2.3744 ± 0.19342
0.5	Ω_p	-1.67/-3.0	-1.6338 ± 0.00002	-2.1398 ± 0.19242
0.5	$\omega_{p,i}$	-1.67/-3.0	-1.6294 ± 0.00002	-1.7162 ± 0.14832
0.2	Ω_p	-1.67/-3.0	-1.6336 ± 0.00002	-2.6023 ± 0.17149
0.2	$\omega_{p,i}$	-1.67/-3.0	-1.6225 ± 0.00002	-2.3364 ± 0.17836

The results for a both breakpoint frequencies are similar (Table 7). Introducing Iroshnikov-Kraichnan (IK) turbulence (spectral index $= -3/2$) into the initialization plasma produces similar results as seen for Kolmogorov turbulence, as seen in Table 8. The ideal spectral index $B = -1.5$, but instead the spectral index is again undervalued by the analysis. Likewise, the dissipation range spectral index magnitudes are also underestimated. See Figure 10

Table 8: Fit Parameters - Slab IK Turbulence

R(AU)	Breakpoint Frequency	Ideal Spectral Indices	Inertial Spectral Index	Dissipative Spectral Index
1.0	Ω_p	-1.5/-3.0	-1.4629 ± 0.00002	-1.9564 ± 0.16225
1.0	$\omega_{p,i}$	-1.5/-3.0	-1.4643 ± 0.00003	-2.0772 ± 0.18736
0.5	Ω_p	-1.5/-3.0	-1.4651 ± 0.00002	-2.5701 ± 0.17724
0.5	$\omega_{p,i}$	-1.5/-3.0	-1.4663 ± 0.00003	-2.0684 ± 0.17340
0.2	Ω_p	-1.5/-3.0	-1.4662 ± 0.00002	-3.3748 ± 0.19117
0.2	$\omega_{p,i}$	-1.5/-3.0	-1.4649 ± 0.00003	-3.1570 ± 0.17533

Likewise, for IK turbulence with an ion inertial length breakpoint frequency, the spectral index fits for the inertial range are lower in magnitude than expected (Table 8).

To isolate the source of the PSD underestimation within the inertial range, the simulated SPC Level 0 data are compared to the input synthetic turbulent plasma. The Level 0 data contains samples taken at 37,500 Hz. The SPC energy windows in the Level 0 data with the highest signal for each input plasma are selected and five windows on either side of the maximum are used for a total of 11 energy windows analyzed. To keep the cadences identical between Level 0 and Level 2 data, the first Level 0 measurement in each window are used. These values are compared to the idealized plasma input at the same measurement time as the maximum modulator value. The results are shown in Table 9.

As the results in Table 9 show, the simulated Level 0 data does not accurately reproduce Kolmogorov or IK parameters. The inaccuracy here is attributed to the Level 0 measurement units, which are in instrument units of current and not in physical units as the synthetic turbulent time series are.

Table 9: Percentages Error: L0 Window-Averaged Data vs. Initial Plasma Sampled

Turbulence	Source	Inertial Index	% Error
IK	L0 Data	-1.1164 ± 0.00249	25.574%
Kolmogorov	L0 Data	-1.1469 ± 0.00238	31.186%
IK	Initial Plasma	-1.4656 ± 0.00003	2.291%
Kolmogorov	Initial Plasma	-1.6323 ± 0.00002	2.061%

When a single point from each L0 data energy window is used to to generate a PSD, the resultant power is grossly in error, 25.6% and 31.2% from the ideal Kolmogorov or IK inertial range value. However, when the corresponding samples from the initial synthetic turbulent time series are analyzed in an identical manner, the resultant power is only 2% in error, which is in agreement with the model's established accuracy (Whittlesey, 2016). Furthermore, the fit values from the window-sampled initialization plasma are remarkably similar to the fit values from the Level 2 data analysis (Table 10).

The resultant spectral index in the inertial range for all values are on the order of 2%. Considering the limitations of the model are within a 2% margin of error (Whittlesey,

Table 10: Percentages Error in Inertial range Fit @ 0.2 AU

Turbulence	Source	Inertial Index	% Error
IK	L2 Data	-1.4662 ± 0.00003	1.5022%
IK	Initial Plasma	-1.4656 ± 0.00003	2.2909%
Kolmogorov	L2 Data	-1.6336 ± 0.00002	1.9840%
Kolmogorov	Initial Plasma	-1.6323 ± 0.00002	2.0606%

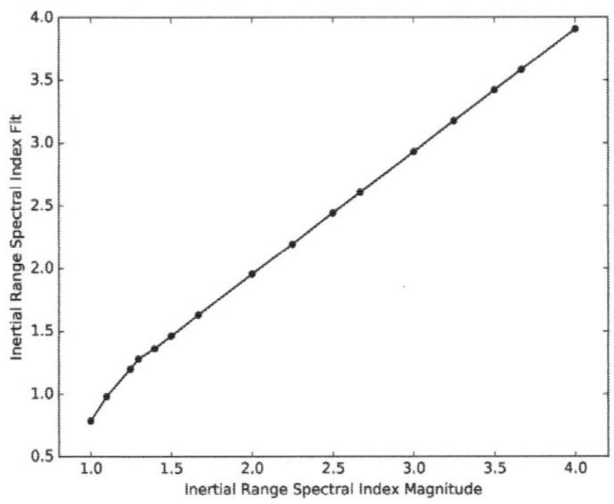

Figure 11: Above: the under-sampled synthetic plasma inertial range spectral index vs. the initial PSD inertial range spectral index.

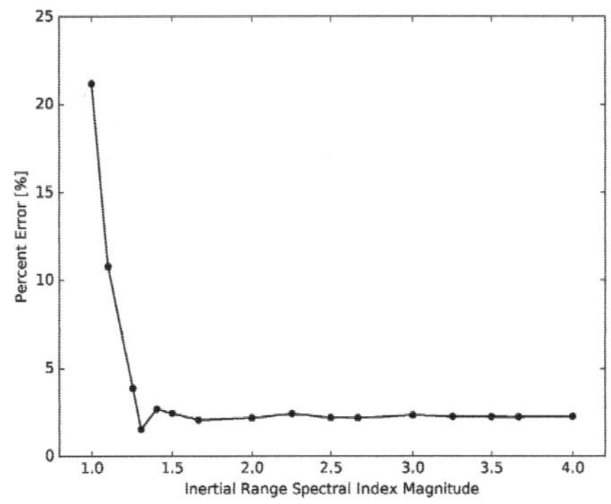

Figure 12: The percentage error vs. the initial PSD inertial range spectral index.

2016), these values are considered a good metric for the accuracy of both the model and instrument's positive ability to resolve the spectral index in both the inertial and dissipation ranges. However, the source of the error in spectral index is yet unfound. To explore the source of error, a set of synthetic turbulent data have been generated for a range of spectral indices ranging between -1.0 and -4.0 for $R = 0.2AU$, with the same density, bulk speed, thermal speed, and ϵ as previously used. The synthetic data are sampled at the same rate as in the comparison to Level 0 SPC data. The results are plotted in Figures 11 and 12. It is apparent that undersampling the synthetic turbulent time series at the SPC Level 2 data rate results in a ~2% error for inertial range spectral indices of magnitude greater than 1.5, but that the error is greater for indices of lesser magnitude.

6. Conclusion: Velocity PSD Spectral Indices are Always Undermeasured

This paper has established that the SPC model and the spectral analysis detailed therein reproduce inertial range spectral indices to a high degree of precision, but a low degree of absolute accuracy, for slab and turbulence. Except for circumstances where the bulk flow is at highly oblique angles to the instrument normal axis, 2D turbulence is not distinguishable from noise in SPC. Examining simulated Level 2 slab turbulence data indicates that the magnitude of the inertial range spectral index is consistently underestimated by 2% of the input value. This underestimation is consistent for both spectral indices chosen. The dissipation range accuracy and precision are considerably less accurate, and with higher variance, likely due to the higher noise levels observed in the dissipation range of the PSD.

Significantly, the spectral index is consistently underestimated in both the inertial and the dissipation range of the inertial range when discretely sampled in a style consistent with thermal plasma instrumentation. This underestimation has been quantified at 2% for SPC's measurement cadence for spectral indices of magnitude greater than 1.5, with possible extension to other thermal plasma instruments of differing cadence.

Finally, note that traditional turbulent spectral analysis typically uses magnetic field measurements instead of velocity measurements (*e.g.* Leamon et al. (1998), Wicks et al. (2010), Bruno and Carbone (2005) and others) due to faster measurement cadences prevalent in magnetic field instrumentation. Thus, it is possible that the higher measurement cadence in measurement by these instruments prevents such a spectral under estimation.

Acknowledgments
We acknowledge the support of people like Justin Kasper, and Tony Case in allowing this work on SWEAP and the Solar Probe Cup (SPC).

References

L. Adhikari, G. P. Zank, R. Bruno, D. Telloni, P. Hunana, a. Dosch, R. Marino, and Q. Hu. the Transport of Low-Frequency Turbulence in Astrophysical Flows. Ii. Solutions for the Super-Alfvénic Solar Wind. *The Astrophysical Journal*, 805(1):63, 2015. ISSN 1538-4357. doi: 10.1088/0004-637X/805/1/63. URL `http://stacks.iop.org/0004-637X/805/i=1/a=63?key=crossref.b29ed47919c2810a56a3b46412d1b`

O. Alexandrova, C. H K Chen, L. Sorriso-Valvo, T. S. Horbury, and S. D. Bale. Solar wind turbulence and the role of ion instabilities. *Space Science Reviews*, 178:101–139, 2013. ISSN 00386308. doi: 10.1007/s11214-013-0004-8.

A Bellomo and A Mavretic. MIT plasma experiment on IMP H and J Earth orbited satellites. *MIT Center for Space Research Technical Report CSR- ...*, 1978. URL `https://scholar.google.com/scholar?oi=gsb95&q=Bellomo%2CA.%2C and A. Mavretic. %22MIT plasma experiment on IMP H and J Earth orbited satellites.%22 MIT Center for Space Research Technical Report CSR-TR-78-2 (1978).&lookup=0&hl=en#1`.

Rb Blackman and Jw Tukey. The measurement of power spectra. *Physics Today*, page 190, 1958. URL `http://www.getcited.org/pub/101189984`.

A. Bonetti, H. S. Bridge, Alan J. Lazarus, B. Rossi, and F. Scherb. Explorer 10 plasma measurements. *Journal of Geophysical Research*, 68(13):4017–4063, jul 1963. ISSN 01480227. doi: 10.1029/JZ068i013p04017.

Joseph E. Borovsky. The velocity and magnetic field fluctuations of the solar wind at 1 AU: Statistical analysis of Fourier spectra and correlations with plasma properties. *Journal of Geophysical Research*, 117(A5):A05104, 2012. ISSN 0148-0227. doi: 10.1029/2011JA017499. URL `http://doi.wiley.com/10.1029/2011JA017499`.

H. S. Bridge, J. W. Belcher, R. J. Butler, A. J. Lazarus, A. M. Mavretic, J. D. Sullivan, G. L. Siscoe, and V. M. Vasyliunas. The plasma experiment on the 1977 Voyager Mission. *Space Science Reviews*, 21(3):259–287, 1977. ISSN 00386308. doi: 10.1007/BF00211542.

R. Bruno, B. Bavassano, and U. Villante. Evidence for long period Alfvén waves in the inner solar system. *Journal of Geophysical Research*, 90(A5):4373, 1985. ISSN 0148-0227. doi: 10.1029/JA090iA05p04373.

Roberto Bruno and V Carbone. The Solar Wind as a Turbulence Laboratory. *Living Reviews in Solar Physics*, 2, 2005. ISSN 16144961. doi: 10.12942/lrsp-2013-2. URL `http://solarphysics.livingreviews.org/Articles/lrsp-2005-4/articlese10.html`.

Benjamin D. G. Chandran and Joseph V. Hollweg. Alfvén Wave Reflection and Turbulent Heating in the Solar Wind From 1 Solar Radius To 1 Au: an Analytical Treatment. *The Astrophysical Journal*, 707(2):1659–1667, 2009. ISSN 0004-637X. doi: 10.1088/0004-637X/707/2/1659. URL `http://stacks.iop.org/0004-637X/707/i=2/a=1659?key=crossref.4ea81e1b7ac6036e64e79341f5a`

T S Horbury, M a Forman, and S Oughton. Spacecraft observations of solar wind turbulence: an overview. *Plasma Physics and Controlled Fusion*, 47(12B): B703–B717, 2005. ISSN 0741-3335. doi: 10.1088/0741-3335/47/12B/S52. URL `http://stacks.iop.org/0741-3335/47/i=12B/a=S52?key=crossref.23ea979751d91ff68014d3594a1`

Justin C. Kasper. *Solar Wind Plasma: Kinetic Properties and Micro-Instabilities.* PhD thesis, MIT, 2003.

Justin C. Kasper, Robert Abiad, Gerry Austin, Marianne Balat-Pichelin, Stuart D. Bale, John W. Belcher, Peter Berg, Henry Bergner, Matthieu Berthomier, Jay Bookbinder, Etienne Brodu, David Caldwell, Anthony W. Case, Benjamin D G Chandran, Peter Cheimets, Jonathan W. Cirtain, Steven R. Cranmer, David W. Curtis, Peter Daigneau, Greg Dalton, Brahmananda Dasgupta, David DeTomaso, Millan Diaz-Aguado, Blagoje Djordjevic, Bill Donaskowski, Michael Effinger, Vladimir Florinski, Nichola Fox, Mark Freeman, Dennis Gallagher, S. Peter Gary, Tom Gauron, Richard Gates, Melvin Goldstein, Leon Golub, Dorothy A. Gordon, Reid Gurnee, Giora Guth, Jasper Halekas, Ken Hatch, Jacob Heerikuisen, George Ho, Qiang Hu, Greg Johnson, Steven P. Jordan, Kelly E. Korreck, Davin Larson, Alan J. Lazarus, Gang Li, Roberto Livi, Michael Ludlam, Milan Maksimovic, James P. McFadden, William Marchant, Bennet A. Maruca, David J. McComas, Luciana Messina, Tony Mercer, Sang Park, Andrew M. Peddie, Nikolai Pogorelov, Matthew J. Reinhart, John D. Richardson, Miles Robinson, Irene Rosen, Ruth M. Skoug, Amanda Slagle, John T. Steinberg, Michael L. Stevens, Adam Szabo, Ellen R. Taylor, Chris Tiu, Paul Turin, Marco Velli, Gary Webb, Phyllis Whittlesey, Ken Wright, S. T. Wu, and Gary Zank. Solar Wind Electrons Alphas and Protons (SWEAP) Investigation: Design of the Solar Wind and Coronal Plasma Instrument Suite for Solar Probe Plus. *Space Science Reviews*, 2015. ISSN 15729672. doi: 10.1007/s11214-015-0206-3.

RJ Leamon, CW Smith, NF Ness, WH Matthaeus, and HK Wong. Observational constraints on the dynamics of the interplanetary magnetic field dissipation range. *... Space Physics (1978 ...*, 103:4775–4787, 1998. URL `http://onlinelibrary.wiley.com/doi/10.1029/97JA03394/full`.

B. A. Maruca. *Instability Driven Limits on Ion Temperature Anisotropy in the Solar Wind: Observations and Linear Vlasov Theory.* PhD thesis, Harvard, 2012.

KW Ogilvie and DJ Chornay. SWE, a comprehensive plasma instrument for the Wind spacecraft. *Space Science ...*, 8:55–77, 1995. URL `http://link.springer.com/article/10.1007/BF00751326`.

Mathew J. Owens and Robert J. Forsyth. The Heliospheric Magnetic Field. *Living Reviews in Solar Physics*, 10, 2013. ISSN 1614-4961. doi: 10.12942/lrsp-2013-5. URL `http://www.livingreviews.org/lrsp-2013-5`.

T. L. Totten, J. W. Freeman, and S. Arya. An empirical determination of the polytropic index for the free-streaming solar wind using Helios 1 data. *Journal of Geophysical Research Atmospheres*, 100(A1):13–17, 1995. ISSN 0148-0227. doi: 10.1029/94JA02420.

FPJ Valero, Jay Herman, and Patrick Minnis. Trianaa Deep Space Earth and Solar Observatory. *NASA background report, ...*, 1999. URL `http://scholar.google.com/scholar?hl=en&btnG=Search&q=intitle:Triana+-+a+Deep+Space+Ear`

Bernard J. Vasquez, Charles W. Smith, Kathleen Hamilton, Benjamin T. MacBride, and Robert J. Leamon. Evaluation of the turbulent energy cascade rates from the upper inertial range in the solar wind at 1 AU. *Journal of Geophysical Research*, 112(A7):A07101, 2007. ISSN 0148-0227. doi: 10.1029/2007JA012305. URL `http://doi.wiley.com/10.1029/2007JA012305`.

M K Verma, D A Roberts, and M L Goldstein. Turbulent heating and temperature in the solar wind plasma. *Journal of Geophysical Research*, 100(A10):839–850, 1995.

P L Whittlesey. *An Estimation of the Solar Probe Cup's Performance & Achievable Science on Solar Probe Plus.* PhD thesis, University of Alabama in Huntsville, 2016.

R. T. Wicks, T. S. Horbury, C. H. K. Chen, and a. a. Schekochihin. Power and spectral index anisotropy of the entire inertial range of turbulence in the fast solar wind. *Monthly Notices of the Royal Astronomical Society*, 5(February):1–5, 2010. doi: 10.1111/j.1745-3933.2010.00898.x. URL `http://arxiv.org/abs/1002.2096` `http://dx.doi.org/10.1111/j.1745-3933.2010.00898.x`.

M Wüest, DS Evans, and R von Steiger. *Calibration of Particle Instruments in Space Physics.* 2007. URL `http://www.issibern.ch/PDF-Files/contents_sr07.pdf`.

Gary P. Zank. *Transport Processes in Space Physics and Astrophysics*, volume 877 of *Lecture Notes in Physics.* Springer New York, New York, NY, 2014. ISBN 978-1-4614-8479-0. doi: 10.1007/978-1-4614-8480-6. URL `http://link.springer.com/10.1007/978-1-4614-8480-6`.

Journal of Physics: Conference Series **767** (2016) 012027 doi:10.1088/1742-6596/767/1/012027

The Heliosphere as Seen in TeV Cosmic Rays

Ming Zhang

Department of Physics and Space Sciences, Florida Institute of Technology, Melbourne, Florida 32901, USA

E-mail: mzhang@fit.edu

Nikolai Pogorelov

Center for Space Plasma and Aeronomic Research and Department of Space Sciences, University of Alabama in Huntsville, Huntsville, Alabama 35899, USA

Abstract. Measurements from several cosmic-ray air shower experiments reveal that the anisotropy of TeV cosmic-ray flux does not agree with a dipole pattern commonly expected from the Compton–Getting effect or from the diffusion of cosmic rays in Galactic magnetic fields. TeV cosmic-ray anisotropy maps often show fine features, some of which are slightly time-dependent. Because the size of the heliosphere is larger than the gyroradius of TeV cosmic rays in the interstellar magnetic field, the electric and magnetic fields of the heliosphere may distort the pattern of cosmic-ray anisotropy that one would see in the local interstellar medium without the presence of the heliosphere. We have developed a method of mapping cosmic-ray anisotropy using Liouville's theorem. In this paper, we show how to use cosmic-ray anisotropy features to determine the direction of the local interstellar magnetic field, the hydrogen deflection plane, the size and shape of the heliotail, and the geometry of the heliosphere bow wave.

1. Introduction

The directional variation (anisotropy) in the intensity of TeV cosmic rays arriving at Earth is routinely measured by a number of air shower experiments, such as Tibet ASγ, IceCube, Super-Kamiokande, Milagro, ARGO-YBG, HAWC and many others [1], [2], [3], [4], [5], [6], [7], [8], [9], [10]. In the last decade, the data quality and event statistics from these experiments have reached a level that can decisively determine the anisotropy and its variation better than 10^{-4}. Surprising features are revealed. At serval TeV energies, although there is an overall large-scale pattern of dipole anisotropy with an amplitude of $\sim 10^{-3}$, small (a few degrees) to medium (tens degrees) scale angular structures are clearly visible. Some small scale features are slightly time-dependent too (e.g., [1]). The small-scale structures, which are often referred to as hot spots [3], [4], become more evident in lower energies of around 1 TeV. The overall dipole orientation and amplitude

Journal of Physics: Conference Series **767** (2016) 012027 doi:10.1088/1742-6596/767/1/012027

do not agree with the Compton–Getting effect from the motion of the heliosphere relative to the local interstellar medium at ~ 25 km/s. What is more puzzling is that the large-scale anisotropy is energy-dependent. At high energies greater than a few hundred TeV, the direction of dipole anisotropy is opposite to that at a few TeV energies ([6], [7], [8]).

A number of authors have tried to interpret TeV cosmic-ray anisotropy measurements using various mechanisms (e.g., [11], [12], [13], [14], [15]). It was also suggested in [16] and [17] that it is the heliosphere that contributes to the observed anisotropy. This is expected because the gyroradius of TeV cosmic rays in an average 3μG interstellar magnetic field is less than the size of the heliosphere, so their trajectories are altered by the heliospheric magnetic and electric fields, resulting in distortion to the pattern in the original interstellar space.

In this paper, we concentrate on using the measurements of TeV cosmic-ray anisotropy to explore the magnetic structures of the heliosphere. After presenting our theory of mapping cosmic-ray anisotropy using Liouville's theorem, we show our calculation results trying to match features in the measured map of cosmic-ray anisotropy obtained by the Tibet ASγ experiment. We identify the features of anisotropy related to our choice of the BV-plane (the plane determined by the magnetic field and velocity vectors in the unperturbed interstellar medium), the plane perpendicular to the local interstellar magnetic field, the heliotail, and the heliospheric bow wave. The bow wave is defined as the region of the interstellar plasma that is affected by the presence of the heliosphere [21]. Note also that our BV-plane is always close to the hydrogen deflection plane (HDP), which is determined by the H-atom velocity vectors in the unperturbed interstellar medium and in the inner heliosphere as determined in the *SOHO* Solar Wind Anisotropy experiment [20].

2. Models

2.1. Liouville mapping of anisotropy

At a fixed energy, the cosmic-ray flux measured by an experiment on Earth is proportional to the particle distribution function in Earth's frame of reference. According to Liouville's theorem, the distribution function is conserved along particle trajectories, and it is also conserved upon Lorentz transformation of reference frames, that is:

$$f(\mathbf{r}_o, \mathbf{p}_o) = f(\mathbf{r}, \mathbf{p}) \tag{1}$$

where $f(\mathbf{r}_o, \mathbf{p}_o)$ is in the Earth frame and $f(\mathbf{r}, \mathbf{p})$ is in the local interstellar medium. Liouville's theorem is valid as far as one can fully determine the particle trajectory.

The distribution function of cosmic rays in the local interstellar medium is not known, although we have a reasonably good knowledge about its energy dependence. This is because the spatial and angular dependence of cosmic-ray distribution in the interstellar medium is sensitive to the history of nearby supernova sources in the past few million years. These dependences are likely to be a function of particle energy, as cosmic-ray interstellar propagation is energy-dependent. Lacking direct measurements, we assume the following distribution function for cosmic rays in the local interstellar medium:

$$f(\mathbf{r}, \mathbf{p}) = f_0 p^{-\gamma} \left[1 + \nabla_\perp \ln f \cdot (\mathbf{R}_g - \mathbf{r}_0) + A_{||1} P_1(\mu) + A_{||2} P_2(\mu)\right] \tag{2}$$

where p is the magnitude of $\mathbf{p}$, $\mathbf{R}_g$ is the location of particle guiding center (one gryoradius vector away from $\mathbf{r}$), $P_1(\mu)$ and $P_2(\mu)$ are the first and second Legendre polynomials of particle pitch angle cosine μ with respect to the magnetic field. Other symbols in Eq (2) are constants: $A_{||1}$ and $A_{||2}$ are the amplitudes of pitch angle anisotropies, $\nabla_\perp \ln f$ is the cosmic-ray density gradient perpendicular to the magnetic field, and $\gamma \approx 4.75$ is the slope of cosmic-ray power-law energy distribution. Since TeV cosmic rays in the interstellar medium are sufficiently magnetized and scattered on the Galactic scale, the distribution function in the interstellar plasma frame of reference is nearly isotropic, so the spatial dependence and pitch-angle dependence is small, thus qualifying it be written in the above perturbation form. The energy dependence gives rise to the Compton–Getting anisotropy due to the transformation of reference frames, and it also includes the effect of particle acceleration during its propagation through the heliosphere. The guiding center dependence contains the b cross gradient anisotropy and the particle drift effect [16]. The pitch-angle dependence is related to cosmic-ray diffusion parallel to the average magnetic field and possible mirroring by the interstellar magnetic field on the Galactic scale. The values of constants $A_{||1}$, $A_{||2}$ and $\nabla_\perp \ln f$ control the contribution from these three types of anisotropy sources. Since we do not know the past history of cosmic rays arriving at Earth, these four constants are free parameters to be determined by fitting cosmic-ray anisotropy measurements with models.

2.2. Heliospheric propagation of cosmic rays

The cosmic-ray trajectory can be calculated using Newton's equation with the Lorentz force. The trajectory can be fully determined once the total magnetic and electric fields are known. We use a multi-fluid MHD heliospheric model implemented in the Multi-Scale Fluid-Kinetic Simulation Suite (MS-FLUKSS) [22, 23]. It addresses the complexity of the charge-exchange processes between solar wind ions and interstellar neutrals and coupling between the interstellar and heliospheric magnetic fields (e.g., [18]). Table 1 lists parameters used to simulate the heliosphere model. Among the outputs of the MHD model are plasma velocity $\mathbf{V}$ and magnetic field $\mathbf{B}$. In ideal MHD, electric field can be calculated with $\mathbf{E} = -\mathbf{V} \times \mathbf{B}$. Note that the fields from MHD solution are average fields, which only reveal large-scale heliospheric structures. In space plasma, there is always a random fast, small-scale field component. For TeV cosmic rays, scattering by magnetic fluctuations is slow, in the order of one scattering every a few years to match their mean free path of a few pc in the interstellar medium, while the particle propagation through the heliosphere of a few thousand AU size is just tens of days. So the transport through the heliosphere is essentially scatter-free in the average fields.

Once the momentum at Earth $\mathbf{p}_o(\mathbf{r}_o)$ is mapped in time backward to the interstellar medium at far enough distance $\mathbf{p}(\mathbf{r})$ for all directions of $\mathbf{p}_o$ viewed from Earth, maps of anisotropy due to all variables in the distribution functions are obtained. According to Eq (2), the composite map is basically a linear combinations of anisotropy maps produced by three anisotropy mechanisms. We do least χ^2 fit to observations of relative cosmic-ray fluxes to determine the free constants in Eq (2).

Journal of Physics: Conference Series **767** (2016) 012027 doi:10.1088/1742-6596/767/1/012027

3. Results

The top panel of Figure 1 shows a map of cosmic-ray relative flux at 5 TeV measured by the Tibet ASγ experiment as a function of declination and right ascension [19]. On the large scale, the cosmic-ray flux shows an enhancement in the tail region and a deficit in the nose region. The middle panel is our model fit with the constants listed in Table 2. In our calculation, Earth is set to be stationary, which corresponds to a yearly average of cosmic-ray anisotropy. A number of features in our model match the observations quite well, except the locations of some boundaries are shifted a little bit due to the uncertainties about the interstellar magnetic field, ion and neutral densities used in our MHD heliosphere simulation. The bottom panel, calculated from Eq (2) without the heliospheric particle propagation, is an inferred map of cosmic-ray anisotropy as would be seen in the local interstellar medium without the presence of the heliosphere. Notice in Table 2, the amplitude of first order pitch-angle anisotropy $A_{||1}$ is much larger than the other two constants, indicating that the interstellar cosmic-ray anisotropy is dominated by diffusion along the magnetic field. $A_{||1}$ is positive, meaning higher cosmic-ray flux at zero pitch angle to the interstellar magnetic field than in the opposite direction. To the zeroth order, the overall dipole pattern in the observation comes from the original interstellar cosmic-ray anisotropy of the bottom panel. The magnetic and electric fields of the heliosphere distort the original interstellar anisotropy map. The amount of distortion can be seen directly by subtracting the interstellar anisotropy from the observation or from our model fit. Figure 2 shows a comparison between them. Although some boundaries are shifted slightly, the patterns in the observed distortion and the model distortion match very well. This suggest that the heliosphere model we have used is quite good in describing the large-scale heliospheric structures.

Let us exam a few features of cosmic-ray anisotropy in Figure 1 and their heliospheric origin. The cosmic-ray flux is enhanced in the tail direction narrowly aligned with the hydrogen deflection plane [20], which coincides with the plane containing the interstellar flow and magnetic field vectors in our heliosphere model. The surface of the heliopause in the tail region is also elongated along the hydrogen deflection plane. In the directions of flux enhancement, cosmic-ray trajectories are focused by the magnetic field in the heliotail or their pitch angles before entering the heliosphere are close to zero degree from the interstellar magnetic field. Slightly outside of the flux enhancement, particularly on the right-hand side, the cosmic-ray flux is dramatically reduced from its original interstellar level (the blue region on the right side of Figure 2). There, the cosmic rays are defocused by the magnetic field of the outer heliosheath. These cosmic rays originally come in with larger original pitch angles. On the other side centered around the right ascension 300°, cosmic-ray flux is enhanced by the heliospheric distortion (the red region on the left side of Figure 2). Thses cosmic rays drift in the direction opposite to the cosmic-ray density gradient when they pass through the outer heliosheath, and they have higher densities originally in the interstellar medium.

The blue region of cosmic-ray flux deficit in Figure 1 is confined roughly around the nose direction. These cosmic rays originally have pitch angles close to 180 degrees to the interstellar magnetic field, in which direction the original interstellar cosmic-ray flux is near the lowest. In our model simulation, the cosmic-ray flux distribution in this flux deficit (blue) region is relatively smooth. This is because these cosmic rays pass through a thinner part of the heliosheath and solar wind, and their trajectories are not very much complicated by the inhomogeneity of the magnetic field due to the short propagation. To the right, the blue region in Figure 1 is approximately bounded by the plane perpendicular to the unperturbed interstellar magnetic field. Inside the blue region, cosmic rays come from the nose side with large pitch angles $> 90°$ and outside they come from the tail direction with small pitch angle $< 90°$. Apparently, the magnetic field in the solar wind and heliosheath along the trajectories of these particles do not deflect the cosmic-ray trajectories very much. Comparing the location of this boundary in our simulation with what is observed by Tibet ASγ, we probably can say that magnetic field direction we used in our heliosphere simulation should be quite close to the true interstellar magnetic field direction. On the left side, the boundary of the blue region in Figure 1 also separates the cosmic rays coming from the nose direction from those coming from the tail direction. However, it is far away from the plane perpendicular to the interstellar magnetic field on that side. This is because these cosmic-ray trajectories are strongly deflected by the heliosheath and heliotail magnetic field. Detailed examination of the trajectories of particles arriving in the directions on the both sides of this boundary (Figure 3) shows that the trajectories are separated by the bow wave in front of the heliosphere. Particle coming from the nose can quickly arrive at Earth, but the particles from the tail direction are severely deflected by the draped interstellar magnetic field in the outer heliosheath and heliospheric magnetic field in the heliotail, and they eventually arrive in the directions close to the nose. The precise location of this boundary in our model calculation is somewhat off from what is observed. Perhaps, this is due to our boundary conditions, which are essentially taken from [24], or maybe our multi-fluid approach does not describe the formation of the bow wave with enough precision. In order to better fit the observations, the bow wave of future models should either move outward or have a larger angle between the bow wave surface and the interstellar flow. In either case, the Mach number (most likely, fast magnetosonic) should be further reduced, which requires that the (magnetosonic) wave speed should be larger than in the current simulation.

In Figure 1, the middle panel of our model calculation demonstrates some spots of cosmic-ray enhancements at high declinations of $> 60°$. The Tibet ASγ observations in Figure 1 top panel do not show the feature. So the enhancements in the model results must be artifacts of our model. If we reduce the size of MHD simulation box by replacing the magnetic field and plasma parameters outside of 2000 AU (in the second circle in Figure 3) with those in the pristine interstellar medium, we find that the artifacts become more obvious (Figure 4). Examination of the trajectories of particles arriving in the direction of the artifacts shows that cosmic rays originally come from the heliotail and the magnetic field at high northern declinations of the heliosheath deflects them southward. From this analysis we conclude that the heliosphere, particularly the heliotail, has to be long, probably more than the maximum of 4400 AU in our MHD simulation.

4. Conclusion and discussion

The heliosphere can distort the map of cosmic-ray anisotropy originated in the local interstellar medium due to the energy, spatial and pitch-angle dependence of the cosmic-ray distribution function. We have used Liouville's theorem to map the interstellar cosmic-ray anisotropy to Earth. The measured map of 5 TeV cosmic-ray anisotropy obtained by the Tibet ASγ experiment suggests that the cosmic-ray anisotropy in the local interstellar medium is dominated by a dipole resulted from cosmic-ray diffusion parallel to the magnetic field, although there contain weaker contributions from the Computon-Getting effect, cosmic-ray density gradient, mirroring by interstellar magnetic field inhomogeneity. We have identified a few features of cosmic-ray anisotropy caused by heliospheric magnetic field structures. These feature can be used to locate the hydrogen deflection plane, the plane perpendicular to interstellar magnetic field, and the heliospheric bow wave.

The measured map of cosmic-ray anisotropy also contains many small-scale variations. Some of them around the ecliptic could be caused the coronal or inner heliospheric magnetic fields. Those far way from the ecliptic must come either for the large-scale interstellar magnetic turbulence or for the heliospheric magnetic field variation in the heliotail region. It is possible we can use these small-scale variations to probe the properties of these magnetic field fluctuations. The original source of all anisotropies is the variation of cosmic-ray distribution function in the pristine interstellar medium. Only when the original interstellar anisotropy is fully figured out can we uniquely decode out the properties of magnetic field fluctuations.

4.1. Acknowledgments

This work was supported in part by NSF Grant AGS-1156056, NASA Grants NNX14AJ53G, NNX15AN72G, and NNX15AB76G, and DOE Grant DE-SC0008721. This work was also partially supported by the IBEX mission as a part of NASA's Explorer program. We acknowledge NSF PRAC award OCI-1144120 and related computer resources from the Blue Waters sustained-petascale computing project. Supercomputer time allocations were also provided on SGI Pleiades by NASA High-End Computing Program award SMD-15-5860 and on Stampede by NSF XSEDE project MCA07S033.

5. References

[1] Amenomori, M., Ayabe, S., Bi, X. J., et al. 2006, Science, 314, 439
[2] Amenomori, M., Ayabe, S., Bi, X. J., et al. 2010, Astrophys. J., 711,119
[3] Abdo, A. A., Allen, B. T., Aune, T., et al. 2008, Phys. Rev. Lett. 101, 221101
[4] Abdo, A. A., Allen, B. T., Aune, T., et al. 2009, ApJ, 698, 2121
[5] Guillian, G., Hosaka, J., Ishihara, K., et al. 2007, Phys. Rev. D, 75, 062003
[6] Abbasi, R., Abdou, Y., Abu-Zayyad, T., et al. 2010, Astrophys. J., 718, L194
[7] Abbasi, R., Abdou, Y., Abu-Zayyad, T., et al. 2011, Astrophys. J., 740, 16
[8] Abbasi, R., Abdou, Y., Abu-Zayyad, T., et al. 2012, Astrophys. J., 746, 33
[9] Di Sciascio, G., Iuppa, R, Argo-Ybj Collaboration, 2012, J. Phys.: Conf. Ser., 375, 052008
[10] Abeysekara, A. U., Alfaro, R., Alvarez, C., et al. 2014, Astrophys. J., 706, 108
[11] Drury, L. OC. and Aharonian, F. A., 2008, Astroparticle Physics 29, 420.
[12] Lazarian, A., and Desiati, P., 2010, Astrophys. J., 722, 188
[13] Giacinti, G., and G. Sigl, 2012, Phys. Rev. Lett., 109, 071101.
[14] Desiati, P., & A. Lazarian 2013, Astrophys. J., 762, 44.
[15] K. Kotera, M. Perez-Garcia and J. Silk, arXiv:1303.1186
[16] Zhang, M., Zuo, P., Pogorelov, N., 2014, Astrophys. J., 790, 5.
[17] Schwadron, N.A., et al. 2014, Science, 1245026.
[18] Pogorelov, N., et al., 2015, Astrophys. J. Lett., 812, L6.
[19] Amenomori, M., et al. 2011, Proc. 32nd ICRC, 1, 62.

Journal of Physics: Conference Series **767** (2016) 012027 doi:10.1088/1742-6596/767/1/012027

[20] Lallement, R. et al., 2005, Science, 307, 1447.
[21] Zank, G. P., Heerikhuisen, J., Pogorelov, N.V., Burrows, R., McComas, D., 2010, Astrophys. J., 708, 1092
[22] Pogorelov, N. V., Borovikov, S. N., Bedford, M. C., Heerikhuisen, J., Kim, T. K.,Kryukov, I. A., Zank, G. P. 2013, in ASP Conf. Proc. Ser. 474, Numerical Modeling of Space Plasma Flows (ASTRONUM-2012), eds. N.V. Pogorelov, E. Audit, and G. P. Zank, 165 (Astron. Soc. Pacific, San Francisco)
[23] Pogorelov, N. V., Borovikov, S. N., Heerikhuisen, J., Kim, T. K., Kryukov, I. A., Zank, G. P. 2014, in XSEDE'14: Proceedings of the 2014 Annual Conference on Extreme Science and Engineering Discovery Environment (ACM, New York, NY, USA)
[24] Zirnstein, E. J., Heerikhuisen, J., Funsten, H. O., Livadiotis, G., McComas, D. J., Pogorelov, N. V. 2016, Astrophys. J. Lett., 818, L18

Table 1. Solar and interstellar plasma and magnetic field parameters used in multi-fluid MHD heliosphere simulation

Parameter	Value
Interstellar magnetic field strength B_{ism}	3 μG
Interstellar magnetic field direction	see Figure 1
Interstellar flow V_{ism}	26.4 km/s
Interstellar flow direction	see Figure 1
Interstellar ion density n_p	0.06 cm^{-3}
Interstellar neutral hydrogen density n_H	0.18 cm^{-3}
Fast solar wind speed V_f	762 km/s
Fast solar wind density at 1 AU n_f	2.4 cm^{-3}
Slow solar wind speed V_s	450 km/s
Slow solar wind density at 1 AU n_s	6.9 cm^{-3}
Heliospheric magnetic field at 1 AU $B_{1\mathrm{AU}}$	47 μG

Table 2. Derived interstellar parameters for the distribution function of 5 TeV cosmic rays

Parameter	$A_{\|\|1}$	$A_{\|\|2}$	$\|\nabla_\perp \ln f\|$	Direction of $\nabla_\perp \ln f$
Value	0.19%	0.037%	0.028%/R$_g$(370 AU)	see Figure 1

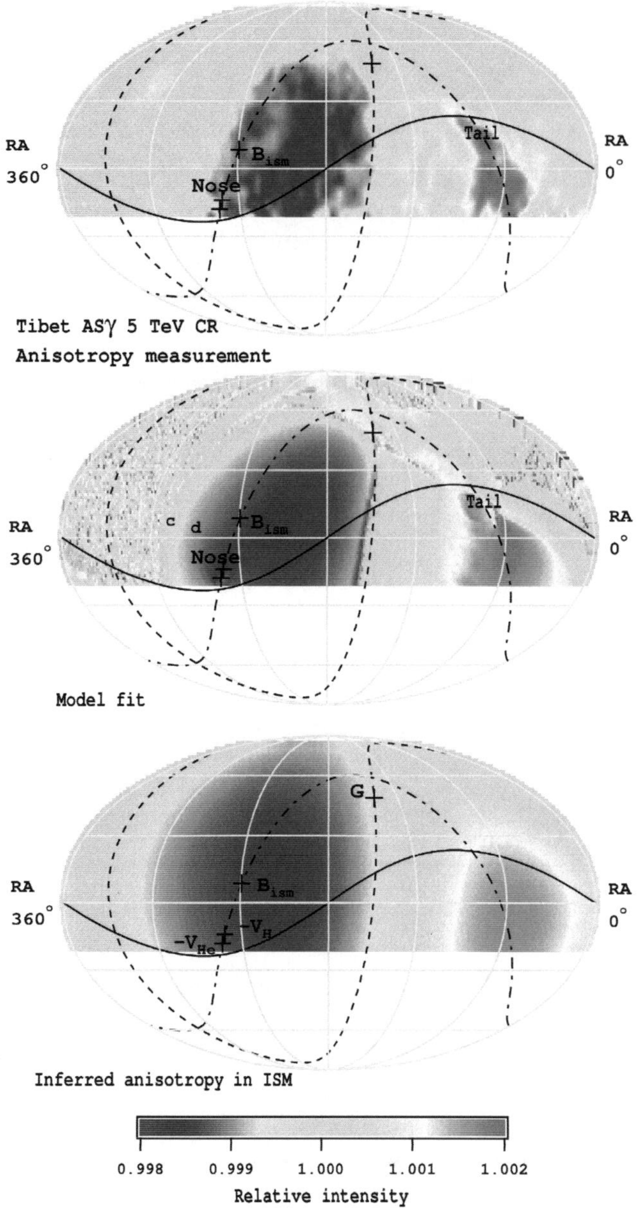

Figure 1. (Top) Tibet ASγ measurements of 5 TeV cosmic-ray relative flux as a function of declination and right ascension. (Middle) least χ^2 fit to Tibet measurements using Liouville mapping of anisotropy and the distribution function in Eq (2). (Bottom) inferred map of cosmic-ray anisotropy in the pristine interstellar anisotropy without the distortion by the heliosphere. The "+" signs indicate the direction of interstellar magnetic field $\mathbf{B}_{ism}$, the directions of interstellar neutral helium ($-\mathbf{V}_{He}$) and hydrogen ($-\mathbf{V}_H$) inflows, and the direction of perpendicular cosmic-ray density gradient G. The dot-dashed curve is the hydrogen deflection plane. The dashed curve is the plane perpendicular to the interstellar magnetic field $\mathbf{B}_{ism}$. The solid curve is the ecliptic. The sun is placed at the nose.

Journal of Physics: Conference Series **767** (2016) 012027 doi:10.1088/1742-6596/767/1/012027

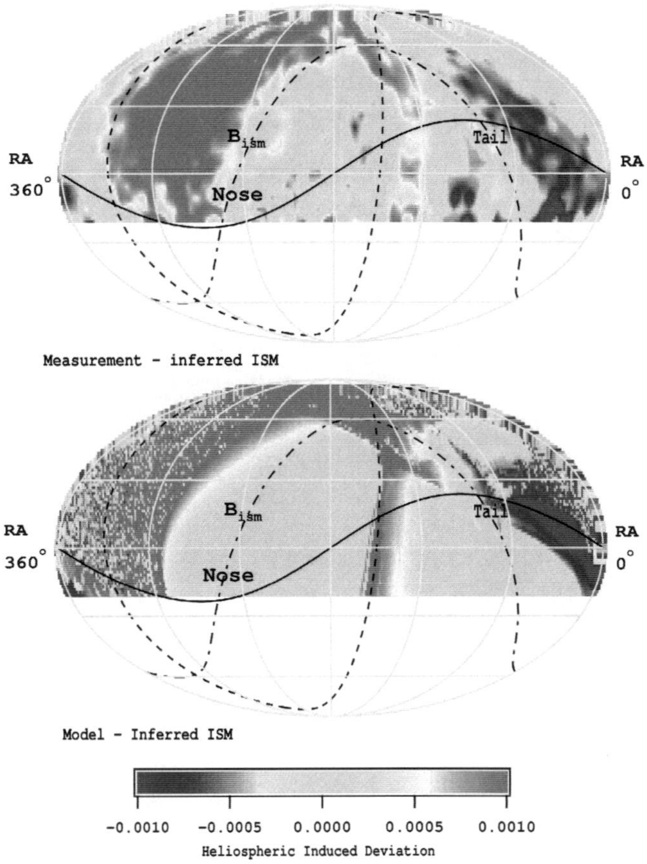

Figure 2. Distortion of cosmic-ray anisotropy caused by the heliosphere. The top is a map of Tibet ASγ measurements subtracted by the inferred interstellar anisotropy and the bottom is our model fit subtracted by the inferred interstellar anisotropy.

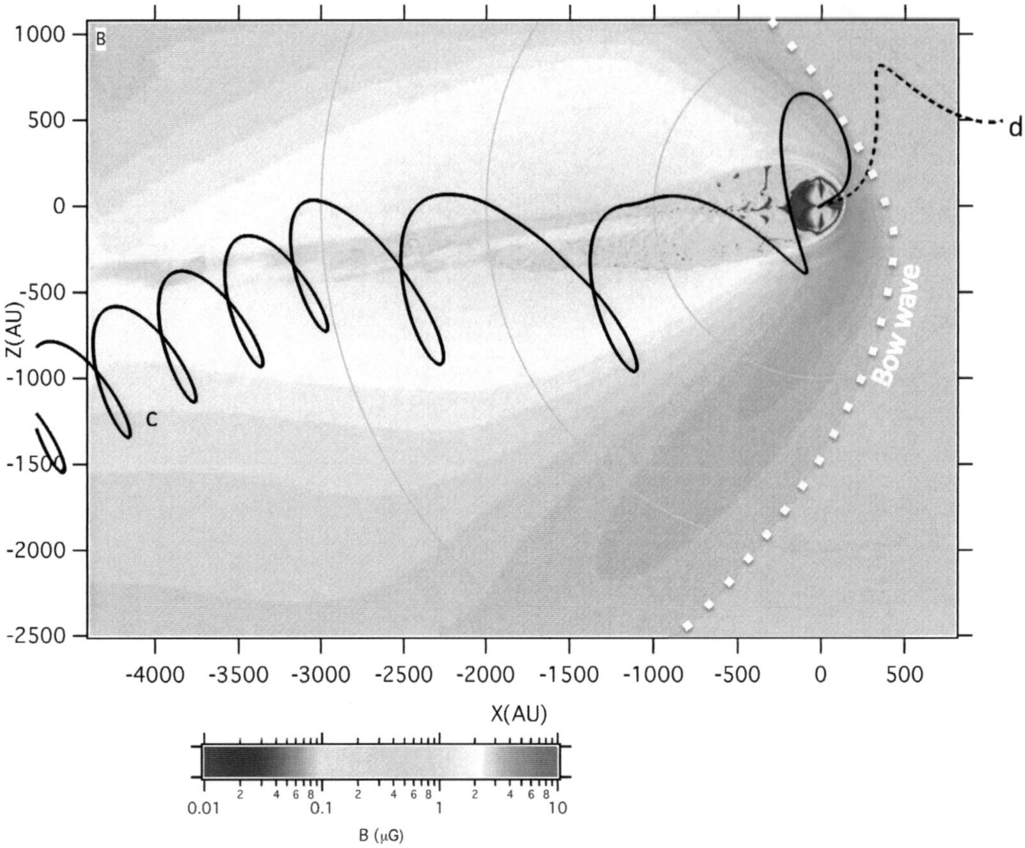

Figure 3. Magnetic field strength in the meridional plane. The two curves are the projections of sample cosmic-ray trajectories arriving in the directions "c" and "d" labelled in Figure 1 middle panel. The white dotted curve shows the location of heliosphere bow wave.

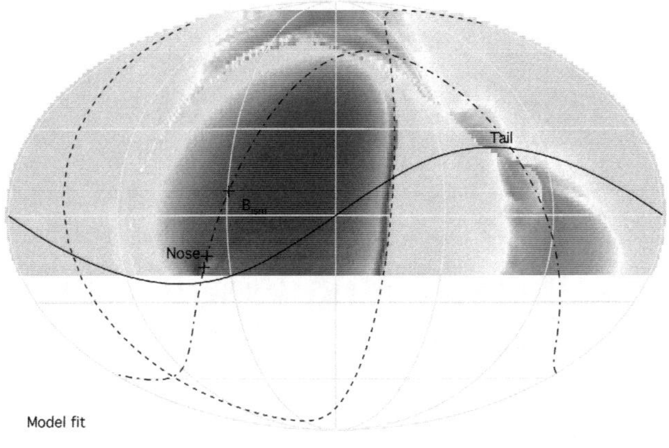

Figure 4. Model calculation of cosmic-ray anisotropy if the heliosphere is truncated and filled with the pristine interstellar magnetic field and flow beyond 2000 AU

Observations and analysis of small-scale magnetic flux ropes in the solar wind

Jinlei Zheng[1], Qiang Hu[1,2]

[1]Department of Space Science, University of Alabama in Huntsville, USA
[2]Center for Space Plasma and Aeronomic Research, University of Alabama in Huntsville, USA

E-mail: [1]jz0006@uah.edu

E-mail: [2]qh0001@uah.edu

Abstract. The small-scale magnetic flux ropes (of duration ranging from a few minutes to a few hours) in the solar wind have the typical topology of winding field lines around a central axis, which is similar to the large-scale flux ropes, i.e., magnetic clouds. However, accumulating evidence suggests that their plasma characteristics, origin, formation mechanism and evolution are different from those of large-scale flux ropes. The small-scale flux ropes are intensively studied in recent years, since they affect particle transport and energization, and are considered as the potential source of local acceleration. The Grad-Shafranov reconstruction technique is a tool to reconstruct the two and a half dimensional field structure based on in-situ measurements captured by an observing platform moving past it. In this study, we reconstruct the flux rope structures in two events using the Grad-Shafranov reconstruction approach. In one event, a twin flux rope structure at 1 AU occurring on 2002 February 1 and two following single flux rope structures are identified behind an interplanetary shock. In the other event, we reconstruct the flux rope structures occurring on 1998 March 25 and 26 at 1 AU in the ambient solar wind. The associated energetic particle signatures and the possible origin of these flux rope structures are discussed.

1. Introduction

The magnetic flux ropes in the solar wind are an important type of transient and coherent structures in the inner heliosphere. Their formation, propagation and evolution shed light on the big picture of the space plasma dynamic processes. The magnetic field of a flux rope has the typical topology of winding field lines around a central axis with distributed currents. According to the scale size of the flux ropes in the solar wind, currently they are empirically classified into two categories: large-scale flux ropes and small-scale flux ropes. In addition to the difference in scale size, they also differ, sometimes, in terms of origination, formation mechanism, plasma profile and evolution process.

The large-scale flux ropes are also called magnetic clouds (MCs). The first magnetic cloud was identified by Burlaga et al. [1] in the solar wind between the heliocentric distances 1 AU and 2 AU. The spacecraft traversed a region with a radial dimension of 0.5 AU (0.25AU in radius), in which the measured magnetic field vectors rotated nearly parallel to a plane. In this region, the magnetic pressure dominated and the total pressure was higher than that of outside. Magnetic clouds are well defined observationally as a magnetic structure possessing these bulk properties based on in-situ magnetic and plasma measurements: enhanced magnetic field strength, smooth rotation in magnetic field direction, and low proton temperature. They have been intensively studied in the past decades. It is widely accepted that the magnetic clouds have the following characteristics: (1) The topology of magnetic field is that of a helical structure, in which the field lines are winding around a central axis, forming the helix shape elongated along the axial dimension. As a spacecraft passes through the helical field lines of a flux rope, the in-situ data of magnetic field vectors show a smooth and continuous rotation. (2) The magnetic field in a magnetic cloud is stronger than the surrounding solar wind, and the magnetic pressure usually dominates. (3) The proton temperature is depressed within the magnetic cloud, resulting in ultra-low proton β value, the ratio between the proton plasma and magnetic pressure. Besides these key properties mentioned above, magnetic clouds have a typical radial scale size reaching a few tenths of an AU at 1 AU [2]. A limited number of studies also examined their radial evolution between 0.3 and 4.2 AU [3].

Observational evidence suggested that the magnetic clouds have a close relation to the coronal mass ejections (CMEs) [4, 5, 6, 7, 8]. The CMEs interact with ambient solar wind when they propagate into the heliosphere. Because the CMEs generally move faster than the ambient solar wind, they form the interplanetary coronal mass ejections (ICMEs), often driving a shock wave. At least a portion of ICMEs has the properties of magnetic clouds as crossed by a single spacecraft [9]. The fraction of MCs among ICMEs has a solar cycle phase dependence [10]. From the studies of 56 magnetic clouds at 0.7AU during 1979-1988, Mulligan and Russell [11] showed that the leading magnetic field in magnetic clouds was controlled by the solar global field. The studies by Lynch et al. [12] suggested that the net accumulative helicity of most magnetic clouds appeared to be modulated by solar activity cycle. The radial scale size of magnetic clouds expands as they move outward from the sun [1, 13, 14, 5, 15, 16]. Besides the radial expansion, a significant poloidal motion was also found in some magnetic clouds [16]. Magnetic flux ropes are also found in the Earth's magnetotail [17] and in other planetary environment. Some non-solar source magnetic flux ropes appear to be created by magnetic reconnection in local plasma environment, such as those by reconnection across the plasma sheet in magnetotail. Some small-scale flux ropes in the solar wind may share the same generation mechanism.

Past research usually employs a cylindrically symmetric, constant-alpha, linear force-free magnetic field model to fit the one-dimensional spacecraft data across a magnetic cloud [18, 19]. A simple solution of such a model was found by [20], in which the axial and transverse components are proportional to the zeroth-order and first-order Bessel functions of the first kind, respectively. Other similar 1D models also exist. However, these models are limited to a pre-defined 1D cross-section geometry and force-free condition. Alternatively, Hu and Sonnerup [21, 22] applied the Grad-Shafranov(GS) reconstruction technique [23, 24] to the in situ data collected when the spacecraft crosses the flux rope structures. In the GS reconstruction, axial symmetric and force-free assumptions are abandoned. The reconstruction results showed the obvious distortion from the axial symmetric geometry. The multiple, non-symmetric flux rope structures were reconstructed by Hu et al. [25] using the GS reconstruction technique. These structures had the double or triple flux rope configurations, in which one flux rope was located immediately next to the other. The magnetic X point/line between adjacent flux ropes implied that they may be relics of the end product of a dynamic evolution process due to magnetic reconnection.

The small-scale flux ropes have the magnetic filed geometry similar to large-scale flux ropes, i.e. MCs. However, they have much smaller scale sizes ranging from $\sim$0.001 to 0.01 AU, and the identification of small-scale flux ropes is less definitive. Unlike MCs which have a clear association with CMEs, originating from solar corona, the origin of small-scale flux ropes is still debatable. Moldwin et al. [26] reported a non-CME flux rope at 5 AU. The scale size of this flux rope is about 0.05 AU, and the plasma profile is different from that of magnetic clouds. They interpreted this flux rope as a result of multiple magnetic reconnection at the heliopspheric current sheet (HCS). The concept of small-scale flux ropes was first introduced by Moldwin et al. [27]. They observed some flux ropes with smaller sizes compared with MCs at 1 AU, while the former had durations of tens of minutes as opposed to a dozen hours to days of MCs. They suggested that these small-scale flux ropes were products of magnetic reconnection in the local solar wind, instead of in the solar corona.

In recent years, small-scale flux ropes have attracted interest of researchers on its different characteristics compared with MCs. Feng et al. [28] identified 144 magnetic flux ropes with radial scale sizes from about 0.004 to 0.6 AU (including both large and small scale flux ropes). Based on the similarity of their energy distributions to solar flares, they suggested that the small-scale and intermediate-scale flux ropes were the interplanetary manifestations of some small CME events. Cartwright and Moldwin [29] identified 68 flux ropes with strong core field using WIND magnetic field data. They excluded the possible mixture of Alfvén waves in the identification. The duration ranged from 39 minutes to 12.2 hours. The comparison showed that these two kinds of flux ropes differed in terms of expansion, whether or not in force-balance, the strengths of core fields, and the association with forward shocks, which implied that they may have different origins. Feng et al. [30] surveyed 125 small-scale and intermediate-scale flux ropes. They suggested that the small-scale and intermediate-scale flux ropes originated from solar eruptions, similar to MCs, since they had similar characteristics. Cartwright et al. [31] did a comprehensive survey of small-scale flux ropes located between 0.3 and 5.5 AU. They found that the occurrence of small-scale flux ropes had a solar cycle dependency, which implies that they may originated from the sun. However, they also found that the lack of depressing proton temperature and the location being near the sector boundary crossings implied that they may be created locally due to magnetic reconnection across the heliospheric current sheet. More recently, Feng et al. [32] examined the counterstreaming suprathermal electron (CSE) signatures of 106 small-scale flux ropes measured by WIND from 1995-2005. They found that the CSE appeared in most small-scale flux ropes located far from the heliospheric current sheets(HCSs), while the CSE appeared in only half of the small-scale flux ropes located near the heliospheric current sheets. A limited number of studies also looked at the corresponding elemental composition and charge state signatures within small-scale flux rope intervals [32]. By using hourly data, those authors found from a subset of small-scale flux rope events identified between 1995 and 2007 that these structures share some common features with MCs. They provided supporting evidence for the origin of small-scale flux ropes from solar corona.

As discussed above, our understanding on small-scale flux ropes is far from comprehensive. Their formation mechanism and origination are still under debate. And the hypotheses of evolution and propagation process are based on insufficient database. The coverage of the latest database is before 2007. The energetic particle profiles related to small-scale flux ropes are less studied. Theoretical studies suggested that small-scale flux rope activities such as contracting, merging (reconnection) can lead to particle acceleration [33, 34]. The observational evidence of the existence of such flux ropes, supported by sound analysis via the GS method, remains to be discovered. The most used model of small-scale flux rope is static constant-alpha force-free cylindrically symmetric field model [18, 19]. This model restricts the fitted small-scale flux ropes to a simple geometry of circular cross section, omitting plasma pressure gradient, which is not consistent with observations. The Grad-Shafranov reconstruction model [21, 22, 25] is able to recover the non-symmetric geometry of the small-scale flux rope in a more general non-force free configuration. So far, it has not been systematically applied to examine the magnetic field configuration of small-scale flux ropes in the solar wind. Furthermore, Sonnerup et al. [35] have developed a basic approach for 2-D time-stationary, ideal, and compressible MHD reconstruction technique that can account for significant remaining plasma flow within the structure. This technique can be implemented in small-scale flux rope reconstructions to recover both the magnetic field and plasma flow structures in dynamic equilibrium.

In this study, we apply the GS reconstruction method to reconstruct a handful of small-scale flux rope structures under different background solar wind conditions. One set of cases is found behind an interplanetary traveling shock and the other set of cases is in the ambient, ordinary solar wind, not associated with an apparent solar source, such as a CME, or a high-speed stream.

2. Grad-Shafranov Reconstruction Technique

The Grad-Shafranov (GS) reconstruction technique is a tool for the reconstruction of $2\frac{1}{2}$-D coherent magnetic field structure from data collected as the structure moves past an observing platform. It was first developed and applied to magnetopause crossings [23, 36, 37]. Hu et al. further developed this technique and applied it to magnetic flux ropes in the solar wind [21, 22, 25]. The GS reconstruction technique is based on the more general non-force-free assumption, and is able to recover the distortion of a flux rope from axial symmetric geometry.

In GS reconstruction technique, the magnetohydrostatic equilibrium assumption is adopted to describe the magnetic field structures. Such a configuration is in static force balance between the plasma pressure gradient and the Lorentz force,

$$\nabla p = \mathbf{j} \times \mathbf{B}. \tag{1}$$

There are two other conditions satisfied by the magnetic field which are $\nabla \cdot \mathbf{B} = 0$ and $\nabla \times \mathbf{B} = \mu_0 \mathbf{j}$. In a cylindrical flux rope, the variation along axial direction is much more gradual than that in the transverse direction. So we can assume that the magnetic field and plasma equilibrium has the translation symmetry along the axial direction (the z axis), which is perpendicular to the transverse plane (the cross-sectional $x-y$ plane). With this assumption, the derivative with respect to z is zero ($\partial/\partial z = 0$). The reconstruction is carried out in a co-moving frame of reference, in which the electric field $\mathbf{E}'$ vanishes. Usually the deHoffmann-Teller (HT) frame is used [38, 39, 40]. With $\mathbf{E}' = 0$, we have $\nabla \times \mathbf{E}' = -\frac{\partial \mathbf{B}}{\partial t'} = 0$, indicating a stationary magnetic field structure. Making use of the magnetic flux function $A(x, y)$ such that the magnetic field vector is

$$\mathbf{B} = [\partial A/\partial y, -\partial A/\partial x, B_z(A)], \tag{2}$$

one can reduce the Equation (1) to a scalar equation

$$\frac{\partial^2 A}{\partial x^2} + \frac{\partial^2 A}{\partial y^2} = -\mu_0 \frac{d}{dA}(p + B_z^2/2\mu_0) = -\mu_0 \frac{dP_t}{dA}. \tag{3}$$

The above equation (3) is the so-called Grad-Shafranov(GS) equation. By solving it numerically [36], the reconstructed map of $A(x, y)$ is obtained. Then the magnetic field $\mathbf{B}$ is acquired by taking derivatives of $A(x, y)$ and a functional fitting, $B_z(A)$.

In practice, we follow the basic reconstruction steps described in the paper by Hu and Sonnerup [22]. The first step is to find a co-moving frame of reference, which is usually the deHoffmann-Teller (HT) frame [38, 39, 40]. This is achieved by minimizing the residual electric field of a frame moving with an unknown velocity. In the ideal case, the residual electric field can be transformed to zero. Then the minimum-variance analysis is applied to find three eigenvalues (the minimum, intermediate, and maximum covariances) and the corresponding eigenvectors. We take the direction along the intermediate eigenvector as the initial trial z axis. A trial-and-error process is carried out to find the optimal z axis orientation, based on the requirement of $P_t(x,0) \sim A(x,0)$ being single-valued. The spacecraft data are projected onto the local reconstruction coordinate (x,y,z), in which the projected spacecraft path is along the positive x direction on the plane perpendicular to z, and the y axis completes the right-handed orthogonal coordinate system. Next we integrate the measured $B_y(x,0)$ to obtain the value $A(x,0)$. By fitting the data pairs $P_t(x,0)$ and $A(x,0)$, we obtain an analytic function for the transverse pressure $P_t(A)$. Using the known function $P_t(A)$ and initial values given by spacecraft measurements along $y=0$, we solve the GS equation [36] to obtain the solution of the flux function $A(x,y)$ in a rectangular domain on the cross section perpendicular to z, which fully characterizes the $2\frac{1}{2}$-D magnetic field configuration of the flux rope.

3. Case Studies

In this section, we present two sets of case studies of small-scale magnetic flux ropes selected from WIND spacecraft data. Cases 1, 2 and 3 belong to an event on Feb. 1st, 2002, all of which are behind an interplanetary shock. Case 4 and 5 are on Mar. 25th, 1998, both of which are in ambient solar wind with no apparent and direct solar source correspondence. Below we present relevant and detailed observations and GS reconstruction results of these cases.

3.1. Flux ropes behind an interplanetary shock

Figure 1 shows that three small-scale flux ropes were identified behind an interplanetary shock. The shock interface was observed on 2002 Jan. 31, 21:38 UT. It is denoted by a vertical dashed line. We can see that the magnetic field magnitude jumps at the shock interface. All magnetic field components change accordingly. Three small-scale flux rope intervals were marked based on mainly magnetic field measurements and the GS reconstruction results to be presented. The magnetic field magnitude kept at a high level (about 12 nT) downstream of the shock, together with elevated solar wind flow speed. The second panel shows that the solar wind speed was enhanced from about 280 km/s to about 360 km/s downstream across the shock interface. These were not typically high speed. The third panel shows that the proton density (in blue color) and the alpha/proton number density ratio changed drastically in the downstream of the shock. It is generally believed that higher alpha/proton ratio implies a coronal source of the plasma. In the last panel, we can see that the proton temperature increased in case 1 and case 2. In case 3, the proton temperature decreased but was still above that in ambient solar wind. This temperature characteristic is different from that of a magnetic cloud, in which the proton temperature is depressed. The average plasma beta values during the time intervals of case 1 and case 3 were 0.09 and 0.14, respectively. These values were close to the typical plasma beta value in MCs ($\lesssim 0.1$). The average plasma beta value during the time interval of case 2 was 0.33, which was above that of MCs.

We also examined the energetic particle profile associated with these flux rope structures. Figure 2 shows the energetic ion dispersion observed by ACE ULEIS instrument. The date range of the top panel is from 2002, Jan. 27 to the end of Jan. 30, which is several days before the flux rope structures were observed. Three flux rope intervals are shown in the bottom panel, which are denoted by the shaded regions in sequence. The dots in this figure represent the energetic ion events, and the colors of dots denote the counts in each energy bin. The inverse slope of each slanted line denotes a nominal length ~1.2 AU traversed by a particle. If the bulk of energetic

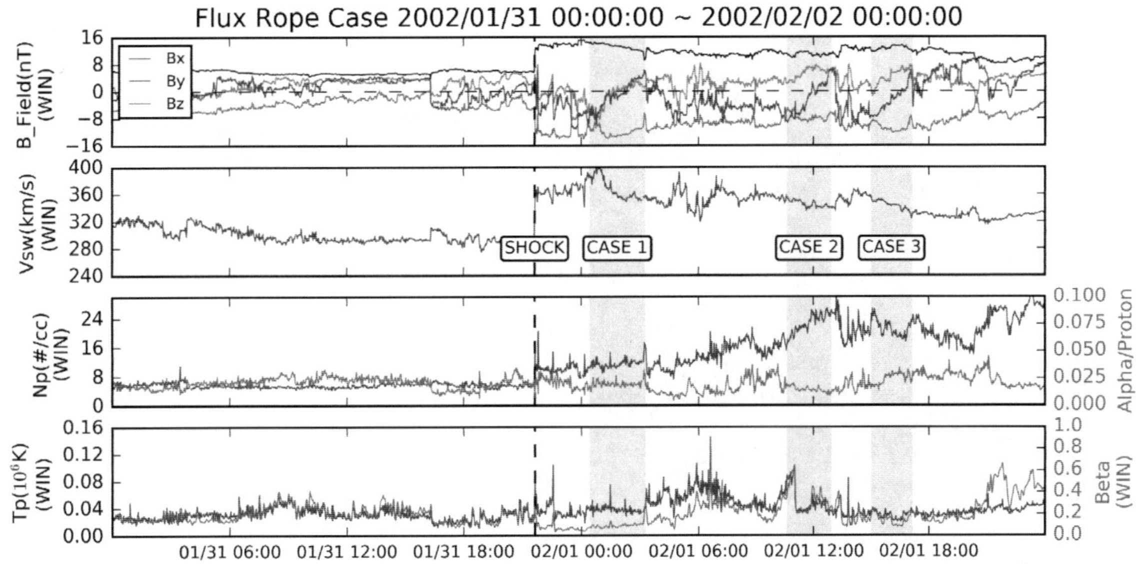

Figure 1. WIND spacecraft data from 2002 Jan. 31, 00:00 UT to 2002 Feb. 2, 00:00 UT. This time interval contains one interplanetary shock and three flux rope structures. The first panel is the magnetic field data from MFI. The black solid line denotes the total magnetic field magnitude. The second panel is the solar wind bulk speed from SWE. In the third panel, the proton number density is shown in blue line and the alpha/proton number density ratio is shown in green line. These two sets of data are from SWE. The fourth panel are the proton temperature (in blue line) and plasma $\beta = N_p k T_p/(B^2/2\mu_0)$ (in green line). The contribution of electrons to plasma β was not included, because the electron temperature was not available. The shock is denoted by a vertical black dashed line. The time of this shock is 2002 Jan. 31, 21:38 UT. The shaded regions represent three small-scale flux ropes, which are labeled by CASE 1, CASE 2 and CASE 3, respectively.

particles was created at the same time and location near the Sun, and if they travel to 1 AU nearly scatter-free, the representing dots will align along the line. The triangular distribution of the enhanced ion population between Jan. 28 and Jan. 29 probably implies significant scattering and other effects on the energetic ions propagating en route from the source to 1 AU within one day. After checking the solar activities, we found there was a halo CME occurred at 2002 Jan. 27, 12:30:05 UT with a speed of 1135 km/s. This was probably the source of those energetic particles. In the bottom panel of Figure 2, the shaded regions represent the case 1, case 2, and case 3 in sequence. The vertical thick black line denotes the location of shock interface. The particle counts were enhanced immediately behind the shock interface, with no clear pattern, which means these particles were created locally, probably energized by the shock. Furthermore, behind the shock, there were three obvious counts enhancement regions. They appear to coincide, partially, with the time intervals of the first two flux ropes identified. Next we are going to present the reconstruction results of case 1, case 2, and case 3 in details.

3.1.1. CASE 1: WIND 2002/2/1 0:25 - 3:15 UT The reconstruction result of case 1 is shown in Figure 3(a). We can see a twin flux rope structure from Figure 3(a). This kind of structure

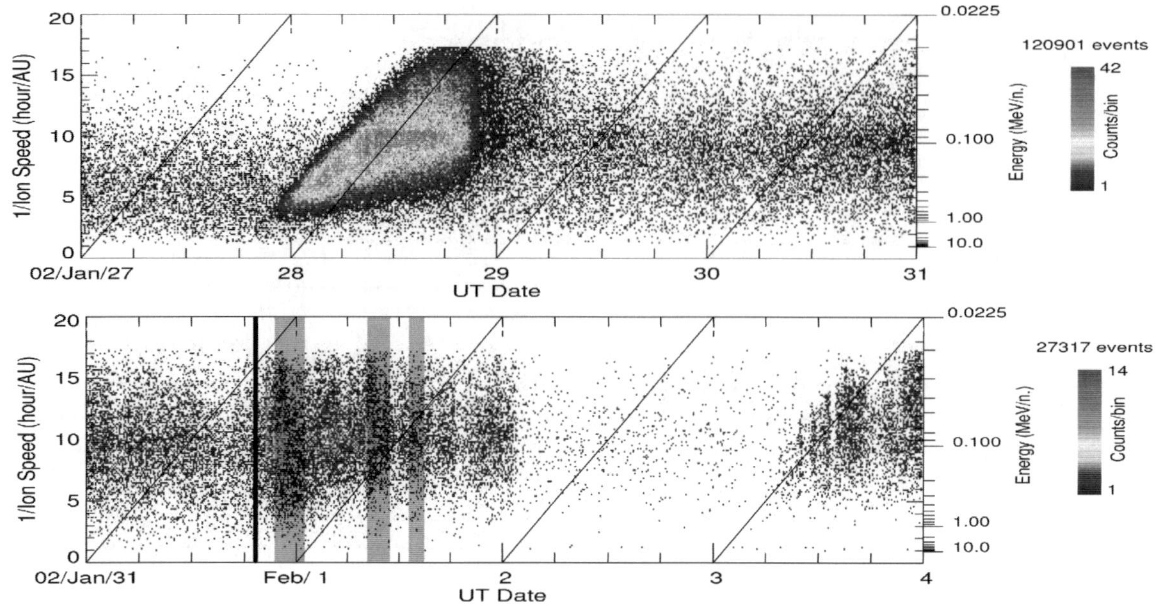

(Courtesy of ACE Science Center.)

Figure 2. The energetic particles profile for Case 1, 2 and 3, which are denoted by three shaded regions in sequence. The data are obtained from ACE ULEIS instrument. The energy range is from 0.0225 MeV to 10 MeV, and the mass range is from 10 to 70 AMU, which includes the Fe, Ca, S, Si, Mg, Ne, O, N, and C elements. The color denotes counts/bin. The left Y axis is the inverse ion speed represented by hour/AU, and the right Y axis is the corresponding energy in MeV/n.

was reconstructed and studied earlier by Hu et al. [25], but for large-scale MCs. The core fields of these two adjacent flux ropes are about 13 nT. The time duration of the entire event was 2 hours and 50 minutes. The flux rope on the left had a scale size of about 0.0075 AU, and the right one had a scale size of about 0.01 AU. From the white arrow we can see that both of them had the right chirality. An X point may be formed by magnetic reconnection yielding this geometry. Figure 3(b) is the fitting curve of $P_t(A)$, which is the transverse plasma pressure, the sum of plasma and axial magnetic pressure. According to the GS reconstruction theory, $P_t(A)$ must be a single valued function of magnetic flux function A. The R_f in the graph is the fitting residue. It is the normalized deviation of the measured data from the fitted curve, which was defined in the paper by Hu et al. [25]. In this case, the residue is 0.15. We consider the fitting with a residue below 0.3 as a reliable fitting. The last panel of Figure 1 shows that the proton temperature was not depressed during the time interval of case 1, which is different from that of MCs.

3.1.2. CASE 2: WIND 2002/2/1 10:36 - 12:55 UT The reconstruction result of case 2 is shown in Figure 4. The time duration was 2 hours and 19 minutes, and the scale size was about 0.01 AU. The core field of this flux rope was about 11nT. The chirality was right-handed, which is the same as that in case 1. This reconstruction has a fitting residue of 0.08 (see Figure 4(b)), which is a very small value, indicating a trustworthy reconstruction. From the third panel of

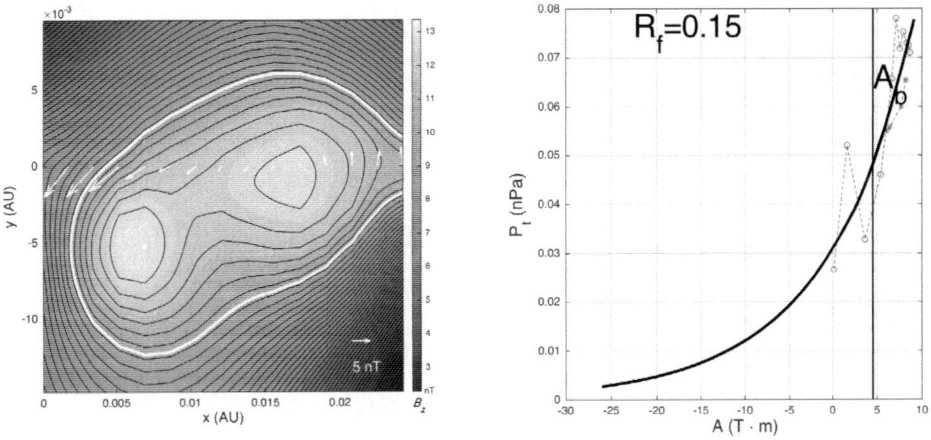

(a) $\hat{\mathbf{z}} = [0.14, -0.99, -0.0089]$ (b)

Figure 3. GS reconstruction result for Case 1 (2002 Feb. 1, 0:25 - 3:15 UT). (a) The cross-section magnetic field configuration of the flux rope. The unit vector $\hat{z}$ represents the z axis orientation of the flux rope in GSE coordinate. The filled color contour represents the axial magnetic field B_z and the overlapped black contour lines represent the transverse magnetic field. The white dot in the center is the location of maximum axial filed B_{zmax}. The white arrows along $y = 0$ are measured transverse magnetic field vectors along the spacecraft path. The white contour line encloses the region $A > A_b$, as denoted in (b). The reconstruction map within this line is reconstructed by double-folded satellite data, and the region beyond this line is reconstructed based on the extrapolated data. (b) The fitting curve (bold solid line) of $P_t(A)$ for Case 1. The fitting is based on the data points denoted by blue circles and red stars, where the circles represent the data along the first half of the spacecraft path and the stars represent those of the second half. The thin vertical solid line denoted by A_b marks the boundary corresponding to the white enclosure line in (a).

Figure 1, we can see that the proton number density was much higher than that of case 1. But the alpha/proton ratio was similar to that of case 1, indicating that the number density of both protons and alpha particles enhanced. The proton temperature was enhanced at the beginning of case 2, then dropped to the same level as that of case 1.

3.1.3. CASE 3: WIND 2002/2/1 14:57 - 17:03 UT The reconstruction result of case 3 is shown in Figure 5. Figure 5(b) shows the fitting curve of function $P_t(A)$ with a low fitting residue 0.07. From Figure 5(a), we can see that the satellite crossed the cross-section of this flux rope near the central axis. The duration of this flux rope was 2 hour and 6 minutes, and its scale size was about 0.015 AU. It had the same right-handed chirality as that of case 1 and case 2. Figure 1 shows that case 3 also had the high proton number density. The average plasma beta value was 0.14, which is close to the typical value in MCs.

From the three cases above, we find that the multiple small-scale flux ropes exist in downstream of a shock. None of them showed the depressed proton temperature. Two cases (case 1 and case 3) had the similar plasma beta values to that of MCs, and one case (case 2) had the

Journal of Physics: Conference Series **767** (2016) 012028 doi:10.1088/1742-6596/767/1/012028

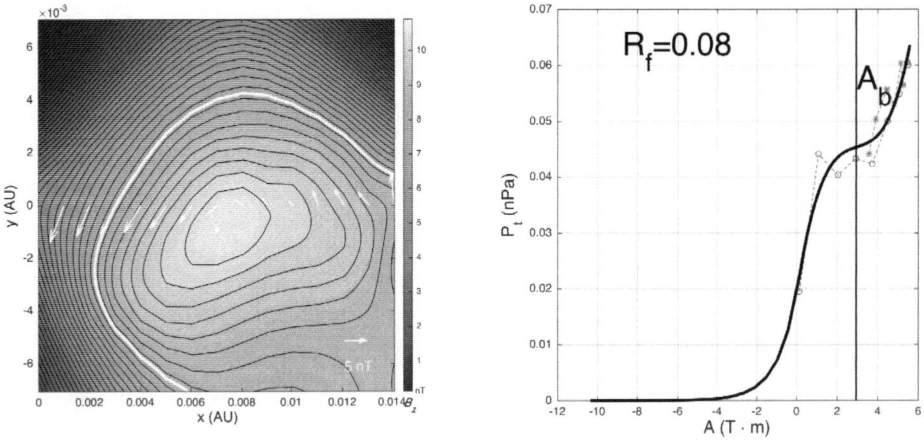

(a) $\hat{\mathbf{z}} = [0.73, -0.67, -0.10]$ (b)

Figure 4. GS reconstruction result for Case 2 (2002 Feb. 1, 10:36 - 12:55 UT). The format is the same as that in Figure 3.

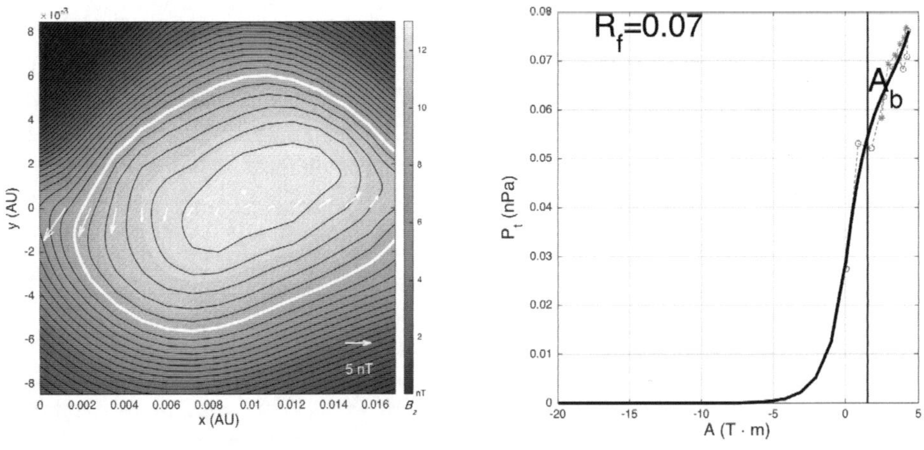

(a) $\hat{\mathbf{z}} = [0.31, -0.93, -0.20]$ (b)

Figure 5. Flux rope reconstruction for Case 3 (2002 Feb. 1, 14:57 - 17:03 UT). The format is the same as that in Figure 3.

higher plasma beta value than that of MCs. There were energetic particle flux enhancements associated with them (see second panel of Figure 2), at least partially. Such associations between particle energization and the small-scale flux rope structures were examined by Trenchi et al.[41] and a mechanism involving both diffusive shock acceleration and the magnetic flux ropes was developed by Zank et al.[42]. A Halo CME occurred at 2002 Jan. 27, 12:30:05 UT was a possible source to drive the shock and generate the trailing complex structure behind the shock. However, given the different plasma profiles from typical MCs, these small-scale flux ropes may not be generated on the Sun by the CME. They may be produced by the shock, or some other processes in the inner heliosphere. The same chirality of the three flux ropes may hint at the same process or the same environment in which they were generated.

3.2. Flux ropes in ambient solar wind

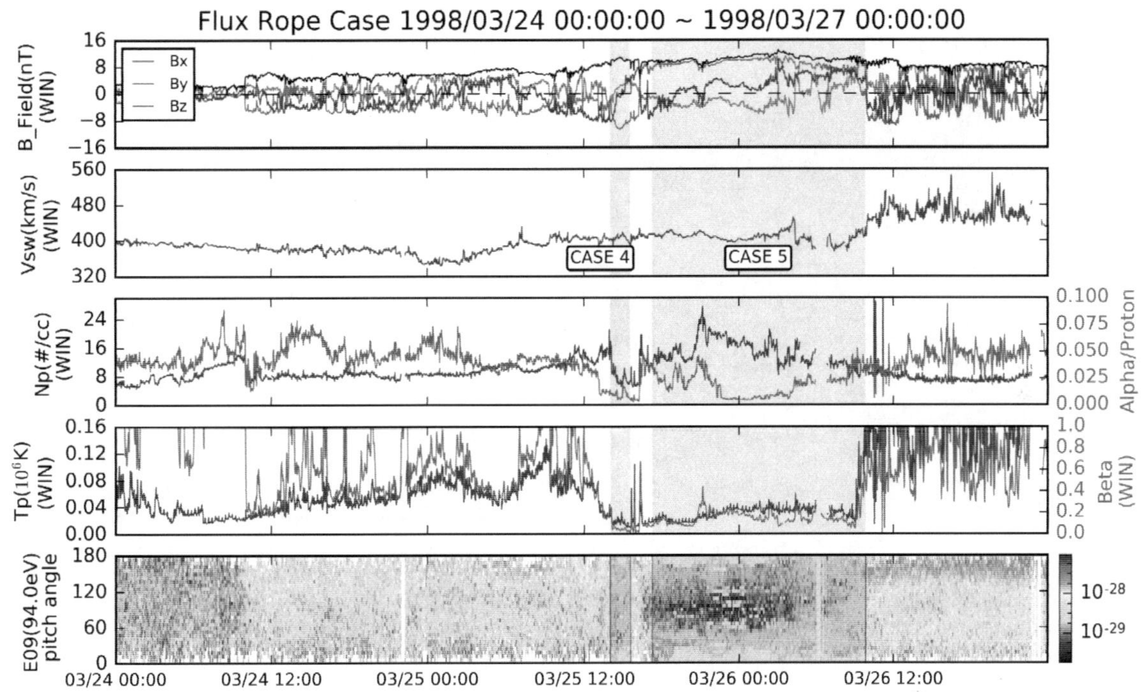

Figure 6. Time series data from 1998 Mar. 24, 00:00 UT to 1998 Mar. 27, 00:00 UT. This time interval contains two flux rope structures. The format of the first four panels are the same as Figure 1. The bottom panel is the suprathermal electron pitch angle spectrogram with the energy of 94.0 eV, obtained by the SWE instrument on WIND.

Figure 6 shows the time series which covers the case 4 and case 5. In this event, the suprathermal electron pitch angle data are available and are shown in the bottom panel. The shaded regions represent the time intervals of case 4 and case 5, respectively. During the interval of case 4, both B_x and B_z components rotated smoothly from negative to positive, while B_y component changed from about -8 nT to 4 nT. The total magnetic field magnitude was about 10 nT across the interval of case 4. During the interval of case 5, there were fluctuations overlapped on the rotation trend, especially in the latter half. The large trend shows that the B_z component rotated from negative to positive, and the B_y component remained uni-directional. The solar wind speed(second panel) was about 400 km/s, which is a bit higher than that of the first three cases. The third panel shows the proton number density (in blue) and the alpha/proton ratio (in green). The proton density was depressed in case 4 and then enhanced in case 5. The fourth panel shows that the proton temperature across case 4 and case 5 was depressed. The average plasma beta (proton temperature only) value in case 4 was about 0.05, which is a very low value, indicating that the magnetic pressure dominates in pressure balance. This value in case 5 was 0.15, indicating a relatively enhanced plasma pressure. From the bottom panel we can see that the flux of suprathermal electrons with large pitch angle(120° - 180°) was enhanced in case 4, while the flux of suprathermal electrons with small pitch angle(0° - 60°) was enhanced in case 5. The gap between case 4 and case 5 is also noteworthy. In this gap, all three magnetic field components changed the sign, indicating a direction reversal of magnetic field. We will discuss this in more details below.

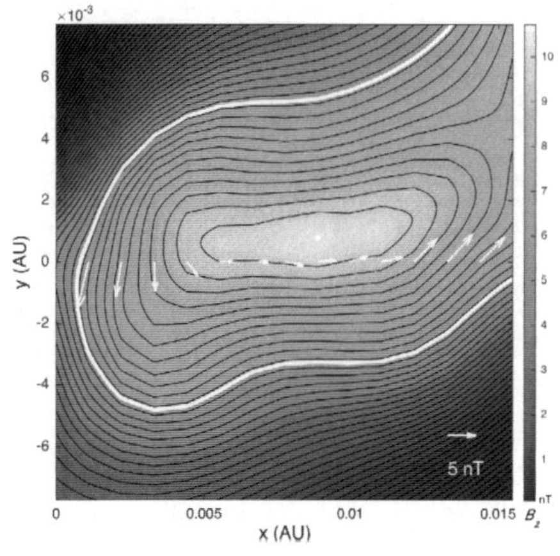

(a) $\hat{\mathbf{z}} = [-0.089, -0.99, -0.084]$

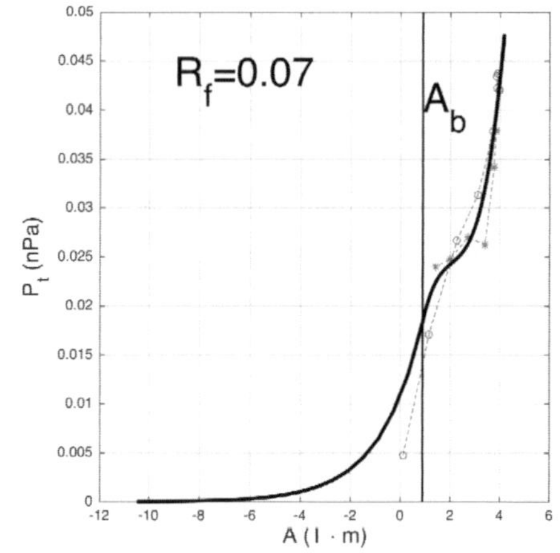

(b)

Figure 7. GS reconstruction result for Case 4 (1998 Mar. 25, 14:03 - 15:39 UT). The format is the same as that in Figure 3.

3.2.1. CASE 4: WIND 1998/3/25 14:03 - 15:39 UT Figure 7(a) shows the GS reconstruction of the small-scale flux rope in case 4. Figure 7(b) shows that the fitting residue is 0.07. This flux rope had a time duration of 1 hour and 36 minutes, and a scale size of 0.01 AU. It had the right-handed chirality. The core field is about 11 nT. According to the bottom panel in Figure 6, the suprathermal electrons with large pitch angles (120° - 180°) was enhanced, which implied the electrons were moving anti-parallel to the magnetic field lines. We will compare it with case 5 below.

3.2.2. CASE 5: WIND 1998/3/25 17:14 - 1998/3/26 09:47 UT Compared with the first four small-scale flux ropes, case 5 had a bigger scale size. Figure 8(a) shows that its scale size was about 0.1 AU, which is ten times as large as the first four flux ropes. This size is approaching the lower limit of the MC scale size. Therefore, we can take it as an intermediate-scale flux rope. The time duration was about 16 hours and 33 minutes, although arguably the interval can be shortened, for example, by excluding the later part showing enhanced fluctuations in magnetic field. Its core field was about 12 nT, which is stronger than that of case 4. Figure 8(b) shows that the fitting residue was 0.09, which is a relatively good fitting. From the fourth panel in Figure 6, we can see that the proton temperature(in blue) across the entire flux rope was depressed. Taking into account the plasma beta value (0.15 on average), this profile is similar to that of MC. The most interesting phenomenon is the suprathermal electrons pitch angle spectrogram(see the bottom panel in Figure 6). At the start of the time interval of case 5, the pitch angle of most suprathermal electrons suddenly changed from 120° - 180° to 0° - 60°, and

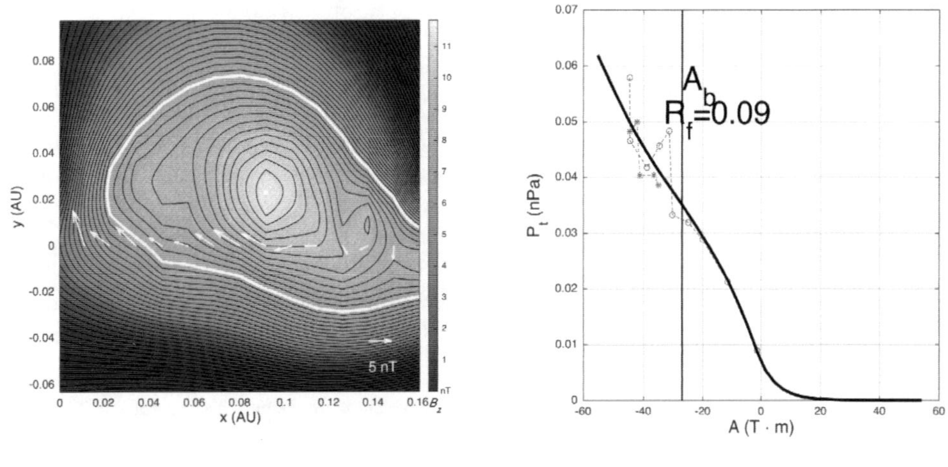

(a) $\hat{\mathbf{z}} = [0.093, 0.90, 0.42]$ (b)

Figure 8. Flux rope reconstruction for Case 5 (1998 Mar. 25, 17:14 - Mar. 26, 09:47 UT). The format is the same as that in Figure 3.

in the end of this interval, the pitch angle of most suprathermal electrons changed back. This flux rope had the left-handed chirality, which is opposite to that of case 4. After checking the z axis directions of these two flux ropes, we found that they had opposite orientation. In case 4, the unit vector of z axis is $[-0.089, -0.99, -0.084]$ in GSE, while in case 5, the unit vector of z axis is $[0.093, 0.90, 0.42]$ in GSE. The angle between them is about $160°$, which means that they are approximately anti-parallel. The hypothesis we proposed in Section 3.2 is supported by this result. Now we are ready to explain the behavior of the suprathermal electrons. Actually, the moving direction of these electrons had not changed. Because the flux rope in case 4 flipped the axial direction compared to the flux rope in case 5, the electron pitch angles changed from $120°$ - $180°$ to $0°$ - $60°$ when the electrons were streaming outward along field lines. The flipping is supported by the fact that in the gap between case 4 and case 5, all of the three components of magnetic field flipped the direction(see the first panel in Figure 6). Because of the anti-parallel directions of the z axes, though case 4 and case 5 have different chirality, in fact the transverse field lines in their adjacent region are anti-parallel. This geometry is able to produce magnetic field line reconnection resulting in a possible configuration of magnetic separatrix. Another interesting fact is, the pitch angle spectrogram of the suprathermal electrons remains unchanged across the flux rope interval in case 5. The bottom panel of Figure 6 also shows that the elections with pitch angle near $90°$ were depleted in the flux rope structure.

4. Summary and Discussion

In summary, we have reconstructed four small-scale magnetic flux ropes and one intermediate-scale magnetic flux rope in the solar wind, using the Grad-Shafranov reconstruction technique. Three of these flux ropes were downstream an interplanetary traveling shock, and two of them were in the ambient solar wind. Table 1 summarizes the key parameters obtained for these cases examined in this paper. All of them have relatively strong core magnetic field (see last column in Table 1), and are located in slow solar wind (the average solar wind speed, $\approx V_{HT}$, $\lesssim 400$ km/s). Our study suggeststhat (1) there exists multiple small-scale magnetic flux ropes behind an interplanetary traveling shock. Additional observations showed their possible associations

Table 1. Summary of five magnetic flux ropes.

Case No.	$Duration^{(a)}$	$(\delta, \phi)^{(b)}$	$< \beta >^{(c)}$	$V_{HT}^{(d)}$	$\Phi_t^{(e)}$	$\Phi_p^{(f)}$	$\tau^{(g)}$	$B_{zmax}^{(h)}$
1	2.83	$(82°, 181°)$	0.09	347	7.2E10	6.8E11	9	13
2	2.32	$(43°, 189°)$	0.33	356	2.4E10	3.9E11	16	11
3	2.10	$(72°, 192°)$	0.14	350	3.5E10	4.2E11	12	13
4	1.60	$(95°, 185°)$	0.05	404	2.4E10	4.9E11	20	11
5	16.55	$(85°, 25°)$	0.15	403	2.1E12	4.2E12	2	12

(a)The time duration of a flux rope (hours).
(b)The δ is the polar angle of the axial direction of a flux rope with respect to solar radial direction, and the ϕ is the azimuthal angle of the axial direction of a flux rope, projected onto the plane of sky, with respect to the Z axis in GSE coordinate.
(c)The average plasma beta value (no electron temperature contribution).
(d)The speed of the deHoffmann-Teller(HT) frame (km/s).
(e)The total axial magnetic flux within the boundary $A = A_b$ of a flux rope (Wb).
(f)The poloidal magnetic flux per 1 AU of a magnetic flux rope (Wb).
(g)The ratio of Φ_p to Φ_t, indicative of average magnetic field-line twist.
(h)The maximum axial magnetic field of a flux rope (nT).

with the energetic particle enhancement. A combined and detailed analysis of all available in-situ observations within certain theoretical framework is needed to further elucidate such associations and the roles of magnetic flux ropes in particle transport and energization. (2)The small-scale magnetic flux ropes occurred in different regions tend to have different characteristics. For example, in the region downstream an interplanetary traveling shock, the small-scale magnetic flux ropes didn't show the depressed proton temperature. The number density ratio of alpha to proton kept at a relatively steady ratio. In the region of ambient solar wind, our two cases showed the obviously depressed proton temperature, which is a typical characteristic of MCs. These differences indicate that they may have different origins and formation mechanisms, and also imply their different capabilities to energize particles.

Acknowledgments
We would like to thank NASA's Space Physics Data Facility for the WIND spacecraft data, and ACE Science Center for ACE data and the energetic ion plots. We acknowledge NASA grants NNX12AH50G, NNX14AF41G and NRL contract N00173-14-1-G006 (NNH13ZDA001N) for support.

References
[1] Burlaga L, Sittler E, Mariani F and Schwenn R 1981 *Journal of Geophysical Research: Space Physics* **86** 6673–6684 ISSN 2156-2202 URL `http://dx.doi.org/10.1029/JA086iA08p06673`
[2] Lepping R P, Jones J A and Burlaga L F 1990 *Journal of Geophysical Research: Space Physics* **95** 11957–11965 ISSN 2156-2202 URL `http://dx.doi.org/10.1029/JA095iA08p11957`
[3] Bothmer V and Rust D 1997 *The Field Configuration of Magnetic Clouds and the Solar Cycle* (American Geophysical Union) pp 139–146 ISBN 9781118664377 URL `http://dx.doi.org/10.1029/GM099p0139`
[4] Smith C W and Phillips J L 1997 *Journal of Geophysical Research: Space Physics* **102** 249–261 ISSN 2156-2202 URL `http://dx.doi.org/10.1029/96JA02678`
[5] Bothmer V and Schwenn R 1998 *Annales Geophysicae* **16** 1–24 URL `http://www.ann-geophys.net/16/1/1998/`

[6] Wilson A (ed) 2003 *Proc. Sources of magnetic helicity over the solar cycle*
[7] Leamon R J, Canfield R C, Jones S L, Lambkin K, Lundberg B J and Pevtsov A A 2004 *Journal of Geophysical Research: Space Physics* **109** ISSN 2156-2202 a05106 URL http://dx.doi.org/10.1029/2003JA010324
[8] Innes D E, Lagg A and Solanki S K (eds) 2003 *Proc. Quantitative Link Between Solar Ejecta and Interplanetary Magnetic Clouds: Magnetic Helicity*
[9] Pizzo V J, Holzer T E and Sime D G (eds) 2003 *Proc. The origin and propagation of coronal mass ejections*
[10] Richardson I G and Cane H V 2004 *Geophysical Research Letters* **31** ISSN 1944-8007 l18804 URL http://dx.doi.org/10.1029/2004GL020958
[11] Mulligan T, Russell C T and Luhmann J G 1998 *Geophysical Research Letters* **25** 2959–2962 ISSN 1944-8007 URL http://dx.doi.org/10.1029/98GL01302
[12] Lynch B J, Gruesbeck J R, Zurbuchen T H and Antiochos S K 2005 *Journal of Geophysical Research: Space Physics* **110** ISSN 2156-2202 a08107 URL http://dx.doi.org/10.1029/2005JA011137
[13] Burlaga L F and Behannon K W 1982 *Solar Physics* **81** 181–192 URL http://dx.doi.org/10.1007/BF00151989
[14] Klein L W and Burlaga L F 1982 *Journal of Geophysical Research: Space Physics* **87** 613–624 ISSN 2156-2202 URL http://dx.doi.org/10.1029/JA087iA02p00613
[15] Lepping R P, Berdichevsky D B, Wu C C, Szabo A, Narock T, Mariani F, Lazarus A J and Quivers A J 2006 *Annales Geophysicae* **24** 215–245
[16] Wang Y, Zhou Z, Shen C, Liu R and Wang S 2015 *Journal of Geophysical Research: Space Physics* **120** 1543–1565 ISSN 2169-9402 2014JA020494 URL http://dx.doi.org/10.1002/2014JA020494
[17] Moldwin M B and Hughes W J 1992 *Journal of Geophysical Research: Space Physics* **97** 19259–19282 ISSN 2156-2202 URL http://dx.doi.org/10.1029/92JA01598
[18] Neugebauer B (ed) 1983 *Proc. On the field configuration in magnetic clouds* (NASA Conf. Publ., NASA CP-2280)
[19] Burlaga L F 1988 *Journal of Geophysical Research: Space Physics* **93** 7217–7224 ISSN 2156-2202 URL http://dx.doi.org/10.1029/JA093iA07p07217
[20] Lundquist S 1950 *Ark. Fys.* **2** 361–365
[21] Hu Q and Sonnerup B U 2001 *Geophysical Research Letters* **28** 467–470 ISSN 1944-8007 URL http://dx.doi.org/10.1029/2000GL012232
[22] Hu Q and Sonnerup B U O 2002 *Journal of Geophysical Research: Space Physics* **107** SSH 10–1–SSH 10–15 ISSN 2156-2202 URL http://dx.doi.org/10.1029/2001JA000293
[23] Sonnerup B U O and Guo M 1996 *Geophysical Research Letters* **23** 3679–3682 ISSN 1944-8007 URL http://dx.doi.org/10.1029/96GL03573
[24] Sonnerup B U O, Hasegawa H, Teh W L and Hau L N 2006 *Journal of Geophysical Research: Space Physics* **111** ISSN 2156-2202 a09204 URL http://dx.doi.org/10.1029/2006JA011717
[25] Hu Q, Smith C W, Ness N F and Skoug R M 2004 *Journal of Geophysical Research: Space Physics* **109** ISSN 2156-2202 a03102 URL http://dx.doi.org/10.1029/2003JA010101
[26] Moldwin M B, Phillips J L, Gosling J T, Scime E E, McComas D J, Bame S J, Balogh A and Forsyth R J 1995 *Journal of Geophysical Research: Space Physics* **100** 19903–19910 ISSN 2156-2202 URL http://dx.doi.org/10.1029/95JA01123
[27] Moldwin M B, Ford S, Lepping R, Slavin J and Szabo A 2000 *Geophysical Research Letters* **27** 57–60 ISSN 1944-8007 URL http://dx.doi.org/10.1029/1999GL010724
[28] Feng H Q, Wu D J and Chao J K 2007 *Journal of Geophysical Research: Space Physics* **112** ISSN 2156-2202 a02102 URL http://dx.doi.org/10.1029/2006JA011962
[29] Cartwright M L and Moldwin M B 2008 *Journal of Geophysical Research: Space Physics* **113** ISSN 2156-2202 a09105 URL http://dx.doi.org/10.1029/2008JA013389
[30] Feng H Q, Wu D J, Lin C C, Chao J K, Lee L C and Lyu L H 2008 *Journal of Geophysical Research: Space Physics* **113** ISSN 2156-2202 a12105 URL http://dx.doi.org/10.1029/2008JA013103
[31] Cartwright M L and Moldwin M B 2010 *Journal of Geophysical Research: Space Physics* **115** ISSN 2156-2202 a08102 URL http://dx.doi.org/10.1029/2009JA014271
[32] Feng H Q, Zhao G Q and Wang J M 2015 *Journal of Geophysical Research: Space Physics* **120** 10,175–10,184 ISSN 2169-9402 2015JA021643 URL http://dx.doi.org/10.1002/2015JA021643
[33] Zank G P, le Roux J A, Webb G M, Dosch A and Khabarova O 2014 *The Astrophysical Journal* **797** 28 URL http://stacks.iop.org/0004-637X/797/i=1/a=28
[34] le Roux J A, Webb G M, Zank G P and Khabarova O 2015 *Journal of Physics: Conference Series* **642** 012015 URL http://stacks.iop.org/1742-6596/642/i=1/a=012015
[35] Sonnerup B U O and Teh W L 2008 *Journal of Geophysical Research: Space Physics* **113** ISSN 2156-2202 a05202 URL http://dx.doi.org/10.1029/2007JA012718
[36] Hau L N and Sonnerup B U O 1999 *Journal of Geophysical Research: Space Physics* **104** 6899–6917 ISSN

2156-2202 URL `http://dx.doi.org/10.1029/1999JA900002`

[37] Hu Q and Sonnerup B U O 2000 *Geophysical Research Letters* **27** 1443–1446 ISSN 1944-8007 URL `http://dx.doi.org/10.1029/1999GL010751`

[38] deHoffmann F and Teller E 1950 *Phys. Rev.* **80**(4) 692–703 URL `http://link.aps.org/doi/10.1103/PhysRev.80.692`

[39] Sonnerup B U O, Papamastorakis I, Paschmann G and Lühr H 1987 *Journal of Geophysical Research: Space Physics* **92** 12137–12159 ISSN 2156-2202 URL `http://dx.doi.org/10.1029/JA092iA11p12137`

[40] Khrabrov A V and Sonnerup B U O 1998 *Journal of Geophysical Research: Space Physics* **103** 6641–6651 ISSN 2156-2202 URL `http://dx.doi.org/10.1029/97JA03731`

[41] Trenchi L, Bruno R, D'Amicis R, Marcucci M F and Telloni D 2013 *Annales Geophysicae* **31** 1333–1341

[42] Zank G P, Hunana P, Mostafavi P, Le Roux J A, Li G, Webb G M, Khabarova O, Cummings A, Stone E and Decker R 2015 *The Astrophysical Journal* **814** 137